P9-DGP-628

Designing for Situation Awareness

Designing for Situation Awareness

An Approach to User-Centered Design

Mica R. Endsley, Betty Bolté, and Debra G. Jones

SA Technologies, Georgia, USA

London and New York

First published 2003 by Taylor & Francis
11 New Fetter Lane, London EC4P 4EE

Simultaneously published in the USA and Canada
by Taylor & Francis Inc.
29 West 35th Street, New York NY 10001

Taylor & Francis is an imprint of the Taylor & Francis Group

Publisher's Note
This book has been produced from camera-ready copy supplied by the authors.
Printed and bound in Great Britain
by Biddles Ltd, Guildford, Surrey

British Library Cataloguing in Publication Data
A catalogue record for this book is available from the British Library

Library of Congress Cataloging in Publication Data
Endsley, Mica R.
Designing for situation awareness: an approach to user-centered design / Mica R. Endsley, Betty Bolte, and Debra G. Jones.
p.cm.
Includes bibliographical references and index.
ISBN 0-7484-0966-1 (hbk.) – ISBN 0-7484-0967-X (pbk.)
1.Human-machine systems—Design. 2. Work design. I. Bolté, Betty. II. Jones, Debra G. III. Title

TS167.E53 2003
658.5'036—dc21

2003042670

ISBN 0-748-40966-1 (Hbk)
ISBN 0-748-40967-X (Pbk)

Contents

Preface

'Our Age of Anxiety is, in great part, the result of trying to do today's jobs with yesterday's tools.' – *Marshall McLuhan*

While a clear understanding of one's situation is undoubtedly the critical trigger that allows the knowledge, skills, and creativity of the human mind to be successfully brought to bear in shaping our environment, very often people must work uphill, against systems and technologies that block rather than enhance their ability to ascertain the information they need. Knowledge in a vacuum is meaningless. Its use in overcoming human problems and achieving human goals requires the successful application of that knowledge in ways that are contextually appropriate. Yet, across a wide variety of engineered systems, people face an ever-widening information gap—the gulf between the data that is available and the information that they really need to know.

This book addresses the information gap through system design. It presents a core set of principles and a methodology for engineers and designers who are seeking to nourish the situation awareness of their system's users. Operators of power plants, aircraft, automobiles, ships, command and control centers for military and large-scale commercial enterprises, intelligence operations, medical systems, and information management systems are all in need of technologies that allow people to effectively manage the information available to gain a high level of understanding of what is happening. They need systems designed to support *situation awareness*.

We are indebted to many people. First, we are grateful for the support and patience of our families during the writing of this book—may it lead to a better world for our children. We are thankful for the support of many of our colleagues, whose dedicated research has created the foundation upon which this book is based. Finally, we are indebted to the countless number of pilots, air traffic controllers, soldiers, and system operators who have volunteered their insights, experiences, time, and expertise in the hopes of bettering the tools of their profession. We also appreciate and acknowledge the support of our sponsors, the U.S. Army Research Laboratory Collaborative Technology Alliance and the NASA Langley Research Center, without whose assistance this book would not have been possible.

Mica R. Endsley, Betty Bolté, and Debra G. Jones

Although we are committed to scientific truth, there comes a point where this truth is not enough, where the application of truth to human objectives comes into play. Once we start to think in terms of utility, we must leave the placid environment of the laboratory, take off our white coats, and roll up our sleeves...." *Samuel Florman—'The Civilized Engineer'*

Preface

Part One: Understanding Situation Awareness in System Design

CHAPTER ONE

User-centered Design

1.1 WHO IS THIS BOOK FOR?

We are living in the 'information age'. All around us are signs of the changes induced by the explosion of information technologies. Whether you are working on the shop floor, in the business world, or are just trying to purchase a new computer for your home, the dizzying pace of technological change and the vast amount of information present can be daunting (Figure 1.1). We are constantly barraged with information through TV, radio, mailings, and hundreds of magazines and journals. Within companies, reports and forms have multiplied and every aspect of the business is recorded somewhere.

Figure 1.1 Challenge of the information age.

Widespread communications networks allow us to communicate with colleagues in other cities and other continents as easily as we once communicated with our neighbors—whether they are at home, in the office, flying over the Atlantic ocean, or hiking through the Andes. Matching the access for voice communication are fax machines, e-mail, and the World Wide Web, which bring text and pictures just as easily. And the computerization of information is only the most recent offshoot of this information explosion. A rapid proliferation in the publishing world has seen a deluge of magazines, journals, and books crowding our mailboxes and the shelves of our libraries. The world's great libraries are doubling in size every 14 years; 1000 new books are published internationally every day, and the number of scientific journals has increased to the extent that surveys show that the vast majority of articles go virtually unread. More new information has been produced in the last 30 years than the previous 5000 (Wurman, 1989).

These challenges are even more daunting in complex, large-scale systems: power plant control rooms, aircraft cockpits, centralized command and control centers for military and large-scale commercial enterprises, intelligence operations, and information management systems for medical, insurance, and financial systems. Add to this list environments that were once fitted only with the basics that are now being targeted for introduction of new information technologies. For example, cell phones, electronic navigation systems coupled with global positioning system locational devices, and automation to perform tasks from traffic conflict detection to automated spacing and lane following are being incorporated into the automobile. In each of these environments, the human operator can be severely challenged in rapidly bringing all of the available information together in a form that is manageable for making accurate decisions in a timely manner. There is simply more information than any one person can handle.

In the face of this torrent of 'information,' many of us feel less informed than ever before. This is because there is a huge gap between the tons of data being produced and disseminated, and our ability to find the bits that are needed and process them together with the other bits to arrive at the actual needed information. That is, it seems to be even harder to find out what we really want or need to know. This problem is real and ongoing, whether you sit in a cockpit or behind a desk. It is becoming widely recognized that more data does not equal more information (Figure 1.2).

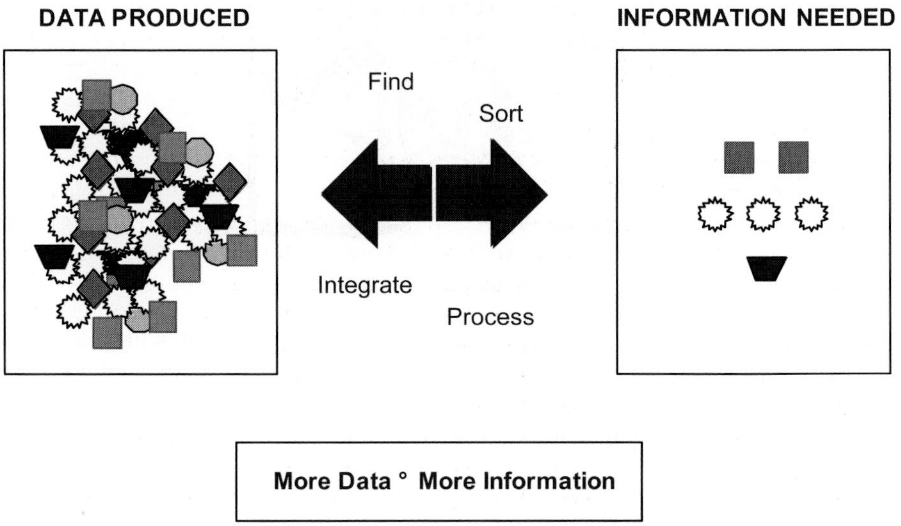

Figure 1.2 The information gap (from Endsley, 2000b).

This book is for those involved in solving this *information gap* through better system design. Engineers, computer scientists, designers, human factors engineers, and psychologists are all working on developing systems to harness information in new ways. These efforts focus on everything from consumer software products to the most sophisticated intelligence systems for detecting and responding to hostile actions in the military arena. Medical systems, information analysis networks with distributed sensors and information input, aircraft and air traffic control systems, offices, automobiles, power plants, and the Internet—all of these systems are in need of user interfaces that allow people to effectively manage the information available to gain a high level of understanding of what is happening. They need systems designed to support *situation awareness*.

1.2 WHY DO WE NEED USER-CENTERED DESIGN?

The problems of information overload are not inevitable nor insurmountable. The most effective way to confront the information gap is by applying a philosophy known as *user-centered design*. The best way to explain user-centered design is to first discuss the pitfalls of current design approaches, which we'll call *technology-centered design*.

1.2.1 Technology-centered design

Traditionally, systems have been designed and developed from a technology-centered perspective. Engineers developed the sensors and systems that were needed to perform each function. They then provided a display for each system that informed the operator of how well that particular system was operating or its present status. So, for example, in the aircraft cockpit a separate display was provided for altitude, airspeed, engine temperature, etc. As technology improved, more and more displays were added. People were left with the job of trying to keep up with the exponential growth of data created by this process. The number of displays in aircraft cockpits increased from none for the early Wright Flyer to hundreds by the 1970s, only to be replaced by multifunction displays which decreased the number of physical displays, but allowed the number of display pages and displayed information to continue to grow exponentially (Sexton, 1988). In the face of changing tasks and situations, the operator is called upon to find, sort, integrate, and process the information that is needed from all that which is available, leading inevitably to an information gap.

Unfortunately, the human has certain information processing bottlenecks. People can only pay attention to a certain amount of information at once. As the display of data in these systems is centered around the technologies producing them, it is often scattered and not ideally suited to support human tasks (Figure 1.3). A considerable amount of additional work is required to find what is needed and extra mental processing is required to calculate the information the operator really wants to know. This inevitably leads to higher than necessary workload and error. Keeping up has become harder and harder to do.

Across major classes of systems, the operator is generally cited as being a causal factor in between 60% and 85% of all accidents (Nagel, 1988; Sanders &

McCormick, 1992). These problems are not just limited to a few select industries or occupations, like power plant operators or pilots. The National Academy of Science has estimated that over 44 000 deaths in the United States annually can be attributed to human error in the medical field, at least two thirds of which are preventable (Kohn, Corrigan, & Donaldson, 1999). Much of this so-called human error is not the result of a faulty human, but rather the direct result of technology-centered designs that have been ill-suited for promoting high levels of sustained human performance over the wide range of conditions and circumstances called for in the real world. Rather than human error, a more accurate term would be *design-induced error.*

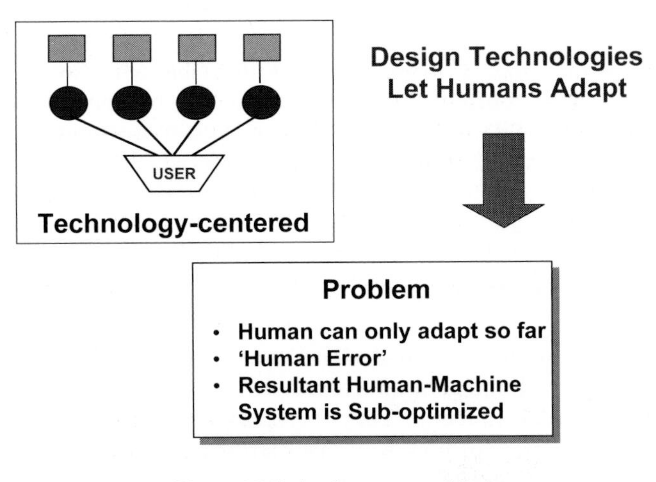

Figure 1.3 Technology-centered design.

As many as 2500 died in the tragic accident at the Union Carbide plant in Bhopal, India in 1984. Consider the following excerpt from an account of one of the century's major accidents: '*Unknown to Dey, the pressure inside the tank was 2 psig only 40 minutes before at 10:20. But the buildup was not apparent because no historical trace of the pressure was shown within the control room, and the operator on the previous shift had not entered this into the log*' (Casey, 1993). Information needed by the power plant operator was not visible from the control room or was not presented by the system at all. The design of the system's interface did not support the operator in detecting significant cues of the building problem or in preventing the events that led to the accident.

Efforts to resolve the perceived problem with error by human operators through the increased use of automation have not necessarily been effective. For example, Boeing Aircraft Corporation reports the flight crew as a primary factor in 76% of hull loss accidents between 1959 and 1981. This number was only reduced to 71.1% in the years 1982 to 1991, when far more aircraft with automated systems were flying. While some significant utility occurred for certain classes of automation, along with that utility came other types of errors induced by problems of increased system complexity, loss of situation awareness, system brittleness, and workload increases at inopportune times (Billings, 1997; Parasuraman, Molloy, Mouloua, & Hilburn, 1996; Wiener & Curry, 1980). The attempt to automate our way out of so-called human error has only led to more complexity, more cognitive load and catastrophic errors associated with losses of situation awareness (Endsley & Kiris, 1995a). While we have seen many of these problems played out historically in the aviation arena, the increasing use of automation in many other types of systems will find widespread reverberation of these problems.

1.2.2 User-centered design

As an alternative to the downward spiral of complexity and error induced by a technology-centered design philosophy, the philosophy of user-centered design is a way of achieving more effective systems. User-centered design challenges designers to mold the interface around the capabilities and needs of the operators. Rather than displaying information that is centered around the sensors and technologies that produce it, a user-centered design integrates this information in ways that fit the goals, tasks, and needs of the users. This philosophy is not borne primarily from a humanistic or altruistic desire, but rather from a desire to obtain optimal functioning of the overall human–machine system (Figure 1.4).

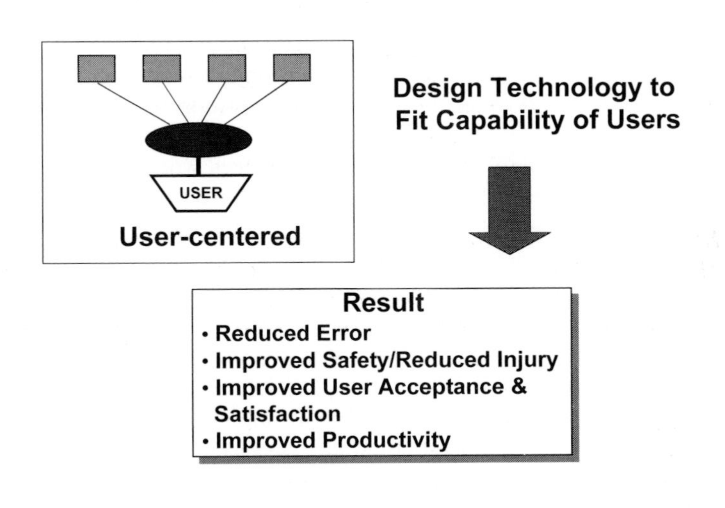

Figure 1.4 User-centered design.

As a result of user-centered design, we can greatly reduce errors and improve productivity without requiring significant new technological capabilities. Along with user-centered design also comes improved user acceptance and satisfaction as a side benefit, by removing much of the frustration common to today's technologies. User-centered design provides a means for better harnessing information technologies to support human work.

1.3 WHAT DOES USER-CENTERED DESIGN MEAN?

While the concept of user-centered design has been bandied about since the mid-1990s, just how to go about achieving this goal has not been exactly clear to many designers. Indeed, many misconceptions abound regarding how to create a user-centered system. It is worthwhile, therefore, to first have a brief discussion on what user-centered design is *not.*

1.3.1 What user-centered design does not mean

User-centered design does not mean asking users what they want and then giving it to them. While this sounds like a perfectly reasonable approach, it has been found to be fraught with pitfalls. Users often have only partial ideas about what might be better than what they are used to. Furthermore, they generally have very limited knowledge on how to effectively present information and design human interactions with complex systems. These issues are compounded by the fact that most systems must be used by many different individuals, each of whom can have significantly different ideas on what they would like to see implemented in a new design effort. The result of this approach is an endless and costly cycle of implementing new ideas, only to have the next team of users decide they want something different. Design solutions tend to be sporadic and inconsistent across features of the interface and many design problems are not recognized. While users are a valuable source of information regarding problems experienced, information and decision needs, working conditions and functions needed, the unfiltered implementation of whatever they want completely neglects the large base of scientific literature regarding which types of interfaces work and which do not. It also discards the significant benefits that come from objective interface testing. Good ideas could be discarded by virtue of group dynamics and bad ideas may stay around as their pitfalls are not known.

User-centered design does not mean presenting users with just the information they need at any given moment. One well-meaning attempt to help manage the information overload problem is the idea of information filtering. Using computers, an assessment is made of the operator's current tasks and then the computer presents just the information and tools needed to support that task. When the task changes, so does the information provided. This too sounds like an ideal solution, however significant problems exist with this approach.

First, it has proven to be very difficult for systems to accurately detect just which tasks and information are needed at any particular time in a dynamic system. Thus, the significant risk of system error exists, whereby the wrong information is presented or removed at any given time. Even if engineers solved this problem adequately, the real problem is that such an approach leaves the operator with significant difficulties. During the course of most operations, operators must rapidly and frequently switch between goals as circumstances dictate, often with very rapid responses required. Individuals do not instantly understand what is happening in a situation by simply looking at instantaneously presented information. It takes a certain amount of time to orient oneself to a situation, to ascertain the key factors and their critical features.

Furthermore, the dynamics of the situation (how much and how rapidly things are changing) are an important feature that is lost with this approach. The operator will also be left with the job of trying to keep up with a system that is rapidly changing. Information will be removed in the middle of a task and new information presented, leaving the operator constantly trying to figure out what is going on with the system. Information-filtering concepts always place the operator in the role of being reactive rather than proactive, which severely limits performance effectiveness. The ability to project ahead and be ready for upcoming situations will be lost.

User-centered design does not mean systems that make decisions for the user. Research on many types of decision support systems have shown that systems that provide users with advice on what decisions to make do not necessarily result in the best performance (Endsley & Kiris, 1994a; Kibbe, 1988; Selcon, 1990). In general, if the system advice is ambiguous, the system is far more likely to reduce human decision quality and speed. For example, if the system provides recommendations and levels of confidence for two options, option A and option B, and it shows option A with a confidence level of 48% and option B with a confidence level of 52%, then the decision maker will be slower to decide what to do than if no information had been provided at all by the system.

If the system advice is wrong, it can lead to decision biasing; the human will be more likely to make a wrong decision than if left alone (Smith *et al.*, 1995). For example, an aviation diagnostic system that provides checklists for the pilot to run through when a problem has been detected can misdiagnose the problem and provide the wrong checklists for the actual problem. Such misdirection can lead the pilot down the garden path, delaying or significantly interfering with the pilot's ability to correct the real problem. If the pilot had been left alone to diagnose the problem, the pilot would have been more likely to diagnose the correct problem in the first place. Even when system advice is correct, it turns out that overall decision making and performance may be slower *without* the expected improvements in decision quality, as the human operator has to consider the decision support system advice in conjunction with other sources of information to arrive at his or her final decision (Endsley & Kiris, 1994a).

Other problems can also be introduced. For example, Microsoft® introduced the Assistant (an animated figure that appears to offer advice) in later versions of its word processing software. Very often, however, this Assistant does not provide useful advice, but rather slows down the work process as users are forced to look away from their task and take steps to respond to the Assistant. Computer users need more effective methods of help.

User-centered design does not mean doing things for the user. Many efforts to add ease of use by performing tasks for the user have backfired. A very fine line exists between doing things *for* the user and doing things *to* the user. Microsoft® Word has an automated function that changes the numbering of paragraphs and lists, but frequently renumbers incorrectly, causing the user to exert considerable effort to fight the system for control of the document's format. This problem is not particular to this product, but rather can be found with automobiles that decide when to turn lights on and when to lock doors, and with a variety of newer systems that boast such automated 'benefits.' The user is 'out-of-the-loop' of such actions and has the additional burden of trying to decide what the system will and won't do in different situations and how to make it complete formerly easy tasks. In general, the automation of tasks can lead to as many problems as it solves and must be addressed very carefully (see Chapter 10).

1.4 PRINCIPLES FOR USER-CENTERED DESIGN

If these things are not user-centered design, then what is? This book offers the perspective that user-centered design can be achieved through several key principles and processes.

1.4.1 Organize technology around the user's goals, tasks, and abilities

Traditionally, the human factors and ergonomics field has sought to design systems so as not to require people to perform tasks that exceed their capabilities perceptually or physically. In recent years, more focus has also been put on the mental abilities of people. Task analyses are completed to insure that the system provides the data needed to support user tasks in the order of their use. User-centered design has much in common with these traditional objectives of human factors. It goes further, however, in that it addresses the far more challenging situations brought about by computers and automation, and the substantially more advanced issues brought about in complex system domains. Whereas traditional human factors approaches are quite suitable for linear, repetitive tasks, user-centered design is more suitable for complex systems in which users need to pursue a variety of (sometimes competing) goals over the course of time, and no set sequence of tasks or actions can be prescribed. In these types of systems, interfaces and capabilities need to be designed to support the changing goals of the operator in a dynamic fashion. For example, a military commander must be able to rapidly switch between offensive and defensive goals, and the information system he uses needs to support both of these goals and seamlessly switch between them. In the information age, goals are the central organizing principle of work performed and interfaces need to be designed to support goal-oriented processing. In *Designing for Situation Awareness*, you will learn about a methodology for user-centered design that is goal centered.

1.4.2 Technology should be organized around the way users process information and make decisions

A considerable amount of research has been conducted on how people make decisions in real-world complex settings. This research has found significant differences from the idealized model of decision making in which optimization across possible alternatives is featured. Instead, researchers in many complex domains find that people will act first to classify and understand a situation. This internal model of the situation triggers the appropriate response from memory, immediately proceeding to action selection (Klein, 1989; Klein, Calderwood, & Clinton-Cirocco, 1986; Lipshitz, 1987; Noble, Boehm-Davis, & Grosz, 1987). In seminal work in this area, Dreyfus (1981) emphasized the role of situational understanding in real-world, expert decision making, building upon the extensive works of deGroot (1965) in chess, Mintzburg (1973) in managerial decision making, and Kuhn (1970) in the field of science. In each of these three diverse areas, they found that experts use pattern-matching mechanisms to draw upon long-term memory structures that allow them to quickly understand a given situation. They then adopt the appropriate course of action corresponding to that type of situation. Rarely do these experts spend much time mulling possible alternative courses of action. For the preponderance of decisions, the majority of their time and effort is spent in assessing and classifying the current situation. Terming this *Naturalistic Decision Making*, Klein (Klein, 1989) has greatly expanded on these findings by developing models of how people make decisions in the real world. A person's situation awareness becomes the key feature dictating the success of the decision process in most real-world decision making.

Decision makers in these complex domains must do more than simply perceive the state of their environment to have good situation awareness. They must understand the integrated meaning of what they are perceiving in light of their goals. Situation awareness, as such, incorporates an operator's *understanding of the situation as a whole*, which forms the basis for decision making. The evidence shows that an integrated picture of the current situation may be matched to prototypical situations in memory, each prototypical situation corresponding to a 'correct' action or decision. Designing to support information processing and decision making, therefore, involves designing to support the operator's ability to gain and maintain situation awareness in a dynamic world environment.

1.4.3 Technology must keep the user in control and aware of the state of the system

Research has shown that high levels of automation for tasks can put users out-of-the-loop, leading to low levels of situation awareness (Endsley & Kiris, 1994b; Pope, Comstock, Bartolome, Bogart, & Burdette, 1994). As long as situation awareness is compromised, the ability of the user to be an effective decision maker is threatened. In 1996, an American Airlines B-757 crashed in the mountains around Cali, Colombia. The crew had lost awareness of the actions of the aircraft's automated flight management system computer that was providing navigation and was unable to regain sufficient situation awareness to resume safe control of the aircraft. A key premise of user-centered design is that while a person does not need to perform every task, the person does need to be in control of managing what the systems are doing in order to maintain the situation awareness needed for successful performance across a wide variety of conditions and situations. Because automation of various types is becoming ubiquitous, user-centered design principles must address the way in which automation must be blended with humans to create effective overall systems. Automation must allow the user to remain in control to optimize situation awareness, and thus performance.

1.5 SITUATION AWARENESS: THE KEY TO USER-CENTERED DESIGN

These principles lead to the realization that situation awareness is key to achieving a user-centered design. If operators can achieve a high level of situation awareness, they will be more effective system components than if situation awareness is denied or hard to achieve. Situation awareness is the engine that drives the train for decision making and performance in complex, dynamic systems (Figure 1.5).

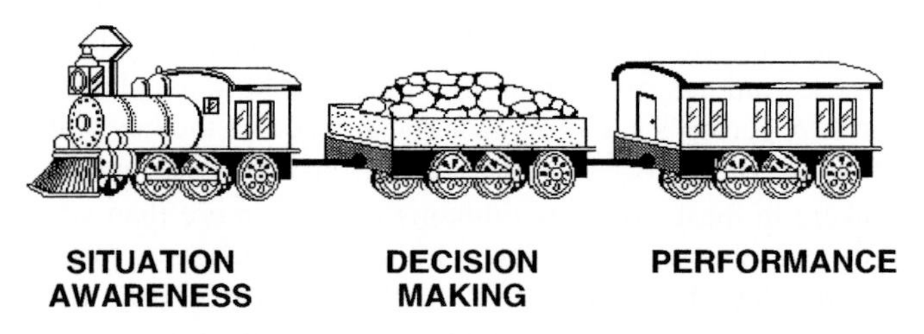

Figure 1.5 Situation awareness drives decision making and performance.

In complex and dynamic environments, decision making is highly dependent on situation awareness—a constantly evolving picture of the state of the environment. In reviewing the three principles of user-centered design, it can be seen that if we can design systems that provide a high level of situation awareness, we will have fulfilled these objectives. First, situation awareness is goal-oriented. By definition, the elements of the environment that people need to be aware of are determined based on the goals associated with that job. A goal-directed task analysis methodology is used to determine which data people need to be aware of, how that data needs to be understood relative to operator goals, and what projections need to be made to reach those goals (see Chapter 5). These analyses provide the basis for understanding how situation awareness maps to goals and for creating goal-driven designs.

Second, supporting situation awareness directly supports the cognitive processes of the operator. If a person's situation awareness is the key factor driving the decision-making process, then the most useful way to support that process will be to create system interfaces that are effective at creating a high level of situation awareness. In at least one study, 88% of human error was found to be due to problems with situation awareness (Endsley, 1995b). That is, in the majority of cases, people do not make bad decisions or execute their actions poorly; they misunderstand the situation they are in. Thus, the best way to support human performance is to better support the development of high levels of situation awareness.

Finally, keeping the user in control is fundamental to good situation awareness. By developing design methods that keep users in control and at high levels of situation awareness, we will create user-centered designs. This book describes a methodology and design principles for achieving these objectives.

CHAPTER TWO

What is Situation Awareness?

Now that we understand what user-centered design is, we need to look more closely at the meaning of the term *situation awareness (SA)*. This chapter may be more theoretical than you would expect in a book about designing systems, but in order to better understand what we are trying to achieve, a thorough understanding of how SA is developed in the mind is necessary. This chapter will provide you with a sound foundation from which we can develop principles for designing systems that will support and enhance SA.

2.1 SA DEFINED

Basically, SA is being aware of what is happening around you and understanding what that information means to you now and in the future. This awareness is usually defined in terms of what information is important for a particular job or goal. The concept of SA is usually applied to operational situations, where people must have SA for a specified reason, for example in order to drive a car, treat a patient, or separate traffic as an air traffic controller. Therefore, SA is normally defined as it relates to the goals and objectives of a specific job or function.

Only those pieces of the situation that are relevant to the task at hand are important for SA. While a doctor needs to know all of a patient's symptoms in order to diagnose an illness or ailment, the doctor does not usually need to know every detail of the patient's history. Likewise, the pilot of an aircraft must be aware of other planes, the weather, and approaching terrain changes, but the pilot does not need to know what the copilot had for lunch.

The formal definition of SA is '*the perception of the elements in the environment within a volume of time and space, the comprehension of their meaning, and the projection of their status in the near future*' (Endsley, 1988). The term *situation awareness* comes from the world of the military pilot, where achieving high levels of situation awareness was found to be both critical and challenging early in aviation history. Situation awareness is also important in many other domains, although it has sometimes received different names. In the world of air traffic controllers, for instance, they generally refer to SA as *the picture,* a mental representation of the situation on which they base their decisions.

While the individual *elements* of SA in this definition can vary greatly from one domain to another, the importance of SA as a foundation for decision making and performance applies to almost every field of endeavor. SA is studied in wide-ranging fields, such as education, driving, train dispatching, maintenance, power plant operations, and weather forecasting, in addition to aviation and military operations. The use of SA as a key driver for decision making and performance also extends beyond these fields to nonwork-related activities, including recreational and professional sport teams, self-protection, even acting. Situation awareness is the real-world changing knowledge that is critical for effective decision making and action.

The formal definition of SA breaks down into three separate levels:

- Level 1 – *perception* of the elements in the environment,
- Level 2 – *comprehension* of the current situation, and
- Level 3 – *projection* of future status.

We will discuss each level of SA as well as the internal and external factors that affect it (shown in Figure 2.1) in order to develop a complete understanding of what SA is and how to more effectively design tools and systems to help people develop and maintain it, even under very challenging circumstances.

2.1.1 Level 1 SA: Perception of elements in the environment

The first step in achieving SA is to perceive the status, attributes, and dynamics of relevant elements in the environment. For each domain and job type, the SA requirements are quite different. A pilot needs to perceive important elements such as other aircraft, terrain, system status and warning lights along with their relevant characteristics. In the cockpit, just keeping up with all of the relevant system and flight data, other aircraft, and navigational data can be quite taxing. An army officer needs to detect enemy, civilian and friendly positions and actions, terrain features, obstacles, and weather. An air traffic controller or automobile driver has a different set of information needed for situation awareness.

Perception of information may come through visual, auditory, tactile, taste, or olfactory senses, or a combination. For instance, a wine maker may collect very critical information on the status of the fermentation process through taste and smell as well as through visual inspection. A physician uses all the senses and information available in assessing the health of a patient. These cues may be very subtle. A trained cardiologist can hear minute differences in the rhythm of a heartbeat and can see significant patterns in an ECG printout that the untrained observer would miss. An experienced pilot can know something is wrong just by hearing the pitch of the engine or seeing the pattern of lights on an air field. In many complex systems, a strong emphasis is placed on the electronic displays and read-outs that are provided, but the reality is that much of Level 1 SA also comes from the individual directly perceiving the environment—looking out the window or feeling the vibration. Verbal and nonverbal communications with others form an additional information source that is drawn upon and contributes to Level 1 SA.

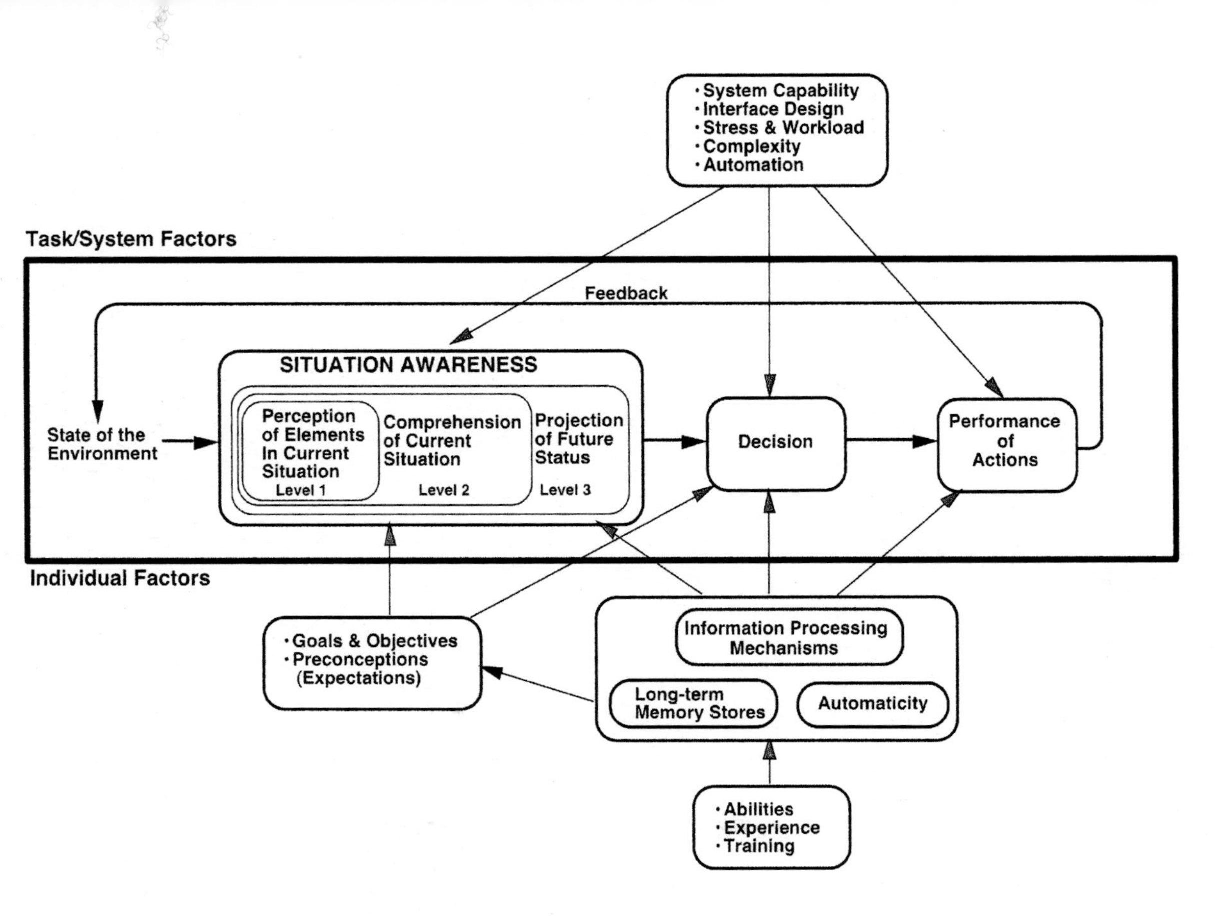

Figure 2.1 Model of situation awareness in dynamic decision making (from Endsley, 1995c). Reprinted with permission from *Human Factors*, Vol. 37, No. 1, 1995.

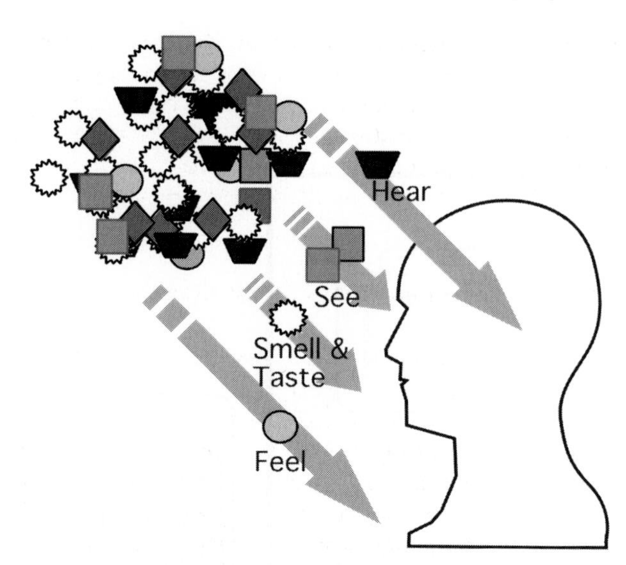

Figure 2.2 Level 1 SA – Perception of needed data.

Each of these sources of information is associated with different levels of reliability. Confidence in information (based on the sensor, organization, or individual providing it), as well as the information itself, forms a critical part of Level 1 SA in most domains.

In many domains it can be quite challenging just to detect all the needed Level 1 data. In military operations it may often be difficult to assess all the needed aspects of the situation due to obscured vision, noise, smoke, confusion, and the dynamics of a rapidly changing situation. The army officer must try to determine the current state of affairs while enemy forces are actively working to withhold that information or to supply false information. In the aviation domain, runway markings may be poor, and relevant information may not have been passed along to the pilot. In very complex systems, such as power plants or advanced aircraft, there is a great deal of competing information to wade through, making the perception of needed information quite challenging. In almost all of these domains, a great deal of data is constantly vying for the person's attention (Figure 2.2).

Most problems with SA in the aviation domain occur at Level 1. Jones and Endsley (1996) found that 76% of SA errors in pilots were related to not perceiving needed information. In some cases (roughly two-fifths) this occurred because the needed information was not provided to the person who needed it or was not provided clearly due to system limitations or shortcomings. For example, the runway lines had become faded and obscured, a gauge was cloudy, or static obscured a radio transmission. In some cases (roughly one-fifth), they did detect the needed information, but then later forgot about it after they took in other new information. In other cases (approximately one-third), all the information was present, but key information was not detected. This may have happened when they were distracted by outside factors (e.g., a phone call or nonwork-related conversation), but more often when they were dealing with other information relevant to competing tasks within their job. In some cases, they may not have brought up the window or display where the information they needed was present because they were viewing information on other displays. Designing for SA means ensuring that the necessary information is obtained by the system and presented in a way that makes it easily processed by the system users who can have many competing pieces of information contending for their attention.

2.1.2 Level 2 SA: Comprehension of the current situation

The second step in achieving good SA is understanding what the data and cues perceived mean in relation to relevant goals and objectives. Comprehension (Level 2 SA) is based on a synthesis of disjointed Level 1 elements, and a comparison of that information to one's goals (Figure 2.3). It involves integrating many pieces of data to form information, and prioritizing that combined information's importance and meaning as it relates to achieving the present goals. Level 2 SA is analogous to having a high level of reading comprehension as opposed to just reading words.

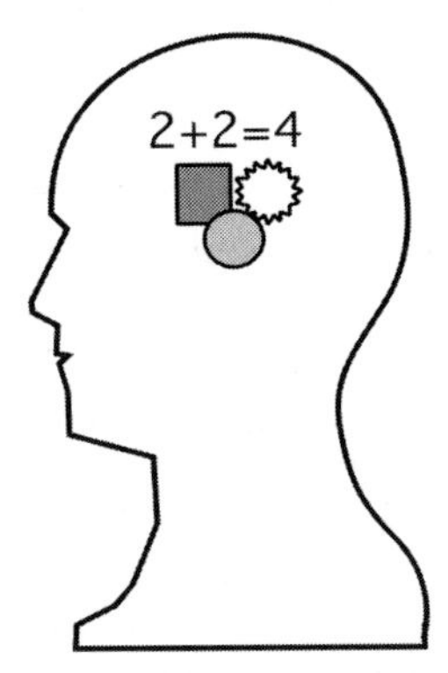

Figure 2.3 Level 2 SA – Comprehension of information.

Consider the driver approaching an intersection. The driver sees a yellow light, so she understands she needs to proceed with caution, based on the distance from the intersection. Her perception of the rate of deceleration of the car in front of her allows her to determine whether it is stopping or proceeding through the intersection and its impact on the rate of closure with that car. The driver's understanding of how the situation impacts on her goals defines what is needed to have Level 2 SA.

Upon seeing warning lights indicating a problem during takeoff, a pilot must quickly determine the seriousness of the problem in terms of the immediate air worthiness of the aircraft and combine this with knowledge of the amount of runway remaining in order to know whether it is an abort situation. A novice pilot may be capable of achieving the same Level 1 SA as more experienced pilots, but may fall far short of being able to integrate various data elements along with pertinent goals in order to comprehend the situation as well.

Level 2 SA for the military commander may involve comprehending that a report of movement in a given location means that enemy troops are massing nearby. Or it may mean seeing vehicle tracks along the road and from that ascertaining what types of troops and units are ahead of the officer's own troops.

By understanding the importance of the pieces of data, the individual with Level 2 SA has associated a specific goal-related meaning and significance to the information at hand. The importance of goals and how they affect SA will be discussed more fully later in this chapter.

Approximately 19% of SA errors in aviation involve problems with Level 2 SA (Jones & Endsley, 1996). In these cases people are able to see or hear the necessary data (Level 1 SA), but are not able to correctly understand the meaning of that information. For example, a pilot may have realized his aircraft's altitude was at 10 000 feet, but did not realize that the clearance to the terrain below was insufficient or that he had deviated from the level assigned by air traffic control. Developing understanding from the many pieces of data perceived is actually fairly demanding and requires a good knowledge base or mental model in order to put together and interpret disparate pieces of data. A novice, or someone in a new type of situation, may not have these knowledge bases to draw upon, and therefore will be at a distinct disadvantage when it comes to developing Level 2 SA. We will talk more about the role of mental models later on in this chapter.

2.1.3 Level 3 SA: Projection of future status

Once the person knows what the elements are and what they mean in relation to the current goal, it is the ability to predict what those elements will do in the future (at least in the short term) that constitutes Level 3 SA. A person can only achieve Level 3 SA by having a good understanding of the situation (Level 2 SA) and the functioning and dynamics of the system they are working with (Figure 2.4).

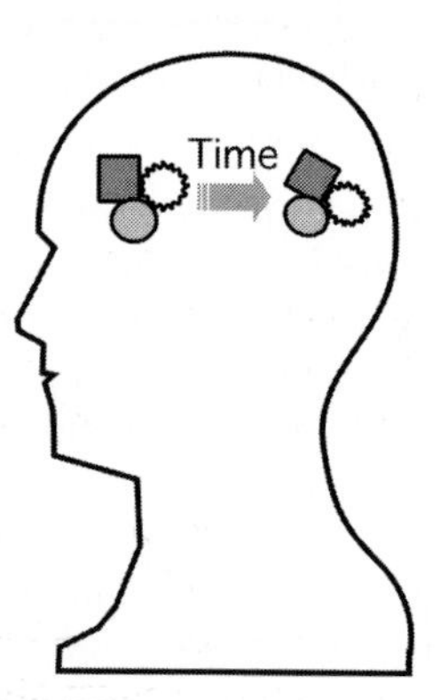

Figure 2.4 Level 3 SA – Projection of future status.

With Level 3 SA, a driver knows that if she proceeds into the intersection, she is likely to be struck by cars on the crossing street. This projection allows her to be proactive in making decisions. Army commanders can project in which direction enemy troops will approach and the likely effects of their own actions based on the Level 2 SA they have developed. Pilots and air traffic controllers work actively to project the movements of other aircraft and anticipate problems well in advance.

The use of current situation understanding to form projections requires a very good understanding of the domain (a highly developed mental model) and can be quite demanding mentally. Experts in many domains devote significant amounts of their time to forming Level 3 SA, taking advantage of any downtime to develop these projections. By constantly projecting ahead, they are able to develop a ready set of strategies and responses to events. This allows them to be proactive, avoiding many undesirable situations, and also very fast to respond when various events do occur.

A failure to accurately project (form Level 3 SA) from Level 2 SA may be due to insufficient mental resources (if the person is overloaded with other information processing, for example), or due to insufficient knowledge of the domain. Often it is also due to overprojecting current trends, for example, assuming that an aircraft will continue to descend at its current rate instead of flattening its descent rate. Only 6% of SA errors were found to fall into this category in an examination of SA errors in aviation (Jones & Endsley, 1996). This is probably due to significant difficulties in obtaining Level 1 and Level 2 SA in this domain, rather than any ease in developing good Level 3 SA. Without sufficient expertise or well-designed information systems and user interfaces, people may fail at the early stages of SA, never progressing to Level 3.

2.2 TIME AS A PART OF SA

Both the perception of time and the temporal dynamics of various elements play an important role in SA. Time, in general, has been found to play a significant part in SA in many domains (Endsley, 1993b; Endsley, Farley, Jones, Midkiff, & Hansman, 1998; Endsley & Kiris, 1994c; Endsley & Robertson, 1996a; Endsley & Rodgers, 1994). An often critical part of SA is understanding how much time is available until some event occurs or some action must be taken.

Operators in many domains filter the parts of the world (or situation) that are of interest to them based not only on space (how far away some element is), but also how soon that element will have an impact on their goals and tasks. Time is a strong part of Level 2 SA (comprehension) and Level 3 SA (projection of future events).

The dynamic aspect of real-world situations is another important temporal aspect of SA. An understanding of the rate at which information is changing allows for projection of future situations (Endsley, 1988; 1995c). Since the situation is always changing, the person's situation awareness must also constantly change or become outdated and thus inaccurate. In highly dynamic environments, this forces the human operator to adapt many cognitive strategies for maintaining SA.

2.3 SITUATION AWARENESS AS A PRODUCT OF THE PROCESS

People have a number of different cognitive mechanisms upon which SA is built in their working memory. While the things that are important for SA in any given domain are quite different, the basic cognitive processes are quite consistent across domains as they are a function of the mental equipment that all people possess. Situation awareness is the product of these processes (Figure 2.5).

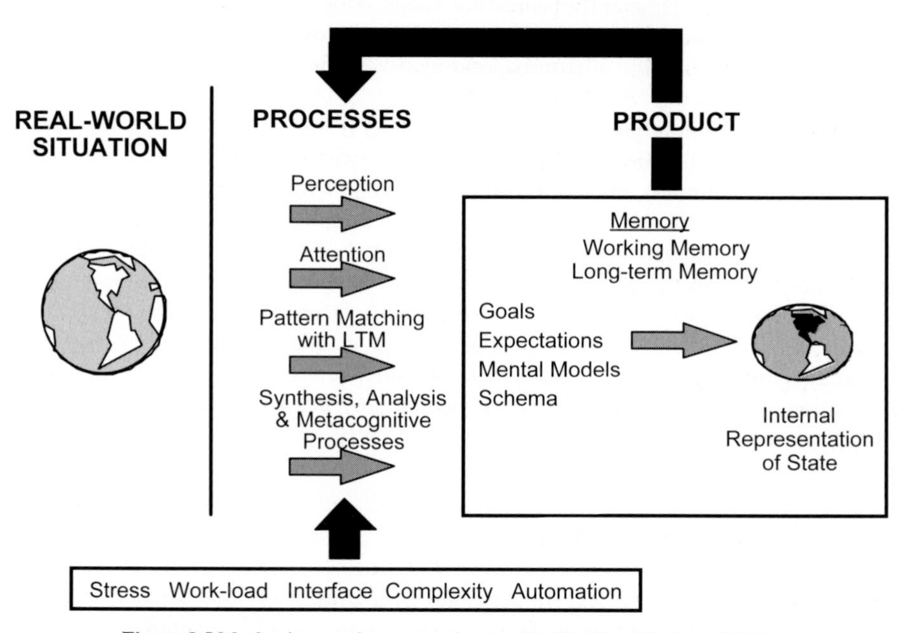

Figure 2.5 Mechanisms and processes involved in SA (from Endsley, 2000b).

A person's SA will in turn have an effect on what information is searched out and attended to, with product affecting process in a circular fashion. By developing a good understanding of how people develop SA and the challenges and limitations on that process, strategies for designing to enhance SA will be more apparent. We'll look in more detail at each of the processes and mechanisms that are important for developing SA (Endsley, 1995c).

2.4 PERCEPTION AND ATTENTION

Objects and their characteristics are initially processed simultaneously through pre-attentive sensory stores. People cannot attend to all information at once, however. A person's ability to perceive multiple items simultaneously is limited by their attention, which is finite. This, in turn, greatly limits the amount of SA an individual can achieve. For example, when driving down the road, a driver might monitor the traffic ahead, the music on the radio, and perhaps the feeling of bumpiness of the road. However, gauges, conversations, and traffic behind the driver may not all be perceived simultaneously and therefore are not attended to as often. The driver must periodically scan for this information. Those elements requiring frequent attention due to their importance or rate of change are typically monitored most often. For example, the speed gauge will be monitored more frequently than the fuel gauge in an automobile.

While some information can be more easily processed simultaneously (for instance, audio and visual information), it is difficult to process other types of information simultaneously. If information comes through the same modality (visual, auditory, tactile, or smell), draws on the same resources (e.g., central processing), or requires the same response mechanism (e.g., speaking or writing), it is more difficult to attend to simultaneously (Wickens, 1992). The limit of how many elements one person can pay attention to at one time affects how much information a person can process, forming a central bottleneck for SA.

2.5 WORKING MEMORY

Working memory and long-term memory play essential roles in helping achieve SA. Individuals can store information in their working memory on a temporary basis. Only a limited amount of unrelated information can be held and manipulated in working memory (7 plus or minus 2 'chunks') and a person must actively work to keep information there or it will decay.

Newly perceived information is combined with existing knowledge in working memory to create a new or updated mental picture of the changing situation. These pieces of information are also processed and used to create projections of what may happen in the future. These projections, in turn, help a person decide what actions to take as a result. All of this must take place in working memory. The combination of processing information to achieve high levels of SA and deciding on future actions can tax working memory to a great extent. Working memory is very limited, and forms the second major bottleneck for SA (Fracker, 1987).

2.6 MENTAL MODELS, SCHEMA, AND SCRIPTS

Luckily, working memory is not the only form of memory that individuals possess. Long-term memory actually allows individuals to work around the limitations that working memory places on SA. Long-term memory structures known as schema and mental models play a significant role in improving a person's SA (Figure 2.6). Mental models were defined by Rouse and Morris (1985) as '*mechanisms whereby humans are able to generate descriptions of system purpose and form, explanations of system functioning and observed system states, and predictions of future states.*' Mental models are complex structures people use to model the behavior of specific systems, whether the systems are related to air traffic control, medical procedures and surgeries, or highway driving.

A *mental model* is a systematic understanding of how something works. For example, people can develop a mental model of an appliance that would include an understanding of where information is located, the order in which buttons need to be pressed for certain desired actions, and expected system behaviors based on user inputs. These mental models may be formed for not only physical objects, say an engine or a power plant, but also for organizational systems. Drivers develop a mental model of how traffic is expected to flow and how other cars behave; they will only drive on one side of the road, and will yield to oncoming traffic at intersections, for example. Pilots, air traffic controllers, football players, and university students similarly form mental models of how their systems work, including not only formal rules, but also very detailed considerations of the expected behaviors of others.

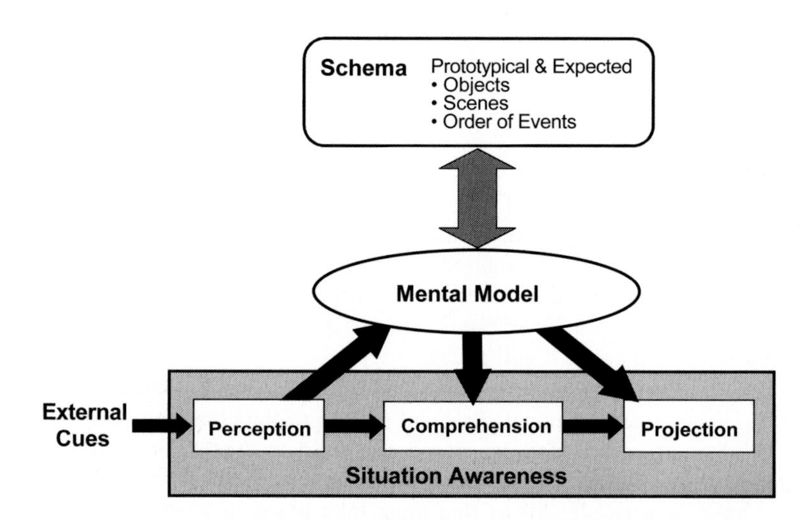

Figure 2.6 Schema, mental models and situation awareness (from Jones & Endsley, 2000b). Reprinted with permission from *Human Factors*, Vol. 42, No. 3, 2000. Copyright 2000 by the Human Factors and Ergonomics Society. All rights reserved.

Mental models are based on both semantic knowledge and system knowledge. Semantic knowledge typically includes knowing *what* as opposed to *how*. It might include, for example, knowing the names of the capitals of all the states or the names of various species of trees. System knowledge, which is critical for mental models, is formed not only on semantic knowledge, but also by understanding how these systems function.

Physicians, for instance, have a well-developed understanding of how the human body works that allows them to understand how a particular cardio-pulmonary resuscitation (CPR) maneuver will affect the lungs and heart and what would occur based on various abnormalities in the body. They can accurately project things they have never even seen, based on well-developed mental models of the body and pulmonary functioning.

Mental models help a person determine what information is important to attend to (e.g., if temperature is changing they should look to see what is happening to pressure), and help to form expectations (if temperature is dropping and pressure is constant, volume should also be dropping). Mental models also enable the higher levels of SA (comprehension and projection) without straining working memory's capabilities. Without a mental model, a person would be very poor at understanding what is happening (why are the lips turning blue even though CPR is being administered?), or what is likely to happen in future (CPR administered to a person with a wound to a main artery will lead to additional blood loss). Mental models are the key enablers of Levels 2 and 3 SA.

While it is possible for a person without a good mental model to mentally work out understanding and projection, it can be very difficult and quite mentally demanding, frequently exceeding what is possible in very dynamic situations. For example, to the uninitiated an air traffic control display can be quite daunting. It will take considerable time to understand how even two aircraft might interact in the airspace. The experienced air traffic controller, however, equipped with detailed knowledge about the characteristics and capabilities of the particular aircraft involved, as well as a detailed knowledge of how different types of traffic tend to flow within a particular airspace sector, can sort out and separate as many as 20 or 30 aircraft with relative ease. All in all, processing novel cues in situations where good mental models don't exist taxes limited working memory, and makes achieving SA much harder and more prone to error.

Mental models also provide default information (expected characteristics of elements) that helps form higher levels of SA even when needed data is missing or incomplete. Thus, an experienced soldier can reasonably fill in the speed a particular convoy is traveling just by knowing the types of vehicles in the convoy. Default information can provide an important coping mechanism for people in forming SA in many situations where they do not have enough information or cannot acquire all the information they need.

Schema or schemata (the plural of schema) are prototypical states of the mental model that provide further processing efficiencies. A doctor has a schema of what people with heart disease will look like in terms of the symptoms and physical characteristics. A different schema will exist that defines what a person with the flu or encephalitis will look like. These schemata or prototypical patterns are retained in memory. People use a process called pattern matching to link cues taken in from the current situation to schemata to pick the best match from those available.

Schemata allow the person to very rapidly classify and understand information perceived (i.e., 'I've seen this situation before!'). Schemata function like shortcuts. People don't have to exercise the mental model each time in order to understand and project based on situation information. The schema basically provides comprehension and projection as a single step, already preloaded in memory for well-known and recognized classes of situations. Therefore, they are very important mechanisms that individuals can use to gain high levels of SA very rapidly, even when confronted with large volumes of data.

Schemata are formed by direct experience, or can be formed vicariously based on reading or hearing about similar cases. Doctors and pilots commonly pass on war stories or case histories to help build this type of knowledge store. In order for schemata to be effective, the individual must be good at pattern matching between situation cues and schemata in memory (and some people appear to be better at this than others).

One significant advantage of mental models and schemata is that the present situation need not be exactly like a previous one in order to recognize it, because people can use *categorization mapping* (a best fit between the characteristics of the situation and the characteristics of known categories or prototypical schemata). This ability is why people can generalize from their experiences to new ones. For example, an army captain may have led his troops into mock battles while on maneuvers, practicing tactics and battle plans in the woods of southern Georgia. But when he leads his soldiers into a real battle in another country, the reality is frequently different from the practice. Thus, he will use pattern matching to find the best match between existing conditions and the learned classes of situations in memory, even if they are not exactly the same, to help develop comprehension and projection. To the degree that this match is close enough on key parameters, comprehension and projection will be good. If it is not, however, comprehension and projection may be off.

As a final aid to SA, people also may develop *scripts* associated with each schema. Scripts are set sequences of actions on what to do in each case that a schema represents. These scripts may have been developed through experience, or may be dictated within the domain. For instance, a doctor has a well-developed script of what to do for a person with a particular disease. A pilot has a well-developed set of procedures to follow for different types of situations. Military commanders follow doctrine. While what to do may be well-defined, the individual must still use mental models and schema to understand the situation well enough to know when to apply a particular procedure or script. When these scripts are present, the load on working memory to decide on the best course of action for a particular situation is even further reduced.

The role of these various mechanisms can be illustrated in another example. An auto mechanic listens to a car and makes various observations or tests when it is brought in for service. He uses his mental model of how the car (based on the year, make, model, etc.) should sound, feel, and behave to interpret what he hears and sees and to direct what he should examine. When he hears something amiss (perhaps a whine, or a clinking), the schema (linked to the mental model) that best matches that cue is activated. This schema provides a classification for the situation that allows him to correctly understand why the car is making that particular noise, its origin, and its impact on the proper functioning and safety of the vehicle (Level 2 SA) and what is likely to happen if the car continues to be driven (Level 3 SA). He also invokes a script from memory associated with that schema that tells him what repairs are needed for that type of problem. This whole process can occur in seconds due to the presence of the mental model. Without it, it could take hours or days for a person to figure out the state of the vehicle.

Considering that mental models, schema, and scripts are developed over time with experience in a particular domain, less experienced individuals will need to spend significantly more mental processing time sorting out what is happening to develop SA and will likely overload working memory, leading to significant SA gaps. Designing for novice operators (whether pilots, railroad engineers, or surgeons) requires designing to give them more assistance with these processes than is required for experienced operators. Designing for experienced operators requires that we facilitate the great utility of these mechanisms by enhancing the pattern-matching process.

In summary, mental models, schema, and scripts constitute very important mechanisms for SA, providing a significant advantage to those with the experience to have developed them. When people have a fully developed mental model for a specific system, or domain, the mental model provides three things:

- dynamic direction of attention to critical environmental cues,
- expectations regarding future states of the environment based on the projection mechanism of the model, and
- a direct, single-step link between recognized situation classifications and typical actions, enabling very rapid decisions to be made.

2.7 GOALS AND SA

Individuals know what they are trying to accomplish within their task or job, i.e., the specific goals for the tasks they perform. In what is known as *top-down information processing* (Casson, 1983), these goals help determine which environmental elements to pay attention to as people perform their tasks. For example, a doctor seeks information from the patient regarding the symptoms the patient is experiencing and conducts a detailed examination of the patient's physical characteristics. By actively gathering information, the doctor can work to accomplish the goal of alleviating physical problems. A driver seeks information on road conditions and traffic problems that are relevant to the goal of reaching a particular destination. These are examples of goal-driven (or top-down) processes and are common forms of information processing. Specific information is searched for in the environment that is relevant to the individual's goals.

Conversely, data-driven (or bottom-up) processing is when information 'catches' a person's attention, completely independently of their goals. Data-driven processing is when incoming information is processed based on the priority of its inherent perceptual characteristics. For example, flashing icons on a computer screen will grab a person's attention, even though they may be trying to read other information. Loud noises (e.g., alarms) and certain colors (e.g., red) will similarly tend to grab attention. These cues, if indicative of something important, may lead people to reprioritize their goals. For instance, if a driver sees an accident about to happen, his goal will change from navigating to avoiding the accident. A pilot may abort the goal of landing in order to avoid another plane that appears on the flight path.

Goal-driven processing is when a person's SA is driven by current goals and expectations which affect how attention is directed, how information is perceived, and how it is interpreted. The person's goals and plans dictate what aspects of the surroundings are noticed. The perceived information is then blended and interpreted in light of these goals to form Level 2 SA. A pilot, for example, will focus on the landing instrumentation, out-the-window view, and directions from air traffic control when landing an aircraft. The information perceived (e.g., airspeed of 180 knots) will have meaning in relation to the current goal (e.g., keep the airspeed between 170 and 190 knots). The pilot will not focus on takeoff information or maintenance requirements, keeping attention and thus SA devoted to the goal of landing the aircraft safely. A different goal (e.g., go-around) would lead to an entirely different interpretation of the same data.

If the pilot stays locked into a given goal, however, and is not responsive to cues that indicate a new goal is more important (e.g., a mechanical failure or an aircraft incursion on the runway), a significant SA failure can occur. People must be able to quickly switch between data-driven and goal-driven processing in order to perform successfully at their jobs. The pilot must be able to land the airplane, but if the path is in jeopardy due to an incursion of another airplane in that airspace, the pilot will need to alter the goal of landing to one of avoiding a collision. Thus the pilot will switch between goal-driven processing (trying to land the airplane) to data-driven processing that changed the goal (aborting the landing to avoid a collision), to goal-directed processing associated with the new goal.

Because the active goal(s) has an effect on SA, the effective juggling of these multiple, often competing goals as priorities change is very important for SA. Each job environment contains multiple goals, which change in priority over a period of time and which may be in direct competition (e.g., production vs. safety). Generally, a person only works on a selected subset of these goals at one time. As shown in Figure 2.7, SA will help determine which goal(s) should be active (have the highest priority). The goals selected as active or primary determine several critical functions.

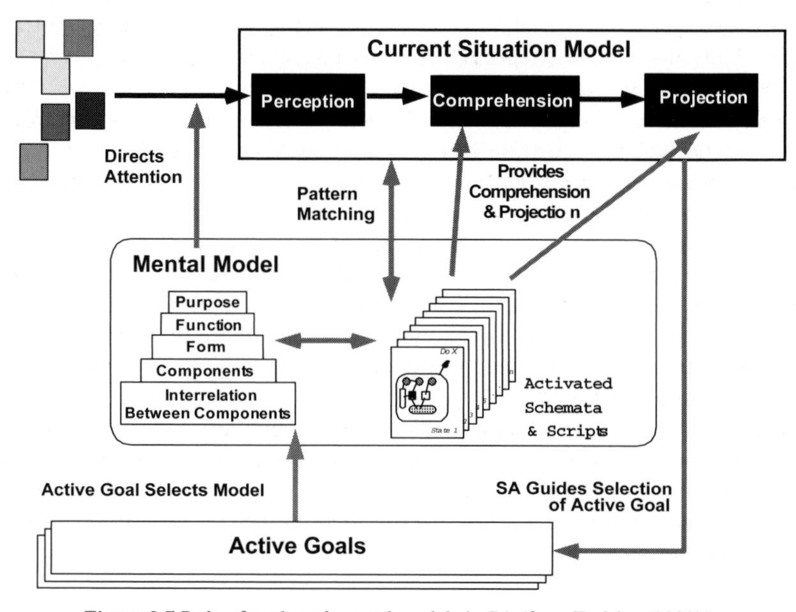

Figure 2.7 Role of goals and mental models in SA (from Endsley, 2000b).

First, the active goal guides which mental models are chosen. For example, an Army commander's goal may be to ambush a supply caravan, so he'll choose the ambush mental model. A businesswoman may have a goal of increasing the revenue of one department for the next year, so she'll activate her mental model of a marketing plan for that department. Because the mental model plays an important role in directing attention to cues and information in the environment and interpreting those cues, this selection is critical.

Second, prioritizing goals is critical to successfully achieving SA. If people focus on the wrong goals, they may not be receptive to the correct information, or may not seek out needed information at all. In many cases, they may actually be working toward achieving a goal that is not so important, while all the time thinking they are achieving what the task at hand calls for. For example, an air traffic controller who is working on the goal of providing assistance to a pilot who asks for a new clearance may miss cues that he should be working on the goal of separating two other aircraft. While working on the wrong goal, he neglects to scan the display for other aircraft not related to that goal. If one is not working on the right goals, important information may not be perceived or properly interpreted. For example, if a doctor is working on the goal of obtaining tests for a patient for a particular disease, she may miss information associated with other goals (such as insuring the patient has home care).

Third, both the selected goal(s) and associated mental models will be used to interpret and integrate perceived information in order to understand what the information means. The specific goal helps determine whether the information is important. A pilot who realizes her altitude is at 10 000 feet must interpret this information in light of the goal for the aircraft to be at 11 000 feet to comply with air traffic controller instructions or the goal to clear all terrain. The goal provides a basis for understanding the importance or significance of information perceived.

Alternating between bottom-up data-driven and top-down goal-directed processing of information is one of the vital mechanisms supporting SA. If this process fails, SA can be severely damaged. Being locked into only goal-driven processing can result in significant SA failures. If the person does not process the information needed to make the proper goal selection, he or she will often not notice other needed information (i.e., *attentional narrowing*). Conversely, relying on data-driven processing alone is highly inefficient and overly taxing. The individual will be awash with many competing bits of data. To be successful, people must be receptive to critical cues that will allow them to efficiently prioritize competing goals and information (bottom-up processing), while simultaneously using goal-driven processing to direct attention and to process the most important information (top-down processing). This constant interplay between these two processing modes is critical to SA.

By understanding how a person—whether a pilot, a driver, a physician, or a power plant operator—selects and uses goals, designers can better understand how the information perceived has meaning to people in specific situations. Without understanding effect of goals on SA, the information presented to a person through a system has no meaning.

2.8 EXPECTATIONS

Like goals, a person's predetermined expectations for a given situation can change the way SA is formed. People often have ideas of what they expect to see, hear, or taste in a given situation. Expectations are based on mental models, prior experiences, instructions, and communication from other sources. Pilots, for example, often get detailed preflight briefings that they use to form expectations of what they will encounter inflight. In a manufacturing plant, a bill of lading forms expectations regarding what is to be found in a particular shipment that will be inspected.

Expectations guide how attention is directed and how the person absorbs the information perceived. For example, if drivers expect a freeway on-ramp to be on one side of the road, they will look for it there. This is efficient because they don't have to look everywhere for the needed information. Expectations also provide a vital function as a shortcut in mental processing in information perceived. People have developed this efficiency mechanism to enable them to process the large volume of data in the world. Without this shortcut, working memory would be overloaded in acquiring and interpreting information from the world around us. Because people can help the process of interpreting the world by using expectations, they have more mental resources left for other cognitively challenging processes.

Expectations are a double-edged sword, however. When wrong, false expectations can also lead to misinterpreting the data. If another road is in a location near where the on-ramp is expected, people may 'mis-fit' that road to their expectations and falsely believe they are turning onto the on-ramp. In many cases, important cues will be missed if they are not where they are expected to be or if other objects can be fitted to those expectations.

For instance, in certain displays that showed pictures of outdoor scenes, expected locations of targets were automatically cued on a computer screen (via a circle or pointer). People searching for those targets were much more likely to misidentify other items in these cued locations as targets when the real targets were actually elsewhere, as compared to their performance with displays where expected locations were not indicated (Yeh & Wickens, 2001). Pilots have been known to go seriously off course by incorrectly matching landmarks to their expectations with sometimes tragic results (Endsley, 1995b). Thus, expectations can be good or bad for SA. They help people to find things more quickly when they are right, but can lead them astray when they are wrong.

2.9 AUTOMATICITY AND SA

Experience contributes significantly to SA by allowing people to develop the mental models, schema, and goal-directed processing that are critical for SA in most domains. Experience can also lead to a certain level of automaticity in mental processing, however, where a person's behavior and reactions become somewhat automatic. The pattern-recognition/action-selection sequence (stimulus/response pairing) can become routinized to the point of becoming automatic (Figure 2.8). This can create a positive effect on SA as it frees up mental effort for more demanding tasks. For instance, a person riding a bike is often unaware of where her feet are in their cycle and can instead concentrate on where she wants to go.

We often think of automaticity in terms of physical tasks. Automaticity for mental tasks as well as physical tasks is also possible (Logan, 1988). In this case, incoming cues (e.g., red traffic lights) are directly connected to actions (e.g., pressing the brakes) in a tightly coupled sequence. Automaticity can be observed in many tasks from driving a car to flying an aircraft. Automaticity is advantageous to SA because it requires very little conscious attention to process information, freeing up mental resources for other tasks. This is particularly true for automaticity of physical tasks (like steering the car or balancing on the bicycle).

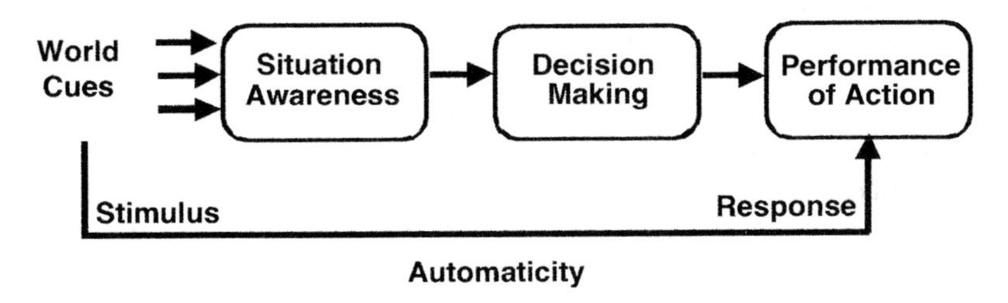

Figure 2.8 Automaticity in cognitive processes (from Endsley, 2000b).

It can also be a significant problem for SA, however, when automaticity of mental tasks is involved. Under these circumstances, information outside the scope of the routinized sequence might not be attended to, thus SA suffers if that information is important. For example, when a new stop sign is erected on a well-traveled route home, many drivers will drive right past it, not even noticing this new and significant piece of information. Checklists are heavily emphasized in the aviation domain to help protect against the shortcomings of processes that can become too automatic and thus prone to certain errors.

2.10 SUMMARY

To summarize this model (Figure 2.1), SA is largely influenced by the limitations of human working memory and attention in complex and dynamic systems. People can use several mechanisms to help overcome these limitations, including the use of goals and goal-directed processing, automaticity, mental models, and pattern matching to schemata of known situations, and expectations. The development of these mechanisms comes with experience and training in particular domains. Good SA in one domain, therefore, rarely translates directly into good SA in another. These powerful mechanisms make high levels of SA possible in very demanding situations, however, they also come with some costs—a built-in propensity to err in some situations.

Situation awareness can vary widely from one person to another, even given the same information and circumstances. As much as a tenfold difference in SA has been noted between pilots, for instance (Endsley & Bolstad, 1994). Conversely, the degree to which any person can achieve a high level of SA is greatly affected by the environment and by the design of the system they operate. Both of these classes of influences are significant in determining how good SA will be in a given situation. We will next describe a few key challenges, both within the person and within the environment, that act together to undermine SA in many different domains and situations.

CHAPTER THREE

SA Demons: The Enemies of Situation Awareness

Building and maintaining SA can be a difficult process for people in many different jobs and environments. Pilots report that the majority of their time is generally spent trying to ensure that their mental picture of what is happening is current and correct. The same can be said for people in many other domains where systems are complex and there is a great deal of information to keep up with, where information changes rapidly, and where it is hard to obtain.

The reason why good SA is so challenging can be laid to rest on both features of the human information processing system and features of complex domains that interact to form what we will call SA demons. SA demons are factors that work to undermine SA in many systems and environments. By bringing these demons to light, we will take the first step toward building a foundation for SA-Oriented Design. We will discuss eight major SA demons:

- Attentional Tunneling
- Requisite Memory Trap
- Workload, Anxiety, Fatigue, and Other Stressors (WAFOS)
- Data Overload
- Misplaced Salience
- Complexity Creep
- Errant Mental Models
- Out-of-the-loop Syndrome

3.1 ATTENTIONAL TUNNELING

Situation awareness within complex domains involves being aware of what is happening across many aspects of the environment. Pilots must instantaneously keep up with where they are in space, the status of the aircraft systems, the effect of turbulence on passenger comfort and safety, other traffic around them, and air traffic control directives and clearances, to name a few elements. Air traffic controllers must concurrently monitor separation between many different pairs of aircraft (there may be as many as 30 or 40 aircraft under their control at any one time), process the information required to manage aircraft flows and pilot requests, and keep up with aircraft seeking to enter or leave their sector. A stock car driver must monitor the engine status, the fuel status, the other race cars on the track, and the pit crew signals.

Successful SA is highly dependent upon constantly juggling different aspects of the environment. Sometimes multiple pieces of information are simultaneously processed in order to perform one or more tasks; for example, monitoring the road while driving and monitoring the radio for traffic information. This is called *attention sharing*. People face numerous bottlenecks in attention sharing, however, particularly within a single modality, like vision or sound, and thus it can only occur to a limited extent (Wickens, 1992).

As they cannot access all the needed information simultaneously, people also set up systematic scans or information sampling strategies to insure that they stay up-to-date in their knowledge of what is happening. A scan across needed information may occur over a period of seconds or minutes, as in the case of the pilots and the air traffic controllers discussed here, or it may take place over a period of hours, as in the case of a power plant operator required to log the status of hundreds of different systems several times over the course of the day.

In all of these cases, and for systems with any level of complexity, good SA is highly dependent on switching attention between different information sources. Unfortunately, people often can get trapped in a phenomenon called attentional narrowing or tunneling (Baddeley, 1972; Bartlett, 1943; Broadbent, 1954). When succumbing to attentional tunneling, they lock in on certain aspects or features of the environment they are trying to process, and will either intentionally or inadvertently drop their scanning behavior (Figure 3.1). In this case, their SA may be very good on the part of the environment they are concentrating on, but will quickly become outdated on the aspects they have stopped attending to.

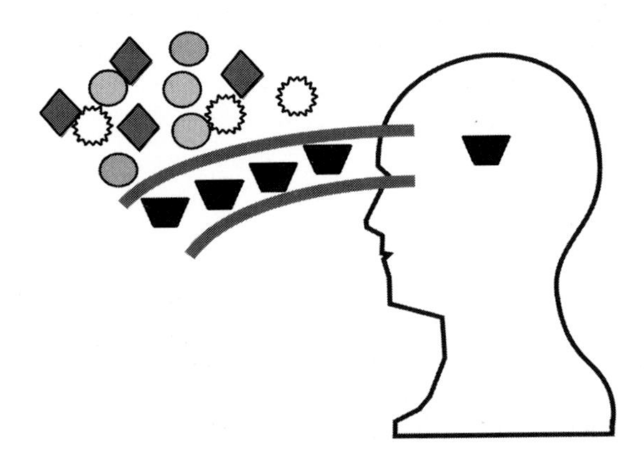

Figure 3.1 Attentional tunneling demon: fixating on one set of information to the exclusion of others.

In many cases, people will believe that this limited focus is okay, because the aspect of the situation they are attending to is most important in their minds. In other cases, they simply fixate on certain information and forget to reinstate their information scan. Either situation can result in a critical loss of SA. The reality is that keeping at least a high-level understanding of what is happening across the board is a prerequisite to being able to know that certain factors are indeed still more important than others. Otherwise it is often the neglected aspects of the situation that prove to be the fatal factors in loss of SA.

In the most well-known example of attentional tunneling, an Eastern Airlines aircraft crashed into the Florida Everglades, killing all aboard. All three crew members had become fixated on a problem with an indicator light and neglected to monitor the aircraft's flight path, which, as it turned out, was not correctly set in the autopilot (National Transportation Safety Board, 1973).

While the consequences are not always so severe, this problem is actually fairly common. The most common type of SA failure involves situations where all the needed information is present, yet is not attended to by the person monitoring the situation. In studying aircraft and air traffic control accidents and incidents, Jones and Endsley (1996) found that 35% of all SA errors fell into this category. While various factors contribute to this problem, most often the people were simply attending to other task-relevant information and had lost SA on important aspects of the situation.

Attentional tunneling is not just a problem in aviation. It must be guarded against in many other arenas. As we see a greater use of technology in automobiles (from cell phones to the use of computerized navigation systems), a significant problem with attentional tunneling is rearing its ugly head. Use of a cell phone while driving as much as quadruples the risk of an accident according to one study, and that appears to be independent of whether the phone is handheld or not (Redelmeier & Tibshirani, 1997). The problem is not a function of physical interference, but attentional distraction. Dynamically switching attention between these devices and the driving task can be a challenge. Similarly, the greater use of technologies, such as helmet-mounted displays, for soldiers may result in problems of attentional narrowing on displays to the neglect of what is going on around them. Future design efforts in these domains and others need to explicitly consider the effects of this attentional tunneling SA demon and take steps to counteract it.

3.2 REQUISITE MEMORY TRAP

Human memory remains central to SA. Here we are not referring to long-term memory (the ability to recall information or events from the distant past), but rather short-term or working memory. This can be thought of as a central repository where features of the current situation are brought together and processed into a meaningful picture of what is happening (fed by knowledge stored in long-term memory as well as current information taken in). This memory bank is essentially limited. Miller (1956) formalized the proposition that people can hold approximately seven plus or minus two *chunks* (related pieces) of information in working memory. This has significant implications for SA. While we can develop the ability to hold quite a bit of situational information in memory through the use of a process called *chunking*, in essence working memory is a limited cache for storing information. SA failures can result from insufficient space in that cache and from a natural decay of information in the cache over time. With experience, people do learn to condense or combine multiple pieces of information into more compact and easy to remember chunks. So, for instance, the air traffic controller does not need to track 30 separate aircraft, but perhaps five or six different groups of related aircraft, which is cognitively more manageable. The development of rich mental models of the environment over time contributes significantly to people's ability to form meaningful chunks of information for more efficient storage.

Even so, information does not stay in this memory cache indefinitely. Unless people actively work to keep it there (by repeating or revisiting the information, for example), it will rapidly fade from memory (Baddeley, 1986). This loss may occur as quickly as 20 to 30 seconds for abstract information (e.g., a phone number or aircraft call sign), or the information may remain accessible for a while longer if connected to other information or mental models in long-term memory (Endsley, 1990a).

With respect to SA, memory plays a critical role (Figure 3.2). Many features of the situation may need to reside in memory. As the person moves on to scan different information from the environment, previously accessed information must be remembered and combined with new information. Auditory information must also be remembered as it usually cannot be revisited the way visual displays can. Given the complexity and sheer volume of information required for SA in many systems, it is no wonder that memory limits create a significant SA bottleneck.

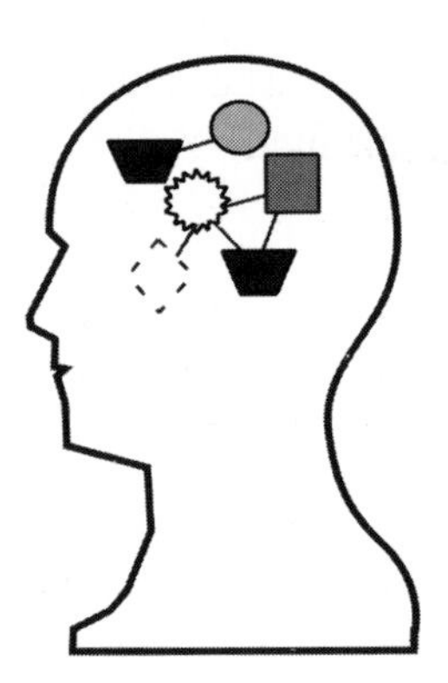

Figure 3.2 Requisite memory trap: relying on limited short-term memory.

In many cases, significant error can result from systems that rely heavily on a person's memory for performance. A major aircraft accident occurred at Los Angeles International Airport when a taxed air traffic controller forgot an aircraft had been moved to a runway and assigned another aircraft to land on the same runway. She could not see the runway and had to rely on memory for her picture of what was happening there (National Transportation Safety Board, 1991).

While it is easy to fault the individual in cases like this, a more appropriate response is to fault system designs that necessitate overreliance on a person's memory. Surprisingly, many systems do just this. Pilots must routinely remember complex air traffic control instructions and clearances, drivers attempt to remember verbal directions, mechanics are expected to remember tolerance limits and what other activities are taking place in the system, and military commanders must assimilate and remember where many different soldiers are on the battlefield and what they are doing based on a steady stream of radio transmissions. It is no wonder that SA is so sorely taxed in these situations because memory failures are bound to occur.

3.3 WORKLOAD, ANXIETY, FATIGUE, AND OTHER STRESSORS (WAFOS)

In many environments, SA is taxed by the conditions under which people must operate. These stressors can take several forms. In many situations people may be under a significant amount of stress or anxiety. This can occur on a battlefield or in the office. Understandably, stress or anxiety can be an issue when one's own well-being is at stake, but also when factors like self-esteem, career advancement, or high-consequence events are involved (i.e., where lives are at stake). Other significant psychological stress factors include time pressure, mental workload, and uncertainty.

Stressors can also be physical in nature. Many environments have high levels of noise or vibration, excessive heat or cold, or poor lighting. Physical fatigue and working against one's circadian rhythms can also be a major problem for many people. Pilots of long-haul aircraft, for instance, often fly for long durations and at night. Soldiers are routinely called upon to function on small amounts of sleep and after heavy physical exertion.

Each of these stressors can significantly strain SA (Figure 3.3). First, they can act to reduce an already limited working memory by using up a portion of it. There are essentially fewer cognitive resources available for processing and holding information in memory to form SA. As reliance on working memory can be a problem anyway, stressors such as these only exacerbate the problem. Second, people are less able to gather information efficiently under stress. They may pay less attention to peripheral information, become more disorganized in scanning information, and be more likely to succumb to attentional tunneling. People are more likely to arrive at a decision without taking into account all available information (termed *premature closure*). Stressors will undermine SA by making the entire process of taking in information less systematic and more error prone.

Clearly, these types of stressors can undermine SA in many ways and should be avoided or designed out of operational situations whenever feasible. Unfortunately, this is not always possible. A certain degree of personal risk will always be involved in combat, for example. In these cases it will be even more important to design systems to support effective, efficient intake of needed information to maintain high levels of SA.

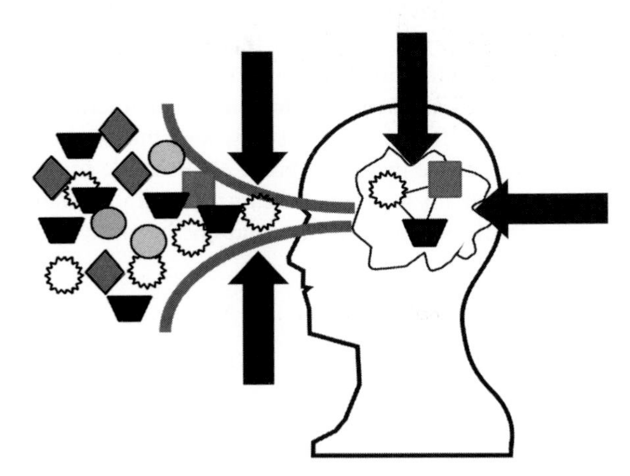

Figure 3.3 Stressor demons: reducing a person's capacity to process information.

3.4 DATA OVERLOAD

Data overload is a significant problem in many arenas. In these situations, the rapid rate at which data changes creates a need for information intake that quickly outpaces the ability of a person's sensory and cognitive system to supply that need. As people can only take in and process a limited amount of information at a time, significant lapses in SA can occur. The human brain becomes the bottleneck.

This demon, quite understandably, creates a significant challenge for SA. If there are more auditory or visual messages than can be processed, the person's SA will be quickly outdated or contain gaps, either of which can be a significant handicap in forming a mental picture of what is happening.

While it is easy to think of this problem as simply a natural occurrence that people are ill-suited to handle, in reality it is often a function of the way that data is processed, stored, and presented in many systems. To think in engineering terms, the problem is not one of volume, but of bandwidth—the bandwidth provided by a person's sensory and information processing mechanisms. While we cannot do much to dramatically alter the size of the pipeline, we can significantly affect the rate that data can flow through the pipeline (Figure 3.4).

Data that is jumbled and disorganized flows through the pipeline very slowly. Data that is presented in certain forms, streams of text for example, also move through the pipeline far more slowly than that which is presented graphically. By designing to enhance SA, significant problems with data overload can be eliminated or at the very least reduced.

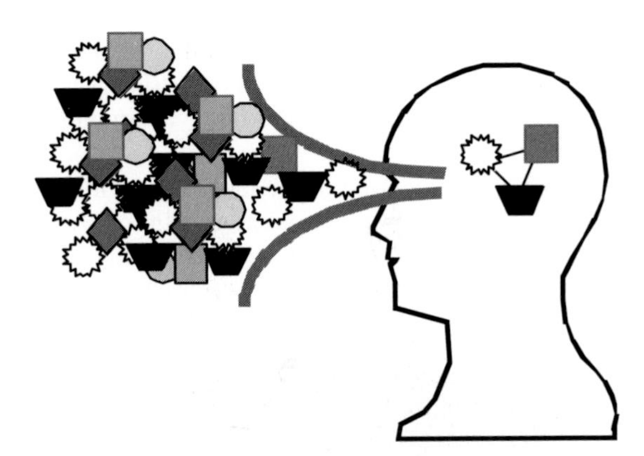

Figure 3.4 Data overload: overwhelming amounts of data can reduce SA.

3.5 MISPLACED SALIENCE

Many pieces of information vie for one's attention in the world. For a driver, this might include billboards, other drivers, roadway signs, pedestrians, dials and gauges, radio, passengers, cell phone conversations, and other onboard technologies. In many complex systems, the situation is analogous with numerous system displays, alarms, and radio or phone calls competing for attention.

People will generally attempt to seek out information relevant to their goals. For example, an automobile driver may search for a particular street sign among competing signs and objects in the world. Simultaneously, however, the driver's attention will be caught by information that is highly salient. *Salience*, the compellingness of certain forms of information, is largely determined by its physical characteristics. The perceptual system is simply more sensitive to certain signal characteristics than others. So, for instance, the color red, movement, and flashing lights are much more likely to catch one's attention than other features. Similarly, loud noises, larger shapes, and things that are physically nearer have the advantage in catching a person's attention. These are generally features that can be thought of as important for survival in evolutionary terms and to which the perceptual system is well adapted. Interestingly, some information content, such as hearing one's name or the word 'fire,' can also share similar salience characteristics.

These natural salient properties can be used to promote SA or to hinder it. When used carefully, properties like movement or color can be used to draw attention to critical and highly important information and are thus important tools for designing to enhance SA. Unfortunately, these tools are often overused or used inappropriately. If less important information is flashing on a display, for example, it will distract the person from more important information. Some air traffic control displays do this, flashing aircraft symbols that it believes are in conflict and drawing the controller's attention to them. If the aircraft are truly in conflict, this would be a good thing. But false alarms are common: such is the case when the controller has already taken actions to separate the aircraft or the aircraft are already scheduled to turn or climb at a certain point taking them out of conflict. In this case, the use of highly salient cues (flashing lights) creates unnecessary distractions that can act to degrade SA of the other information the controller is attempting to attend to.

In many systems, flashing lights, moving icons, and bright colors are overused. This creates the Las Vegas Strip phenomenon (Figure 3.5). With so much information drawing attention, it is difficult to process any of it well. The brain attempts to block out all the competing signals to attend to desired information, using significant cognitive resources in the process.

Figure 3.5 Misplaced salience.

While misplaced salience may be difficult to control in the natural world, in most engineered systems it is entirely in the hands of the designer. Unfortunately, in many systems there is a proliferation of lights, buzzers, alarms, and other signals that actively work to draw people's attention, frequently either misleading or overwhelming them. Less important information can be inadvertently made to look more important. For example, in one aircraft display a large ring of uncertainty was drawn around the location of aircraft symbols whose locations were determined from low reliability sensor data. This led to the unintended consequence of drawing the pilot's attention toward this less certain information and making it seem more important than other aircraft displayed that were based on more certain information and thus had much smaller circles around them, when the opposite effect was actually desired. Misplaced salience is an important SA demon to avoid in system design.

3.6 COMPLEXITY CREEP

Related to the demon of data overload is the demon of complexity creep. Complexity has run rampant in new system development. Many system designers have unwittingly unleashed complexity with the practice of feature escalation. Televisions, VCRs, even telephones, have so many features that it is very difficult for people to form and retain a clear mental model of how the system works. Studies indicate that only 20% of people can properly operate their VCR (Hoffberg, 1991; Verity, 1991). In consumer product usage, this can result in annoyance and frustration on the part of the consumer. In critical systems, it can result in tragic loss of life. Pilots, for instance, report significant problems in understanding what the automated flight management systems onboard their aircraft are doing and what they will do next (Wiener, 1989; Wiener & Curry, 1980). This problem continues to persist, even for pilots who have worked with these systems for a period of years (McClumpha & James, 1994).

At the root of this problem is that complexity makes it difficult for people to form sufficient internal representations of how these systems work. The more features, and the more complicated and branching the rules that govern a system's behavior, the greater the complexity (Figure 3.6).

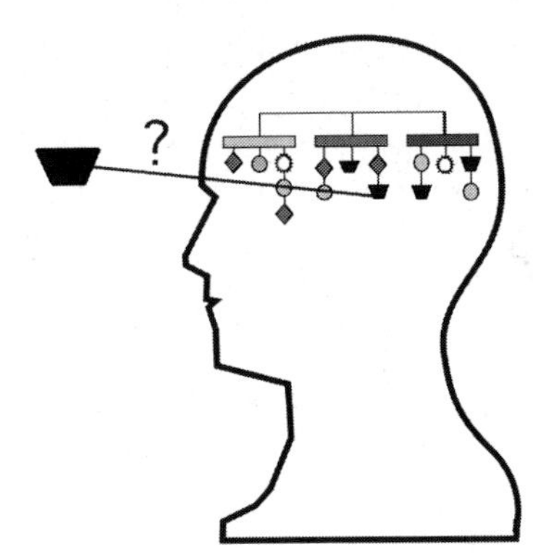

Figure 3.6 Complexity creep: systems with too many features make it difficult for a person to develop an accurate mental model of how the system works.

Complexity is a subtle SA demon. While it can slow down the ability of people to take in information, it primarily works to undermine their ability to correctly interpret the information presented and to project what is likely to happen (Levels 2 and 3 SA). They will not understand all the features of the situations that will dictate some new and unexpected behavior, or subtleties in the system's programs that will cause it to work in different ways. A cue that should indicate something about what is happening with the system will be completely misinterpreted, as the internal mental model will be inadequately developed to encompass the full characteristics of the system.

While training is often prescribed as the solution to this problem, the reality is that with increasing complexity there is a much greater chance that people will have scant experience with system behavior in situations that occur only infrequently, will need significantly more training to learn the system, and will be more susceptible to forgetting subtle system nuances. We will devote Chapter 8 to the problem of overcoming the complexity demon in designing for SA.

3.7 ERRANT MENTAL MODELS

Mental models are important mechanisms for building and maintaining SA in most systems. They form a key interpretation mechanism for information taken in. They tell a person how to combine disparate pieces of information, how to interpret the significance of that information, and how to develop reasonable projections of what will happen in the future. Yet, if an incomplete mental model is used, poor comprehension and projection (Levels 2 and 3 SA) can result (Figure 3.7). Even more insidiously, at times the wrong mental model may be used to interpret information. For example, a pilot who is used to flying one model of aircraft may mistakenly interpret displays of a new aircraft incorrectly by using the mental model that was correct for the previous aircraft. This has led to accidents when important cues are misinterpreted. Similarly, doctors may misinterpret important symptoms in patients when they have been misdiagnosed. The new symptoms will be misinterpreted to fit the earlier diagnosis, significantly delaying correct diagnosis and treatment. Mode errors, in which people misunderstand information because they believe that the system is in one mode when it is really in another, are a special case of this problem.

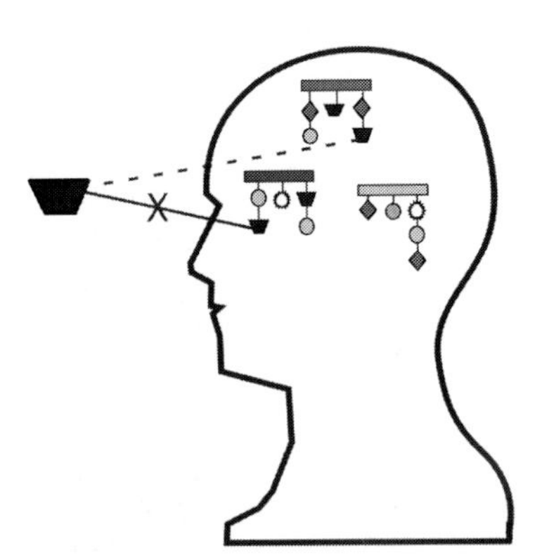

Figure 3.7 Errant mental model: use of wrong mental model leads to misinterpretation of information.

Mode errors are a significant concern in many automated systems in which multiple modes exist. For example, pilots have been known to misinterpret displayed descent rate information because they believe they are in one mode, where the descent rate is in feet per minute, when actually they are in another mode, where it is displayed in degrees.

The errant mental model demon can be very insidious. Also called a *representational error*, it can be very difficult for people to realize they are working on the basis of an errant mental model and break out of it. A study with air traffic controllers found that even very blatant cues that a wrong mental model had been activated were not properly detected and interpreted 66% of the time (Jones & Endsley, 2000b). People tend to explain away conflicting cues to fit the mental model they have selected, even when such explanations are far-fetched, and thus are very slow to catch on to these mistakes, if ever. This not only results in poor SA, it also leads to a situation in which people have difficulty detecting and correcting their own SA errors on the basis of conflicting information.

It is very important, therefore, to avoid situations and designs that lead people into the use of errant mental models. Standardization and the limited use of automation modes are examples of key tenets that can help minimize the occurrence of such errors.

3.8 OUT-OF-THE-LOOP SYNDROME

Automation leads to the final SA demon. While in some cases automation can help SA by eliminating excessive workload, it also can act to lower SA in certain cases. The complexity that comes with many automated systems and mode errors that can result when people mistakenly believe the system is in one mode when it is not, are SA demons that relate to automation. In addition, automation can undermine SA by taking people *out-of-the-loop*. In this state they develop poor SA on both how the automation is performing and the state of the elements the automation is supposed to be controlling (Figure 3.8).

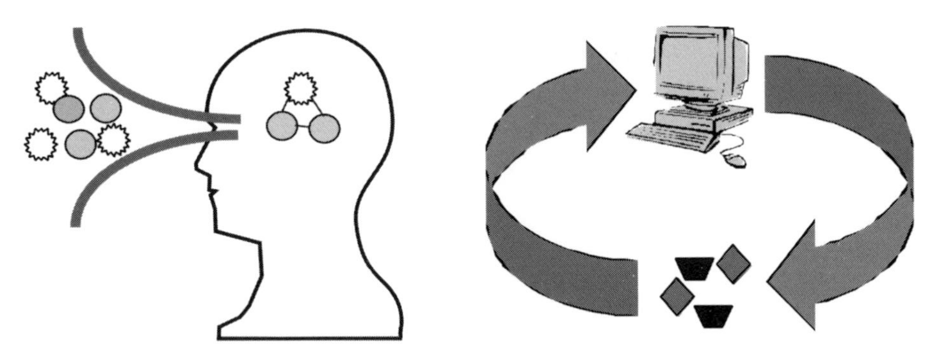

Figure 3.8 Out-of-the-loop syndrome: automation can undermine SA.

In 1987 an aircraft crashed on takeoff at an airport in Detroit, killing all but one person onboard. An investigation into the crash revealed that the automated takeoff configuration and warning system had failed (National Transportation Safety

Board, 1988). The pilots were unaware they had misconfigured the flaps and slats for takeoff, and were unaware that the automated system was not backing them up as expected. While the reasons for any accident are complicated, this is an example of an SA error that can result from automation approaches that leave people out-of-the-loop in controlling system functions.

When the automation is performing well, being out-of-the-loop may not be a problem, but when the automation fails or, more frequently, reaches situational conditions it is not equipped to handle, the person is out-of-the-loop and often unable to detect the problem, properly interpret the information presented, and intervene in a timely manner. The underlying causes and solutions to this problem will be further elaborated in Chapter 10. With the increased employment of automated aids in everything from kitchen appliances to power plants, the appropriate design of automation to avoid the out-of-the-loop syndrome is critical.

3.9 SUMMARY

A number of pitfalls that stem from the inherent limitations of human information processing and the features of many man-made systems can undermine SA. Designing to support SA needs to take these SA demons into account by avoiding them whenever possible. Good design solutions provide support for human limitations and avoid known problems for human processing.

In the next chapter we will discuss an overall approach and methodology for system design. In the remainder of this book we will develop design principles for building systems that enhance SA and minimize the likelihood that their operators will fall prey to these SA demons.

CHAPTER FOUR

The Design Process

4.1 SYSTEMS DEVELOPMENT LIFE CYCLE

In order to better understand how to inject a consideration of situation awareness into the design of systems, we will provide a brief overview of the typical system design process used across many different industries. The names of the phases and exact process can vary quite a bit from industry to industry and from company to company. In general, however, the broad descriptions of the processes and the design cycle presented here have a role in many different types of development efforts.

4.1.1 Waterfall model of design

The *systems development life cycle* is shown in Figure 4.1. It ranges from the initial identification of the need for the product (or product modification) to the deployment and operation of that product. It is depicted here as a series of stages,

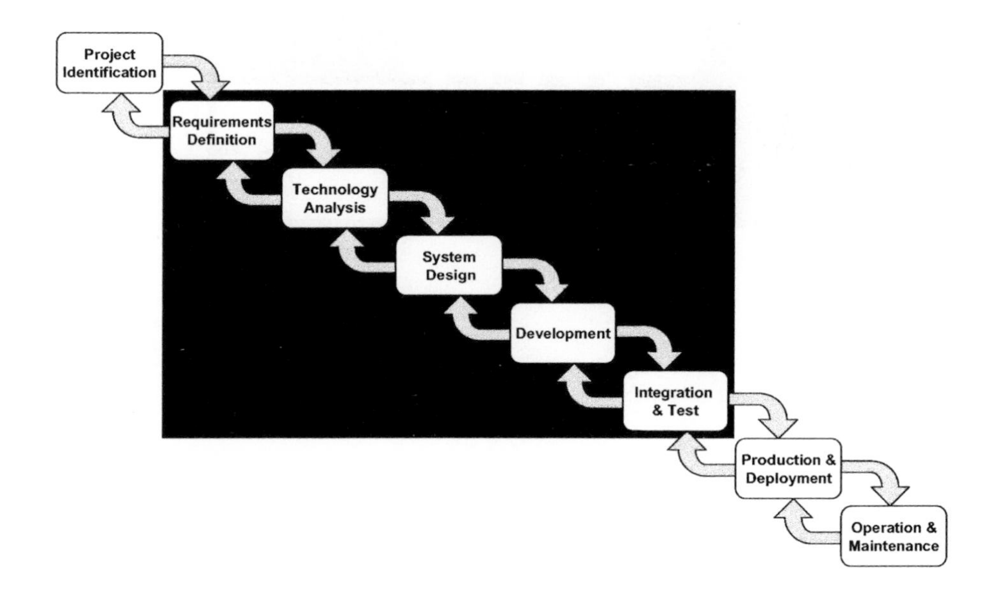

Figure 4.1 Systems development life cycle.

each of which flows into the next. Information from later stages also flows backwards, however, as different stages can be revisited and further work completed based on revisions found necessary in the later stages. This model is often called the *waterfall model of design*. (For more information on this process see Chapanis, 1996; Hoffer, George, & Valacich, 2002; Meister & Enderwick, 2002).

4.1.2 Concurrent engineering model

As a modification to this basic model of design, a *concurrent engineering* model recognizes that in the real world these processes do not occur in neat phases with work on one phase completing before work on the next stage commences. In reality, the systems development life cycle may look more like that depicted in Figure 4.2. Each phase overlaps considerably as engineers work concurrently through analysis, design, testing, and development.

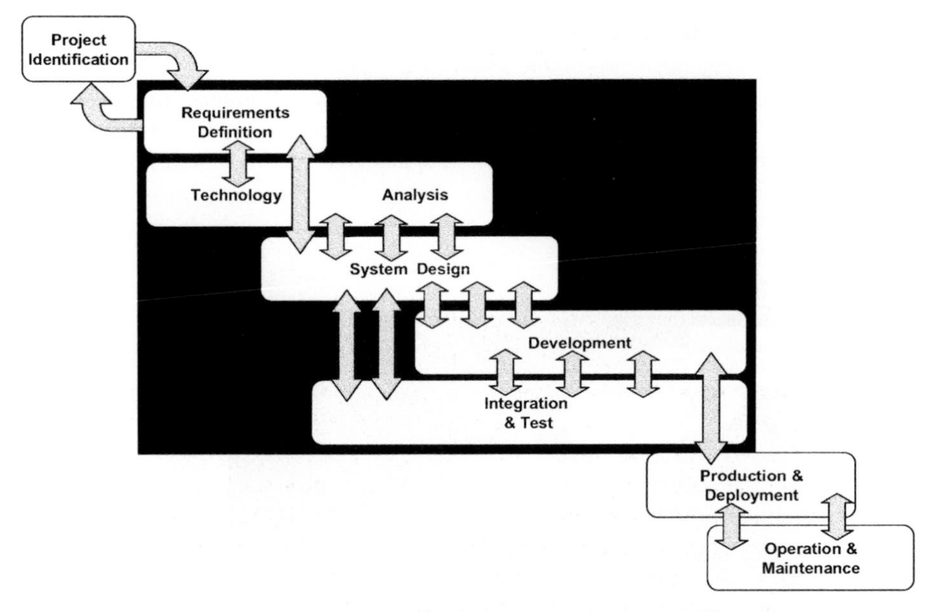

Figure 4.2 Concurrent engineering in the systems development life cycle.

Concurrent engineering allows not only a tightening of schedule as different phases are completed concurrently, but also emphasizes the use of *integrated product teams* (IPTs) composed of individuals from many different specialties (e.g., marketing, manufacturing, sales, and engineering) who bring different issues to bear on the process. This reduces the need for changes to the design of the product later on when changes are much more costly and difficult to make. Companies can eliminate or minimize the need for costly redesigns to make a product easier to manufacture or maintain with the use of IPTs and concurrent engineering.

In many industries, an overlap in technology analysis, system design, integration, testing, and production has become quite common. Generally, information and results from each phase flow to the next phase on a fairly frequent basis. Software and hardware mock-ups and prototypes are created based on early designs for testing. Early prototypes are tested and the results passed back to the ongoing design process for further modifications. Additional technology analysis and research is conducted concurrently as holes and weaknesses are identified or new technologies become available on the market. This work is completed in an ongoing cycle, often called a *spiral design* model (Figure 4.3). In each iterative pass of the cycle, the focus becomes narrower and narrower, weeding out ineffective ideas and answering design questions to hone the final design concept.

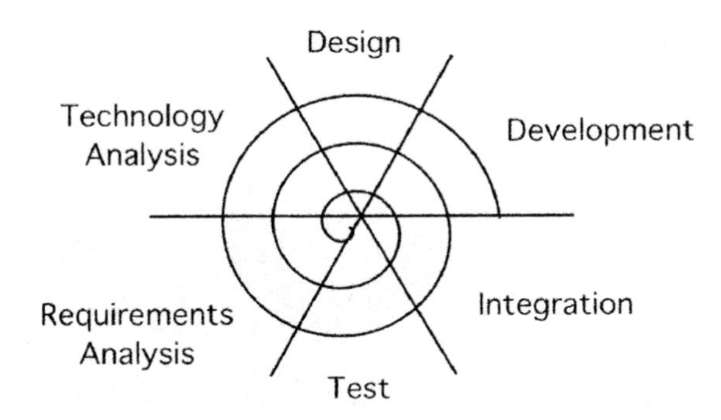

Figure 4.3 Spiral design process.

The software industry is also seeing an increase in the deployment of early versions of software (beta versions or sometimes first release versions), even before the final code is completed. The market then serves as the test bed and the users become the guinea pigs. The product only becomes stable in later versions. Development and testing continue simultaneously with early release of the product. While this has become a too common (and frustrating) practice in software development, it is rarely seen in high-consequence systems, such as aviation, driving, or power plant control, where the costs of failures are quite high.

As an upside to this practice, products can be brought to market more quickly. As a downside, deployment of beta versions seriously neglects system users, leaving them to find the bugs in the programs, cope with system crashes, and generally struggle to even make the software work. It may leave these customers loath to use that product again, even when the problems are fixed. While this practice is becoming all too common in the software industry, it is highly discouraged as not only being not user-friendly but actually quite user-hostile. It completely neglects the costs on the users' side as they suffer the consequences of lost work time, degraded capabilities, and loss of data.

Concurrent engineering poses certain challenges, but can be managed wisely to achieve user-centered design objectives. Simultaneous system deployments should be avoided, however, to prevent frustrating and perhaps alienating current and potential system users.

4.2 USER INTERFACE DESIGN PROCESS

Rather than cover the entire systems development life cycle in depth, we will focus here on only those aspects that are heavily involved in the design and development of the system's capabilities and interface, represented by the shaded area in Figures 4.1 and 4.2. We will discuss how the SA needs of system operators can be addressed within this process. A simplified design model is shown in Figure 4.4.

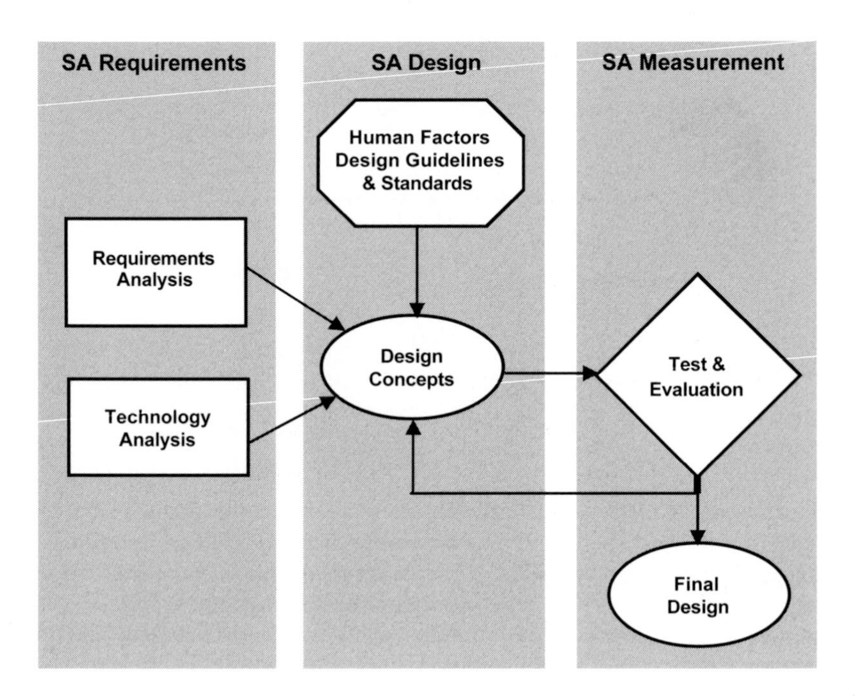

Figure 4.4 User interface design process.

4.2.1 Requirements analysis

Requirements analysis is perhaps one of the most important phases of the design process. In this phase it is necessary to translate the broad goals and objectives of the project into specific systems requirements. This analysis will include a number of different considerations.

Generally, the designers develop an *operational concept* that describes their vision for how the proposed system will be used. Such a concept would include the types of missions or functions that should be performed using the system and the capabilities the system should bestow on the user. Such a description provides important background on the intended use and functions of the system. Functions that should be performed (e.g., navigate from point A to point B) and any operational constraints (e.g., time requirements or reliability requirements) should be carefully documented.

In addition to the operational concept, at this phase it is very important to define the *environmental conditions* in which the system will be used. This will often include a consideration of factors such as:

- Ambient noise levels,
- Lighting levels,
- Susceptibility to weather and temperature variations,
- Vibration,
- Privacy,
- Security requirements,
- Expected pace of operations,
- Positions of use (e.g., sitting, standing, while mobile),
- Workload and stress levels,
- Frequency of use (e.g., occasional, intermittent, frequent, continuous),
- The need to integrate with other current systems, and
- Unusual characteristics in the working environment such as weightlessness or environmental toxicity.

The *user characteristics* should also be identified at this phase. The types of users the system accommodates should be listed along with their pertinent characteristics. This should include a description of:

- Gender (male, female, or both),
- Sizes, including height and weight (percentile of the population to be accommodated),
- Skill levels, training, and background knowledge (including reading capability and experience with similar types of systems),
- Age ranges (with special note of young or aging populations),
- Visual acuity and hearing capabilities,
- Languages to be accommodated,

- Special clothing or other equipment to be accommodated (such as gloves, masks, or backpacks),
- Any physical disabilities or special requirements, and
- The need to accommodate multiple users on the same system.

Finally, the *operational requirements* of the system's users should be specified. This includes a consideration of how they work and operate in their jobs, the types of physical and cognitive processes they employ, their need for interaction with others, and their SA requirements. Typically, this is conducted through a *cognitive task analysis* (CTA). There are many variants of CTA being employed (see Schraagen, Chipman, & Shalin, 2000). Most use one or a combination of observations of operations in representative similar environments, interviews with subject matter experts (SMEs), and reviews of available documentation from the domain of interest. When a brand new system is being developed for which existing users and systems do not exist, typically these analyses are based on users and systems that are as similar as possible to that anticipated for the new system.

In this book, we describe a form of CTA called a *goal-directed task analysis* (GDTA) that has been used to identify the goals, decisions, and SA requirements of operators in a wide variety of systems and domains. Presented in Chapter 5, the development of a comprehensive delineation of the SA requirements of a system's operator forms the fundamental foundation for the creation of systems that will support and enhance SA.

Together the operational concept, environmental constraints, user characteristics, and operational requirements form the basic input to the design process. These documents must then be translated into a set of implementable *system requirements*. The system requirements specify for the engineer what the system hardware or software must do and what characteristics it must have to accommodate each of these environmental, operational, and user constraints and characteristics. For example, if users will have corrected 20/20 vision and will operate the device at a viewing distance of approximately 20 inches, then the system requirement would be for a 10 point or larger font. The development of a clear set of system requirements requires designers to have the ability to apply a wide set of human factors data and design guidelines to the unique combination of circumstances, users, and operations inherent in each project.

It is critical to create a set of engineering criteria that clearly identify the system requirements that result from the human usability and SA requirements identified. *It is the system requirements that are identified at this stage that will drive engineering development and the evaluation process.* Typical engineering requirements for a system might be that data is transferred across the network within 6 seconds and there is a 90% success rate on such transitions. Such requirements completely ignore the human processing component, however. Users could be swamped with data that is poorly presented and the resultant system would still meet all engineering requirements and be judged a success. By translating the human requirements into system requirements, many problems downstream can be avoided. In this example, suitable system requirements that are more user-centric might include a requirement for changes in key information to be detected by users within 2 seconds, or for the user to be able to demonstrate SA of 90% of high priority messages. Requirements of this nature will help ensure that designers understand that their job is not done once the information gets to the system, but that the information also must be displayed effectively for human understanding and use.

4.2.2 Technology analysis

Technology analysis is generally conducted simultaneously with requirements analysis and frequently is an ongoing activity in many companies and organizations. Technology analysis typically incorporates an ongoing survey of products in the market that are relevant to the product line of the company and may include things such as new sensors, computer architectures, communications and networking devices, automation and control software, and controls and displays hardware and software. In addition, many companies have ongoing internal research and development (IR&D) programs to actively develop new technological capabilities in these areas.

From the standpoint of SA, many of these technologies may be employed to improve the quality and quantity of information provided to the system user. An active IR&D program works to identify the best design characteristics for new technologies in order to enhance human performance. For example, it might work to specify the effects of different fields of view and display resolutions of a helmet-mounted display on human performance when carrying out a variety of tasks. This allows information on the relative merits of different technologies for enhancing SA and performance to be established.

The technology analysis forms a second important contribution to the design process by ascertaining the range of devices and their characteristics that are available for gathering and displaying the SA requirements to the user and for allowing the user to interact with the system. These technologies will be considered for selection during the design conceptualization phase.

4.2.3 Design conceptualization

There are several different activities that occur during the design concept development phase.

4.2.3.1 Function analysis/function allocation

Function analysis and allocation are traditionally defined as the first step toward design. A *function analysis* involves determining the basic functions that need to be performed to accomplish the job. It does not specify who will do them or how they will be accomplished. A *function allocation* is then performed that specifies who will do each function (human, machine, or a combination of both).

In reality, function allocation actually occurs as an integral part of the interface design effort. Introducing automation to perform various tasks has a profound effect on the SA of the individual operator overseeing or working in conjunction with the automation. Such decisions cannot be divorced from the design of the interface. Function allocation is most properly performed concurrently with other interface design tasks, and should be subject to the same modeling, simulation, and testing efforts.

Traditionally, lists were developed that provided designers with recommendations on what sort of tasks were best performed by humans and by machines. Such approaches have largely fallen out of favor, however (Sanders & McCormick, 1992). This is because 1) technology has developed to such an extent that computers are capable of many tasks that previously only people could perform, blurring the lines of distinction between the two, and 2) a general recognition that very few tasks are really performed solely by either human or machine. Rather, most tasks are performed by an integral combination of the two, sometimes in combination and sometimes dynamically, as control of a function passes back and forth at different times. In Chapter 10 we provide some guidelines for directing the decisions regarding how to apply automation to system design.

4.2.3.2 User interface design

At its heart, design is a creative process. User interface (UI) design is no different. There can be many viable ways to design any system that will meet the requirements established. Given the myriad of technologies available and the hundreds of permutations of assignments of display and control requirements to those technologies, designers can generate great variety in resultant design concepts.

It is almost never possible to test all potential designs to discover their shortcomings from the standpoint of usability. The time and expense of such a proposition is managed through the application of a well-developed set of *human factors principles and design guidelines* that allow designers to avoid many, many pitfalls in that design process. These guidelines have been developed based on half a century of scientific testing and empirical data collection that establishes how to best design system components to enhance human use. Good examples of resources including such guidelines are:

- Boff, K., Kaufman, L., & Thomas, J. (Eds.) (1986). *Handbook of perception and human performance*. New York: Wiley & Sons.
- Eastman Kodak Company. (1983). *Ergonomic design for people at work*. New York: Van Nostrand Reinhold.
- Sanders, M. S., & McCormick, E. J. (1992). *Human factors in engineering and design* (7th ed). New York: McGraw-Hill.
- Salvendy, G. (Ed.). (1987). *Handbook of human factors*. New York: Wiley & Sons.

- Woodson, W. E., Tilman, B., & Tilman, P. (1992). *Human factors design handbook* (2nd ed). New York: McGraw-Hill.

These guidelines are primarily focused on the perceptual and physical design of systems. They include detailed data on the presentation of different types of visual, auditory, and tactile information to allow successful human perception of information. They provide guidelines for the best design of various types of buttons, knobs, keyboards, and continuous control devices (both for mechanical and computerized systems) to minimize error and maximize human performance in the operation of systems and input of data. They also have considerable information on the physical design of workspaces to support human physical capabilties (e.g., how to best support lifting and carrying requirements, the design of handtools, chairs, and other devices to minimize injuries to users, and where to place equipment for the best viewability and physical accessibility by users), and how to design systems to compensate for various environmental issues such as illumination, climate, noise, and vibration.

At a minimum, these design guidelines form a very necessary foundation for SA. If a person cannot see or hear the information they need, SA cannot exist. Alone, however, they generally do not go far enough to guide the design of systems that support high levels of SA under demanding circumstances. This is for several reasons.

First, these guidelines exist, for the most part, at the level of the component. SA, by its very nature, is highly affected by not only the design of individual system components, but by how those components are integrated. So, while two components may be just fine when viewed individually, when viewed together within a demanding setting, SA may be very poor on certain information and higher on other information. Or perhaps it may not be affected much by the addition of certain displays because the user can already gain the needed information elsewhere in the environment. While some integration guidelines do exist (e.g., provide standardization across the system, compatibility between controls and displays, and methods for arranging multiple controls on a panel), far more is needed to create an overall integrated system that supports SA.

Second, these guidelines are primarily focused on the perceptual and physical attributes of humans, but not on their cognitive attributes. Just as we need to design chairs to fit the human body, we also need to design systems to support the way people take in and process information. This state of affairs is largely due to not only differences in the types of systems that have been developed historically, but also to developments within the field of psychology as approaches have gone from strictly behaviorialist to those that have focused more on human cognition. Fields of endeavor such as cognitive engineering (Hollnagel & Woods, 1983; Norman, 1981, 1986; Woods & Roth, 1988) and naturalistic decision making (Klein, Orasanu, Calderwood, & Zsambock, 1993; Zsambok & Klein, 1997) have done much to help build a foundation of knowledge on human cognition.

In this book we offer a set of guidelines and a methodology for designing to enhance SA. These guidelines do not replace these other human factors standards and guidelines, but rather augment them. They provide the next level of principles for addressing—through system design—the very real problems that people have in obtaining and maintaining high levels of SA in complex and dynamic systems.

4.2.3.3 Design concept products

The products of this design process can be expressed in many ways that serve to both facilitate the process and to document the system design. These include:

- Flow analysis or operational sequence diagrams,
- Task analysis,
- Timeline analysis,
- Link analysis, and
- User interface specification.

Flow analysis or *operational sequence diagrams* graphically represent decision sequences used by the system and/or user, showing the order of information input, decisions, and other operations that need to be performed to accomplish each function.

A *task analysis* lists the actual tasks that the user must perform with the system. While sometimes these analyses are done as precursors to the design, they really are very dependent on the technologies and design of the actual system that is developed, and thus are more appropriate as a documentation of the design. The task analysis is used to ensure that all task sequences can be easily performed.

A *timeline analysis* shows the tasks that must be performed along a timeline. This analysis can highlight the need to perform different tasks simultaneously and potential workload problems, and the fit between the design and task performance requirements. While these are straightforward for jobs in which tasks are always done in a set sequence, they become very difficult or meaningless for jobs in which tasks are done in very different orders depending on the flow of events. Different types of computer modeling tools have been developed to help with this problem, many of which generate possible task sequences based on input assumptions, allowing different possibilities of requirements for simultaneous task performance to be considered.

A *link analysis* shows a network of interrelationships among elements of the design, most typically different controls and displays. The importance of the relationship (e.g., how much they relate to the same function) or the frequency of relationship (e.g., how often they need to be used in sequence) are specified. The link analysis allows a designer to examine the degree to which a design layout facilitates easy access to controls and displays as they relate to the performance of tasks, iterating the design until it better facilitates such linkages. Typically, these analyses are useful when the system includes numerous different physical displays and controls laid out in a room or across a workstation. They are not used as often with the design of computerized displays, although they could be.

The *user interface specification* is a document that includes text, graphics, and often flowcharts showing the displays and controls of the user interface. It documents in detail the resultant design concepts produced in the design effort in such a way that hardware and software engineers can implement the design concept.

Many people think of the user interface as something that comes after the engineers have designed the physical system, and is then designed as a front end to allow human interaction and operation. Because the effectiveness of the overall system is largely determined by not just the physical nature of the displays and controls, but also by the inherent functionality of the system (its features, logic algorithms, and automation decisions), this approach works quite poorly. Needed changes to the system design that become apparent when the user interface is developed at the end of the project are also quite expensive and difficult, and often simply never occur.

Ideally, the user interface needs to be specified very early in the design phase and can successfully serve as a key driver of the engineering design and development process. Because many facets of the system design are not known early on (e.g., exactly which technologies might be used and what their characteristics will be), the user interface specification is actually a living document. It may be revised many times over the course of the design process as both the interface design matures through iterative test and evaluation, and as the hardware and software engineering design progresses.

Initially, the user interface design specification may depict a somewhat ideal design, from the standpoint of enhancing overall user system performance. As the engineering design process progresses, it may often happen that certain features are dropped due to time or budgetary constraints. Nonetheless, such depictions can form a useful goal for the design process and some features may be implemented in later versions of the system, even if they cannot be initially included. These efforts need to be constantly fed with information from the technology analysis and other ongoing engineering design efforts in order to be realistic in terms of what is possible, yet form a target for what the system should be.

4.2.4 Prototype development

The development of models, simulations, and prototypes of design concepts (and variants of design concepts) that are conceptualized in the design phase is very useful. It allows many people (including engineers, programmers, marketers, and potential users) to develop a common understanding of what the system might do and what it might look like. In addition to providing a common frame of reference for discussion, these prototypes provide the basis for detecting design problems and shortcomings early in the design process. Without the development of such prototypes and models, many significant problems and deficiencies will only be detected much later after costly development processes have been undertaken and at a point when any change to the design is much more likely to be resisted as too costly. *Prototype development* can take many forms and generally proceeds from fairly low fidelity models and prototypes up through more elaborate simulations and field tests (Figure 4.5), as many design questions are resolved and the final design is zeroed in on. We will discuss each progressive level of prototyping in more detail.

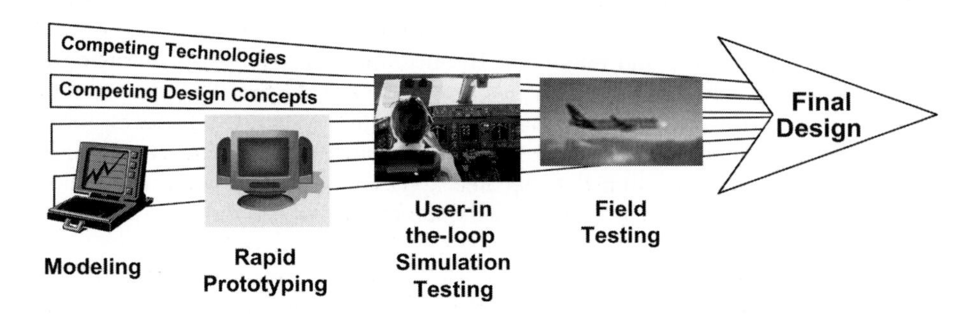

Figure 4.5 Prototyping forms the backbone of the process

4.2.4.1 Modeling

Many types of models exist to help in evaluating user interface design concepts. These include models for:

- Evaluating the *physical layout* of the system with regard to viewability, physical fit, and reach considerations, such as SAMMIE (SAMMIE CAD Ltd., 1990) and COMBIMAN (Evans, 1978),
- Evaluating *workload* issues through tools such as MicroSAINT (Laughery & Corker, 1997) ,
- Evaluating *man-loading, training, and personnel* issues through tools such as IMPRINT (Allender, Kelley, Archer, & Adkins, 1997), and
- Predicting *human information processing characteristics* based on input parameters. Models in this category include MIDAS (Banda *et al.*, 1991; Laughery & Corker, 1997), COGNET (Zachary, Ryder, Ross, & Weiland, 1992; Zachary, 1998), OMAR (Deutsch & Adams, 1995; Deutsch, Macmillan, Cramer, & Chopra, 1997) and ACT-R (Anderson, 1993; Anderson & Lebiere, 1998).

More information on these models can be found in Pew and Mavor (1998). Each of these models may be useful for examining specific aspects of the design early in the design phase, before any prototypes are available. They also are useful in that many variations of design concepts can be evaluated fairly easily. As a downside, it often takes considerable time and effort to adequately model the tasks, design concept characteristics, and features of the system that are needed in order to obtain good output from these models. It also can be quite difficult to validate the quality of the models on which design decisions will be made. Often the models are constructed on the basis of many assumptions regarding the characteristics of the users, the environment, the tasks, and the system design, each of which can markedly affect the output of the model. In order to validate the models developed, usually some type of simulation of the system is required.

Although the physical models available are fairly good, the cognitive models remain fairly limited in terms of the number of components of human cognition considered and the degree to which good predictions of human performance can be made from them. While many of these models continue to be developed to better accomplish their roles in system design, in general it remains necessary to use rapid prototyping and user-in-the-loop simulation to gain good predictions of how well people might perform with a given system design.

4.2.4.2 Rapid prototyping

Rapid and early prototyping of design concepts forms an important bedrock of user interface development. Rapid prototyping is an essential tool that enables designers to make good design decisions and weed out poor ones. By quickly mocking up potential function allocations and display ideas on a computer, a more realistic assessment of their viability can be made. These prototypes may range from fairly simple static drawings and storyboards to computerized depictions of the system displays and controls with some functionality simulated. For most systems it is extremely useful for subject matter experts and representative users to be able to see what the displays will look like and how they will change with different system conditions, automation capabilities, information flows, and user inputs. Audio displays and tactile displays can be similarly prototyped.

What is most important is that these prototypes are created early, allowing needed design input from engineers and users so that design convergence occurs. Multiple display and control options can be considered fairly easily, as software to actually drive the system behavior is usually only crudely simulated at this stage. A 'Wizard of Oz' technique may be used during rapid prototyping in which experimenters behind the scenes input information in order to simulate what the system might do or show to users based on their inputs (Green & Wei-Haas, 1985; Kelley, 1983). Alternately, canned scenarios (animations) can be provided showing how the system might look under certain conditions. Tools as simple as Microsoft® Powerpoint® or Macromind® Director can be used to create these prototypes, or you can use more complicated packages specifically designed for rapid prototyping, such as Emultek's Rapid, Virtual Prototypes' VAPS, or Microsoft® VisualBasic. While designed to allow prototypes to be created fairly easily, these tools do require some programming ability and specialized knowledge to learn how to use them.

Rapid prototyping is best suited for early concept exploration. It allows collection of subjective inputs and some objective data collection regarding potential design concepts. Things such as intuitiveness of displays, labels, and icons, for example, can be examined by looking at user errors in selecting buttons for particular tasks. Rapid prototyping is good for assessment of user interface components (e.g., particular subsystems), but is more limited in terms of assessments of complete integrated systems for more complex situations. In these cases, simulation testing is usually required.

4.2.4.3 Simulation testing

Rapid prototyping of concepts is extremely important in the early stages of design. While much information can be gained from having knowledgeable users review these prototypes and provide subjective comments, it has been found that objective test and evaluation, allowing the collection of more detailed data during simulated use of the design concept, is of paramount importance. Such *simulation testing* is far more likely to provide detailed data on performance problems and the relative costs and benefits of different design options. Based on this information, design concepts will be refined, discarding troublesome components and improving good components, and reevaluated in an iterative fashion until a final design is selected. This use of rapid prototyping and *user-in-the-loop simulation* early and often in the design process is a key risk-reduction factor in user interface design.

As design concepts are refined and the process moves closer to design finalization, the scope and fidelity of the simulation increases. While early testing may examine only subcomponents of the system, later testing tends to involve a fuller range of system components and a broader range of environmental features. The effectiveness of this process is largely influenced by the quality and realism of the simulations, the comprehensiveness of the scenarios, and by the veracity and sensitivity of the measures used to evaluate the concepts.

Simulations can be low fidelity, taking place on a simple desktop computer with simplified versions of the task environment. Microsoft® FlightSim and Wesson International's TRACON II are examples of such programs for the general aviation and air traffic control domains. As a significant limitation in most of these packages, very limited facilities for display modification or data collection are present. Many times, special purpose simulations need to be created to support design testing.

Generally these efforts progress to higher fidelity simulations later in the design process. These may range from PC-based simulations to relatively complex virtual reality simulators developed for a variety of domains and systems. In aviation, full cockpit simulators are often used that include very accurate physical and functional representations of the controls and displays and system behavior of the aircraft models being examined, which can be either fixed-based or include full-motion capabilities. High-fidelity simulators also exist for driving, command and control, air traffic control, and power plant operations, and have recently been developed for areas such as anesthesiology.

Simulations provide the ideal platform for testing integrated concepts (multiple displays and controls across systems needed to perform tasks). Most importantly, a wide variety of objective data can be collected. Direct assessment of operator SA is collected at this level of prototyping, in either low or high-fidelity simulations.

Simulations provide detailed, objective data on the implications of different design decisions early in the design process when changes can be made easily and without the excessive costs that arise from finding problems later in the field.

4.2.4.4 Field tests

In many domains, *field tests* may also be used, sometimes even more so than simulation. This is often due to historical conditions in which suitable simulations were not available. In medicine, for instance, while some manikin-type simulators of patients have recently been developed, much product development and testing is likely to occur in clinical trials in which a small number of physicians may use the device under very controlled circumstances to perform actual cases, usually backed up by other approved technologies and procedures for safety. In these cases, actual working prototypes of the devices are developed and used for testing. In many aviation and military systems, operational test and evaluation of systems is conducted with full working system prototypes in field conditions as a culminating event following simulation testing during design.

Field-testing provides the advantage of greater realism in the environmental conditions and scenarios that can be considered. The data collected often has a high degree of face validity because of this. As a downside, it is much harder to achieve experimental control in a field test. That is, creating scenarios and tests in which users and systems are exposed to exactly the same requirements, conditions, and events may be challenging, as many things, such as what people in the scenarios do, may be out of the designer's control. As such, differences in performance that occur between different system designs may be difficult to attribute to the design, as opposed to differences in the testing conditions.

In addition, another major drawback of field tests is that working functional prototypes of systems are usually required. This generally means that one will be far down the development cycle before such testing can take place. The ability to examine many different design concepts is usually severely restricted. The ability to make needed changes to system designs at that point can also be very limited.

While it is possible to measure operator SA during field-testing in addition to overall performance, it is in many ways much more difficult. The degree of control available to the designer to use many of the methods for SA measurement can be quite restricted and often requires workarounds. In general, a field test is quite appropriate late in the design process to examine how the system will be used in realistic situations and conditions, but should form the capstone to earlier testing accomplished with rapid prototyping and simulation.

4.2.5 Test and evaluation

Test and evaluation of design concepts throughout this process of prototyping, simulation, and field-testing depends largely on having a good battery of measurement tools that will be sensitive to differences in the quality of various design concepts. While there is an important role for subjective evaluation by experienced and knowledgeable representative users, the collection of objective data on the performance of users with the system is crucial. Explicit and objective measurement during this process of iterative test and evaluation insures design objectives will be met. A thorough evaluation of user workload and SA is also critical during system design, as these constructs will largely influence the usability and effectiveness of the system in actual operational circumstances and often provide more sensitive and diagnostic evaluation criteria. In general, evaluation metrics can assess user processes, cognitive states, and performance at different stages (Figure 4.6). More details on measures for test and evaluation can be found in O'Brien and Charlton (1996).

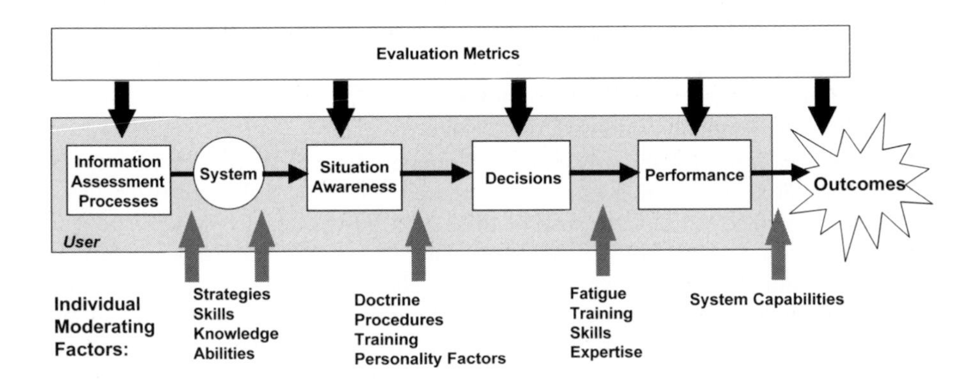

Figure 4.6 Evaluation metrics at different stages of user performance.

4.2.5.1 Subjective measures

Early on, *subjective* input by users can provide useful information. Missing features, confusing labeling, conditions of use not well supported by the design, and problems with implementations can often be identified by users based on even very simple prototypes or simulations. Such subjective data is very limited, however. Many users may be biased toward familiar ideas and concepts. They may initially have negative impressions of design concepts that are new and on which they have often not received much training. Without the opportunity to experience using the system, the ability of users to give a realistic assessment of a potential design concept's utility may be very limited. The biggest limitation of subjective data is that many problems with system designs that can critically affect human performance will not be obvious to naïve users. Designs that have very negative effects on human performance by creating visual illusions or higher error rates, for example, are often not obvious and are only revealed by careful testing and objective data collection.

4.2.5.2 Objective measures of performance

Objective performance measurement usually takes the form of *measures of effectiveness* (MOEs) or *measures of performance* (MOPs). MOEs generally include measures of the overall performance of the combined human/machine system. In military operations, for example, it would typically include measures such as number of kills, number of losses, and number of bullets or missiles fired. For a medical system, they might include number of patient complications and time to perform a case. Measures of performance are usually more granular and focus on particular subsystems. For the human user they might include number of mis-entries and time to perform a wide variety of tasks. Such measures should be collected at every level of prototyping possible.

Performance measures are often not sufficiently granular or diagnostic of differences in system designs, however. While one system design concept may be superior to another in providing users with needed information in a format that is easier to assimilate with their needs, the benefits of this may go unnoticed during the limited conditions of simulation testing or due to extra effort on the part of users to compensate for a design concept's deficiencies.

In many cases, it may be impossible to conduct a sufficient number of test scenarios to examine what would happen to performance under many potential conditions. Without numerous repeated trials and a host of test subjects, it can be difficult to collect enough data to obtain statistical significance of existing differences in performance.

Users also can be quite ingenious. They often will try very hard to achieve good performance and may work harder to compensate for poor designs. Over time, however, in real operations, such design problems will lead to errors and degraded performance. Therefore it is important that these design deficiencies be corrected, rather than relying on users to always be able to compensate for problems.

On the opposite side, oftentimes significant improvements in designs will not be detected based on performance measures alone. For very new types of systems, users may not yet have developed good strategies and tactics for best using new systems to maximize performance objectives. For these reasons, user workload and SA should also be measured directly during test and evaluation activities.

4.2.5.3 Workload measures

The degree to which a system design concept induces high *user workload* indicates a design that has a higher probability of overload and error given multiple events and demands in the operational environment. By directly measuring user workload, design concepts that are less demanding can be selected during design testing.

In other systems, low workload is a concern. Operations requiring long periods of vigilance (where very little happens or changes) can be degrading to SA and user performance. Long-haul transatlantic flights and monitoring radar screens for infrequent targets are classic examples of such situations. A number of different approaches to workload measurement are available. These include:

- *Physiological measures* – such as the p300 component of electro-encephalograms (EEG) and sinus arrhythmia measures obtained from electrocardiograms (ECG),
- *Subjective measures* – of which several well used and validated batteries exist, including the NASA-Task Load Index (Hart & Staveland, 1988), the Subjective Workload Assessment Technique (Reid, 1987), and the Cooper-Harper Scale (Cooper & Harper, 1969), and
- *Performance measures* – including task error measures and measures of spare capacity using concurrent secondary tasks.

For more information on the relative benefits and limitations of each of these approaches see Hancock and Meshkati (1988).

4.2.5.4 Situation awareness measures

By measuring SA directly, it is also possible to select concepts that promote SA, and thus increase the probability that users will make effective decisions and avoid poor ones. Problems with SA, brought on by data overload, nonintegrated data, automation, complex systems that are poorly understood, excess attention demands, and many other factors, can be detected early in the design process and corrective changes made to improve the design.

Direct measurement of SA during prototyping and simulation has been found to be very sensitive to design differences that performance measures alone could not detect during testing (Endsley, 2000a). Measurement of SA also provides good diagnosticity, allowing designers to detect exactly why a given design may have faired well or poorly. They really need to know whether performance shortcomings were due to poor SA (reflecting a system or display problem) or due to poor tactics or task execution (reflecting a training problem). Some SA measures also provide diagnosticity in terms of which aspects of SA (i.e., which specific elements of the situation) may have improved or declined with a particular design concept. This is very useful data to have when deciding how to modify design concepts for further improvement. Several different approaches for SA measurement have been developed, including:

- *Process measures* – including eye tracking, information acquisition, and analysis of communications,
- *Performance measures* – which seek to infer SA from performance on designated tasks,
- *Subjective measures* – which ask the users to rate their SA on a scale, and
- *Objective measures* – which collect data from users on their perceptions of the situation and compare them to what is actually happening to score the accuracy of their SA.

Each of these approaches to situation awareness measurement is discussed in detail in Chapter 12.

4.2.6 Final design

Design conceptualization, prototyping, and test and evaluation are performed in an ongoing iterative fashion throughout the design process. At some point, inevitably the process reaches the stage of a design freeze, moving the product into full-scale development and implementation. While it would be ideal for designers to be able to test many different design options until an optimal solution is identified, in reality this luxury rarely, if ever, occurs. Schedule and cost constraints are usually prohibitive. For this reason early prototyping and simulation testing to find as many design shortcomings as possible are crucial.

In addition, continued monitoring of how users actually use the system in real operational conditions after implementation is very important. Any problems or unforeseen uses of the system can be detected and changes to the system design or operational procedures implemented. For example, after the implementation in cockpits of the Traffic Collision and Avoidance System (TCAS), which displays other aircraft positions to commercial pilots for collision avoidance, it was found that pilots unexpectedly began using the displays to maintain aircraft spacing on final approach, even though the systems were not designed for this purpose (Jones, 2000). Accuracy and resolution of the displays for this use may not be adequate, however, and may lead pilots to develop a false belief of exactly where they are relative to an aircraft they are following in trail. Unexpected uses of new systems and effects on user behaviors should be carefully monitored to detect such problems.

4.3 SITUATION AWARENESS-ORIENTED DESIGN

The design process outlined here is at the heart of user-centered design. Without a systematic analysis of user and operational requirements, an application of sound human factors guidelines and principles, and a systematic and objective testing program, it is almost impossible to create a user-centered system. Many aspects of this design process are commonly employed in the development of complex systems today. Within this context, we have identified three major lines of attack for creating system designs that will support high levels of SA in complex systems (Figure 4.7).

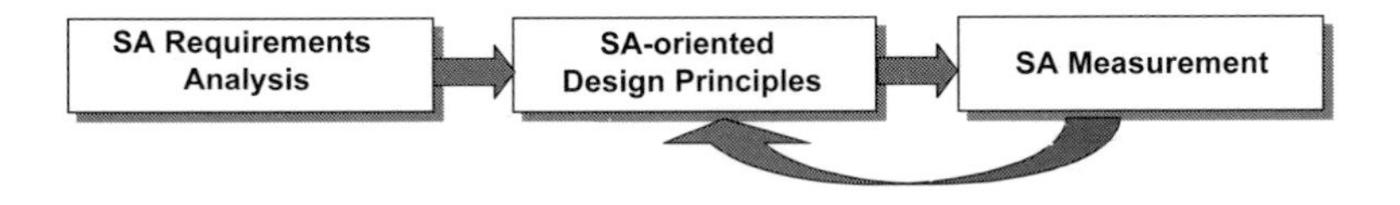

Figure 4.7 SA-oriented Design.

In the next sections of this book, we will discuss each of these techniques in detail. SA requirements analysis, creating the driver for SA-oriented Design, is described in Chapter 5. Principles for designing systems to support SA are discussed in Chapters 6 through 11. Finally, SA measurement (Chapter 12) forms the backbone of this process, providing clear data on the actual effects of design concepts developed based on these design principles and requirements analysis.

Part Two: Creating Situation Awareness-Oriented Designs

CHAPTER FIVE

Determining SA Requirements

5.1 GOAL-DIRECTED TASK ANALYSIS

Before a system can be designed to support SA, a clear understanding must be achieved regarding what 'supporting SA' means in a particular domain. This understanding is accomplished by delineating the operators' dynamic information needs (i.e., their SA requirements). SA requirements focus on the dynamic information requirements relevant in a particular domain rather than the static knowledge the operators must possess, such as system knowledge or rules and procedures (Endsley, 2000a).

SA requirements are delineated through a goal-directed task analysis (GDTA). The GDTA focuses on the basic goals of the operators (which may change dynamically), the major decisions that need to be made to accomplish these goals, and the SA requirements for each decision. The means an operator uses to acquire information are not the focus of this analysis as this can vary considerably from person to person, from system to system, from time to time, and with advances in technology. For instance, at different times information may be obtained through system displays, verbal communications, or internally generated within the operator.

The GDTA seeks to determine what operators would *ideally* like to know to meet each goal, even if that information is not available with current technology. The ideal information is the focus of the analysis; basing the SA requirements only on current technology would induce an artificial ceiling effect and would obscure much of the information the operator would really like to know from design efforts. A completed GDTA is provided in Appendix A as an example for the reader.

5.2 METHODOLOGY OVERVIEW

The GDTA seeks to document what information operators need to perform their job and how the operator integrates or combines information to address a particular decision. This knowledge helps designers determine ways to better present information to operators in order to support situation awareness and, consequently, decision making and performance. Generally operators are interviewed, observed, and recorded individually. The resulting analyses are pooled and then validated by a larger number of operators. The information obtained from these and other methods (e.g., analysis of written materials and documentation, or verbal protocol) are organized into charts depicting a hierarchy of goals, subgoals, decisions relevant to each subgoal, and the associated SA requirements for each decision. Figure 5.1 shows an example of a goal hierarchy.

The number of goals and relevant subgoals vary for different domains. Figure 5.2 illustrates the next level of this hierarchy, which shows the decisions and SA requirements associated with each subgoal. Figure 5.3 shows an alternate delineation of lower level subgoals, allowing greater detail in the decisions and SA requirements specified.

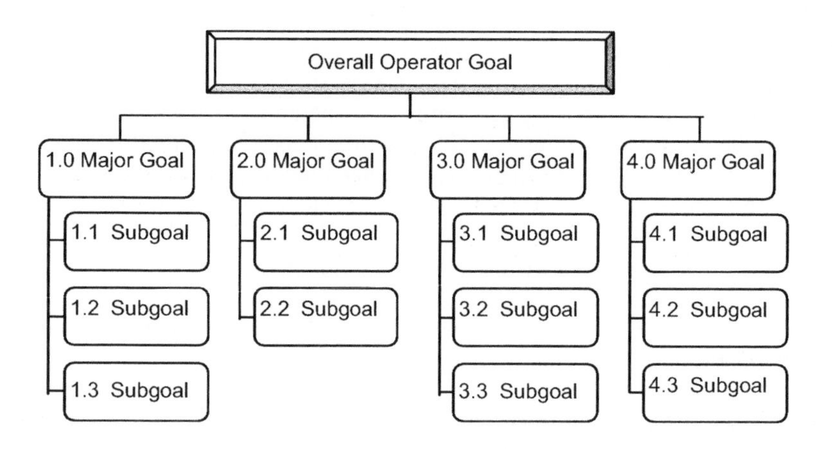

Figure 5.1 Primary goal hierarchy.

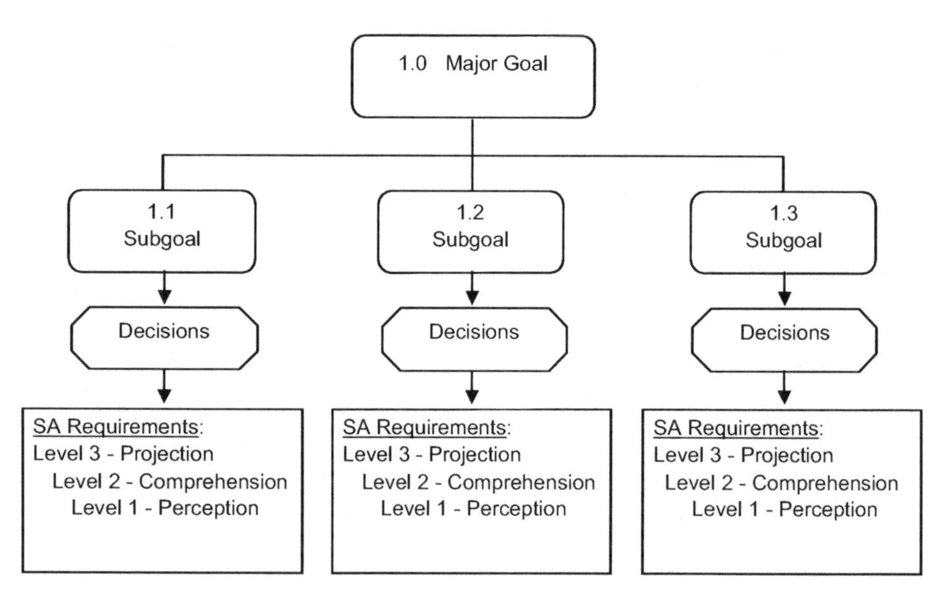

Figure 5.2 Goal-Decision-SA requirement structure.

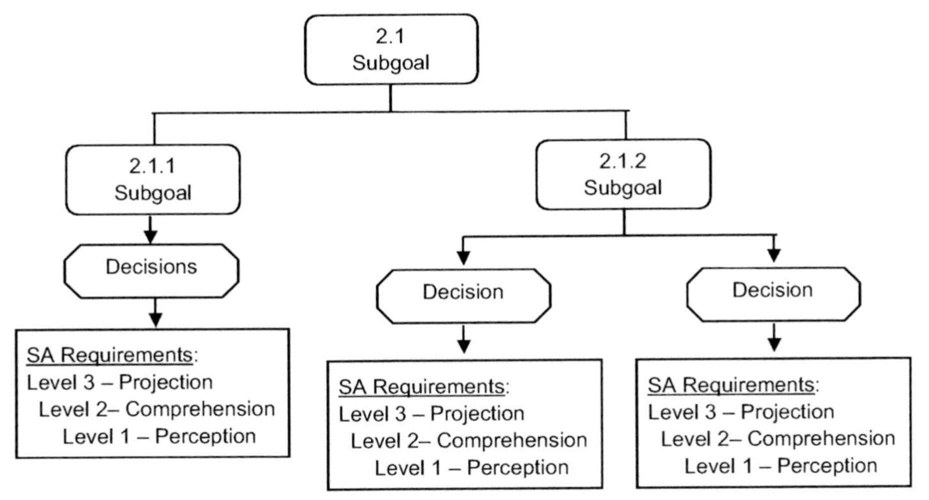

Figure 5.3 Expanded Goal-Decision-SA requirement structure.

5.3 INTERVIEWS

Interviews with subject matter experts (SMEs) are an indispensable source for information gathering for the GDTA. The process begins with an unstructured interview with a domain expert. During this interview, the participants generally are asked what goals they are seeking to achieve while performing their job. Before beginning the GDTA, available materials (e.g., manuals, performance standards, job taxonomies) pertaining to the domain being considered should be reviewed to provide an overview of the domain and the nature of the operator's job. Further, the interview will most likely go more smoothly (and the interviewer's credibility be enhanced) if the interviewer can 'talk the talk' and understand the lingo the interviewee is accustomed to speaking. The interviewer must be cautious, however, not to develop a preconceived notion of what the operator's goals are likely to be and, as a result, to seek only confirming information in the interview.

Typically, each SME should be interviewed individually. Each interview begins with an introduction of the purpose and intent of the data collection effort and a quick review of the interviewee's experience. Typically, the interviewee is then asked about the overall goals relevant to successful job performance, and the participant's responses are used as a starting point for the remainder of the interview. From this initial exchange, the interviewer should be able to identify at least one area to pursue. If more than one potential area for further inquiry is noted, the interviewer should make a note of each area in order to return to each in turn. Consider the following examples and their accompanying analysis.

Example 1:

Interviewer: 'What are the goals of the Fire Support Officer?'

Interviewee: 'Establish fire support liaison with other units. Have to coordinate and communicate with systems outside artillery system.'

Interviewer: 'Why coordinate with other units?'

Analysis: Pick up on the main theme of the participant's response for continued questioning.

Example 2:

Interviewer: 'What is your overall mission goal?'

Interviewee: 'Determine requirements, plan for requirements, and anticipate future requirements to maintain the readiness of the unit.'

Interviewer: 'How do you anticipate future requirements?'

Analysis: Key in on one of the main themes of the participant's response. In this case, the interviewer chose to follow up on the 'anticipation' aspect first. Other possible avenues of approach would have been to follow up on how the requirements are determined or what/how requirements are planned for.

After a statement of initial goals, the interviewee may (1) pause and wait for further questions, or (2) continue the dialog without pause. When the interviewee pauses, one of the identified areas can be selected for further discussion and the conversation continue until all possibilities on the topic are covered. Once this line of questioning is exhausted (sometimes after a lengthy conversation, other times after only minimal conversation), another of the topics can be selected for discussion.

When the interviewee follows a statement of goals with continued conversation, following the interviewee's train of thought is helpful, but the interviewer should also make notes regarding other areas of interest that may be returned to when the conversation pauses. This process of noting potential questions as they come to mind should be followed throughout the entire interview. Relying on memory for questions may work for some, but for most, the questions will be forgotten (but crop up again after the interview when the interviewer is trying to organize the GDTA).

Certain issues will be present in many domains and should be broached at some point in the interview process. One such issue is confidence in information. With the complexity of systems, the uncertainty of information, and the large amounts of data to be sorted, a person's confidence level in information can vary. The way a person utilizes information in order to achieve a goal will be affected by that person's confidence that the information is reliable and valid.

Good questions for the interview include:

- What do you need to know to make that decision?
- What would you ideally like to know?

- How do people do this badly? What do they typically fail to consider
- What would be an example of someone with really good SA?
- How do you use that information?

5.4 DETERMINING THE PRELIMINARY GOAL STRUCTURE

Once the initial interviews have been completed, the task becomes one of organizing all the disparate information collected during the interview into a workable preliminary goal structure that will allow for adequate portrayal of the information requirements. The preliminary goal structure will aid in the collection of additional information, even if it is sketchy or incomplete.

Although each researcher will develop a unique style for accomplishing this task, one approach is to begin by reorganizing the notes from the interviews into similar categories. This categorization can be done by beginning each new category on a page and adding statements from the notes to these categories as appropriate. Sorting the information in this manner helps illuminate areas that constitute goals and may make it easier to outline a preliminary goal structure.

The key elements in the preliminary goal structure are the major goals and subgoals, which are arranged in a relational hierarchy. Figure 5.4 shows an example of a relational goal hierarchy for an Army Brigade Intelligence Officer.

The goals within the hierarchy represent the goals the operator seeks to achieve and are not developed along task or procedural lines. Additionally, the major goals within the hierarchy (e.g., 1.0, 2.0, 3.0) are not sequenced according to any type of timeline (e.g., 1.0 active, then 2.0 active), nor are they arranged in a priority order (e.g., 1.0 is not a higher priority goal than 2.0). The current situation cues the applicable goals, which can change dynamically over time, switching between the goals in any order as needed by the changing environment.

Although defining an adequate goal hierarchy is the foundation of the GDTA, in the early stages this hierarchy will not be perfect, nor should an inordinate amount of time be spent trying to make it so. The main thing at this point is to make a good first attempt. Further interviews with experts will most likely shed new light that will require that the hierarchy be revamped by adding, deleting, or rearranging goals. Developing a goal structure at this point in the process allows for a baseline for future iterations, helps in the process of aggregating information, and in directing information gathering efforts during the next round of interviews. The following sections discuss the three main components of the GDTA: *goals*, *decisions*, and *SA requirements*.

5.4.1 Goal Determination

Goals are higher-order objectives essential to successful job performance. Their labels should be descriptive enough to explain the nature of the subsequent branch and broad enough to encompass all elements related to the goal being described.

The goals in the goal structure should be just that—goals, not tasks nor information needs. Poor goal structuring and delineation will contribute to difficulty in constructing the GDTA.

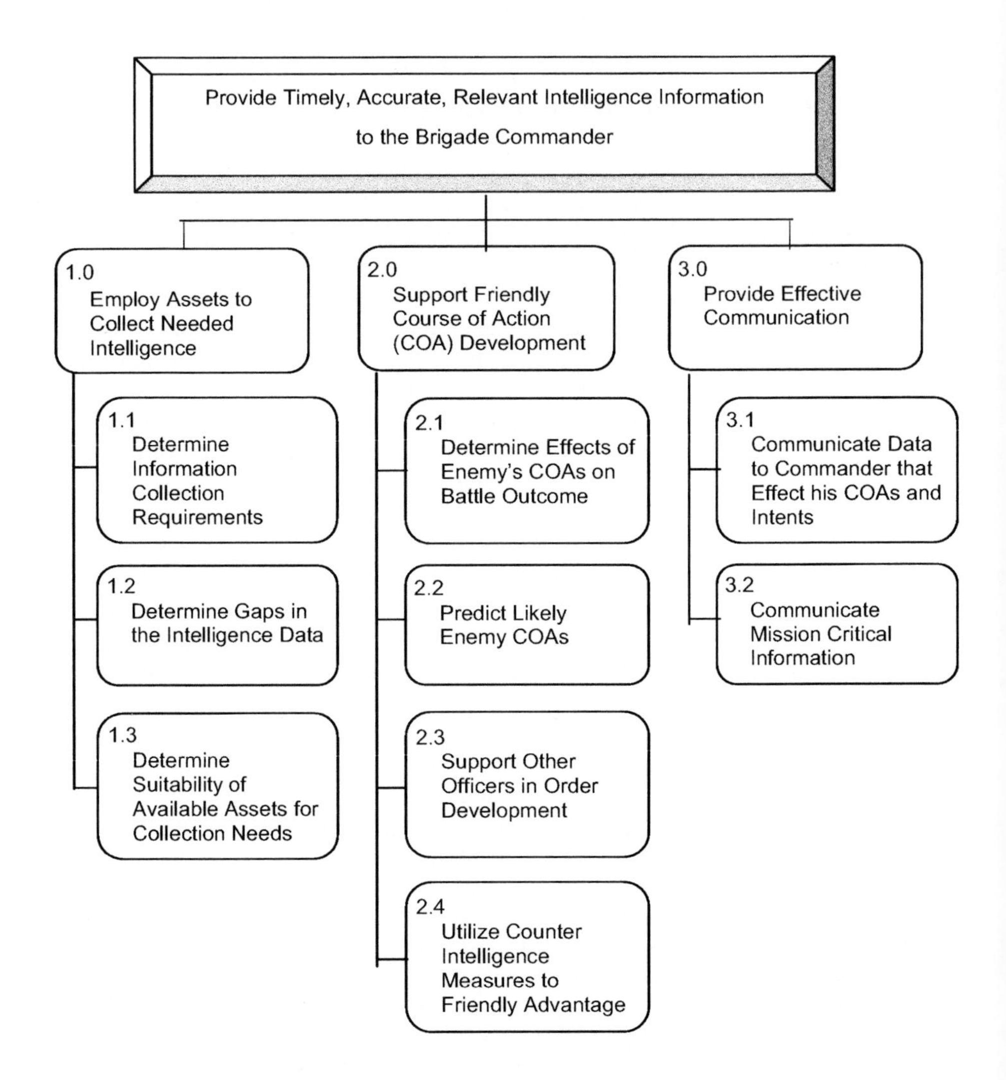

Figure 5.4 Goal hierarchy for Intelligence Officer (US Army).

5.4.1.1 Goals versus tasks

Confusion can result if the interviewer is not careful in distinguishing between two types of activities: physical tasks and cognitive demands. Physical tasks are not goals; they are things the operator must physically accomplish, e.g., fill out a report or call a coworker. Cognitive demands are the items that require expenditure of higher-order cognitive resources, e.g., predict enemy course of action (COA) or determine the effects of an enemy's COA on battle outcome. Performing a task is not a goal, because tasks are technology dependent. A particular goal may be accomplished by means of different tasks depending on the systems involved. For example, navigation may be done very differently in an automated cockpit as compared to a nonautomated cockpit. Yet, the SA needs associated with the goal of navigation are essentially the same (e.g., location or deviation from course).

Although tasks are not the same as goals, the implications of the task should be considered. A goal may be being addressed by a task the SME mentions. For example, consider the following portion of a preliminary draft of the example GDTA (Figure 5.5).

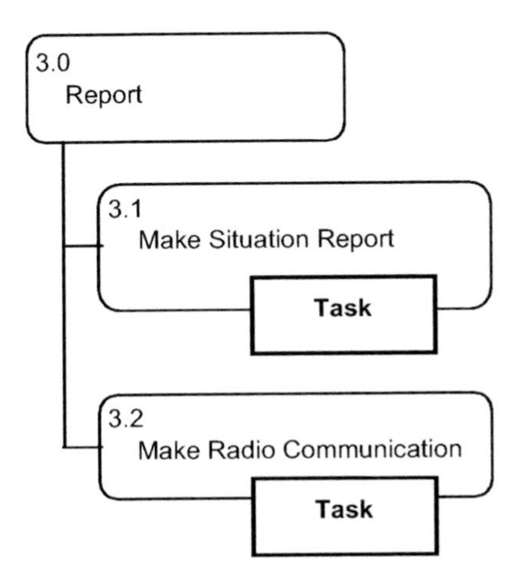

Figure 5.5 Preliminary subset of S2 GDTA.

In this example, both of the items listed as subgoals are tasks rather than goals. However, these actions express real needs: keeping the Commanding Officer and others abreast of critical information. When these needs are rewritten to be more reflective of the actual goal of the task, the following goal structure emerges (Figure 5.6). By focusing on the goals rather than the tasks, future uses of the GDTA will not be constrained. In some future system, this goal may be achieved very differently than by the present task (e.g., the information may be transferred through an electronic network rather than a verbal report).

5.4.1.2. Goals versus information requirements

Another common problem encountered while constructing the GDTA involves appropriately separating goals from information needs. Information requirements can be incorrectly phrased as a subgoal or as a decision. For example, consider subgoal 1.1 from a subset of the preliminary Intelligence Officer GDTA (Figure 5.7). Several items listed as subgoals actually seek to identify a single piece of information, thereby making them an information requirement as opposed to a subgoal. As the process of creating the hierarchy continues, stop often to ask 'What am I trying to do with this?' or 'What am I talking about?' When a goal is listed to 'determine' something, the question needs to be asked 'Why am I trying to determine this? What is the goal?' Items that seem to be a goal because something needs to be determined often turn out to be information requirements rather than actual goals. For example, 'enemy strengths and weaknesses' is an information requirement, not a goal. It is done to meet a higher-level goal of determining the best course of action.

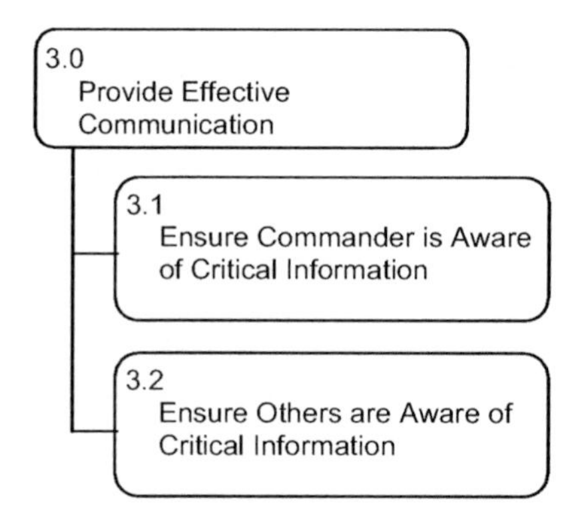

Figure 5.6 Revised subset of Intelligence Officer GDTA.

After defining a preliminary set of goals, you should review the goal structure to determine if goals that originally seemed distinct, albeit similar, can actually be combined under the auspices of a single goal category. This will save a lot of repetition in the analysis. For example, goals involving planning and replanning rarely need to be listed as separate major goals; often representing them as branches under the same major goal is sufficient to adequately delineate any distinct decisions or SA requirements. When combining goals that are similar, all of the information requirements associated with the goals should be retained. If the information requirements are very similar but not quite the same, separate subgoals may be in order, or perhaps future interviews will resolve the inconsistency.

If several goals seem to go together, this clustering might be an indication that the goals should be kept together under the umbrella of a goal one level higher in the hierarchy.

5.4.1.3 Callouts

At times, a subgoal will be essential to more than one goal. In these cases, the subgoal should be described in depth at one place in the GDTA, and then in subsequent utilizations, it can be 'called out,' referencing the earlier listing rather than repeating it in its entirety. Cross-referencing these callouts helps maintain the clarity of the GDTA, makes revisions easier, and minimizes redundancies (see Figure 5.8).

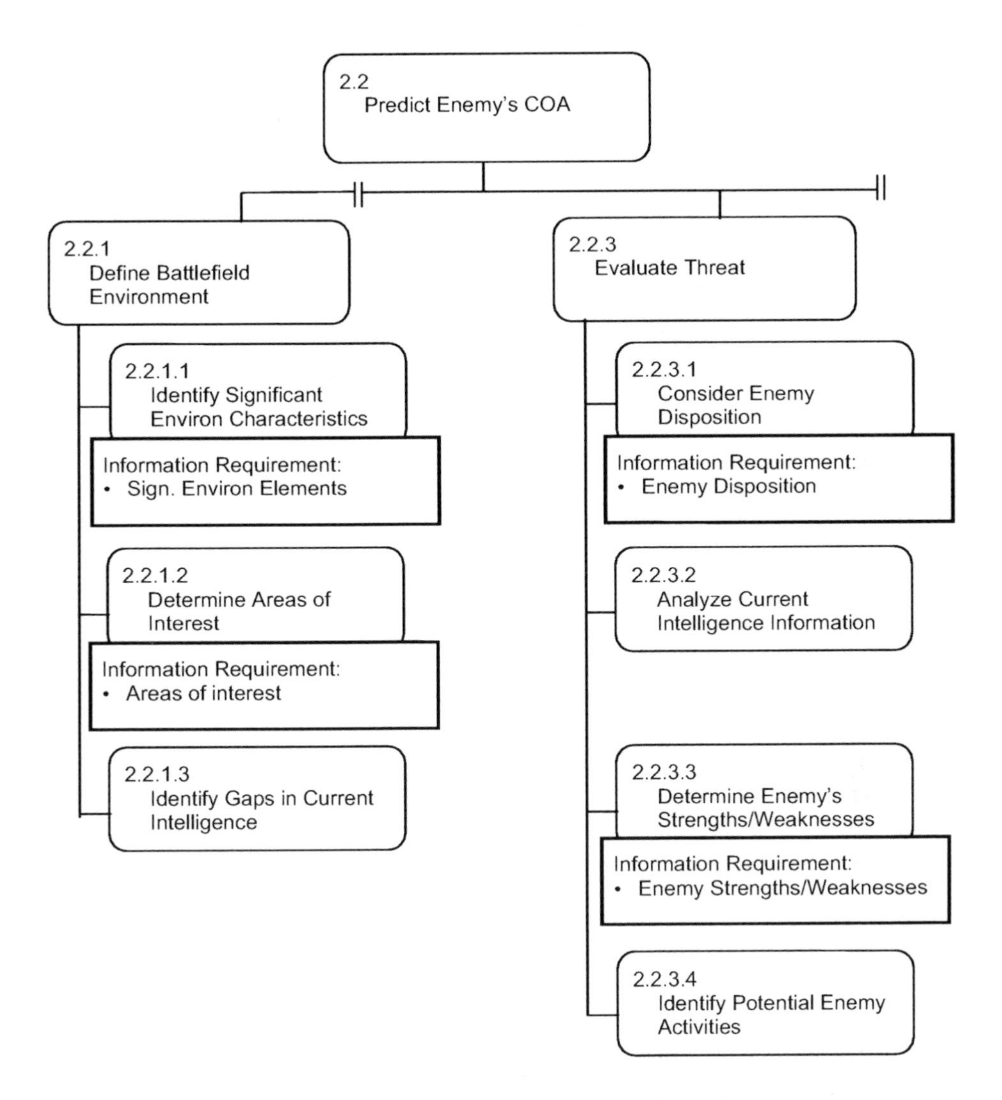

Figure 5.7 Subset of S2 preliminary subgoal hierarchy.

Using callouts keeps the hierarchy manageable and improves the cohesiveness of the document. If a particular goal cannot be accomplished without the output from other blocks or subgoals, the block can be referenced where it is needed and should be positioned in the hierarchy beneath the goal needing the output of that subgoal.

Utilizing callouts instead of repeating information results in an additional benefit during the interviews as these subgoals can be discussed at one point rather than inadvertently taking up valuable interview time reviewing the same subgoals.

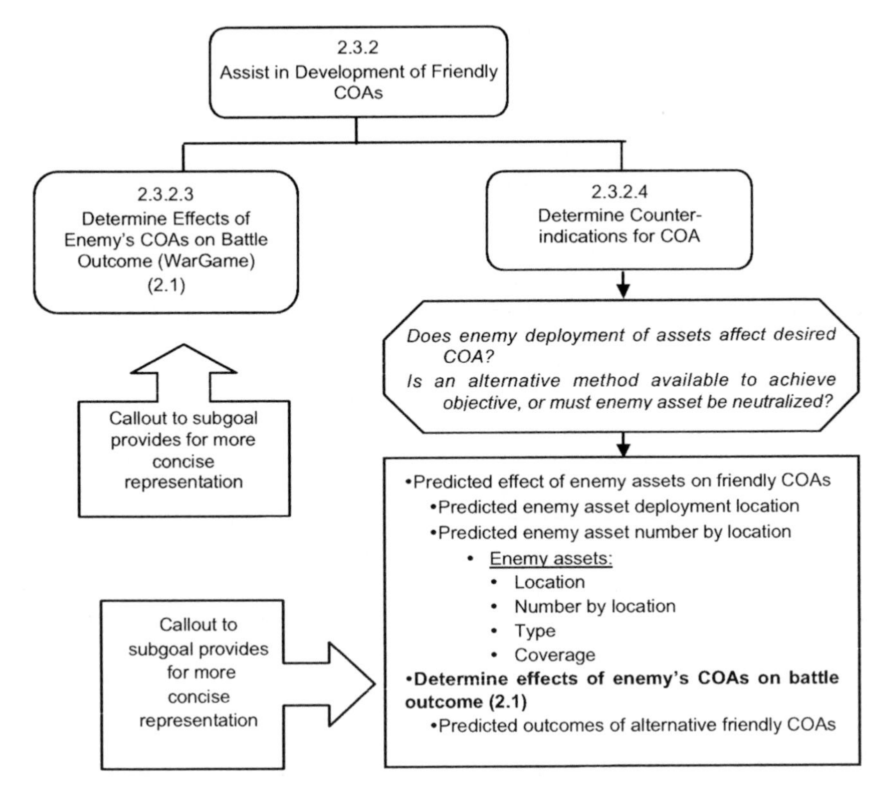

Figure 5.8 Subgoal hierarchy with callout (cross-reference).

5.4.1.4 Summary

Items that are appropriate for designation as goals are those items requiring operator cognitive effort and are essential to successful task completion. They are higher-level items as opposed to basic information requirements. They are not decisions that need to be made; this definition would too narrowly define the goal. They are a combination of sometimes competing subgoals and objectives which must be accomplished in order to reach the person's overall goal. The goals themselves are not decisions that need to be made, although reaching them will generally require that a set of decisions and a corresponding set of SA requirements be achieved. It can take some effort, but determining the actual goals of a particular position is a critical first step upon which the remainder of the analysis depends.

5.4.2 Decisions

The decisions that are needed to effectively meet each goal in the goal hierarchy are listed beneath the goals to which they correspond. These decisions (and their corresponding SA requirements) can be listed separately or bunched in the hierarchy, depending on personal preference and the size of the goal hierarchy (Figure 5.9). One advantage for listing them separately is that the SA requirements for each decision can be easily discerned from each other. This aids in the process of insuring that all information needed for a decision is present. If after final analysis the SA requirements for several decisions show complete overlap, they can then be combined in the representation for conciseness.

Decisions are posed in the form of questions, and the subsequent SA requirements provide the information needed to answer the questions. Although decisions are posed as questions, not all questions qualify as a decision. Questions that can be answered 'yes/no' are not typically considered appropriate decisions in the GDTA. For example, the question 'Does the commander have all the information needed on enemy courses of action?' can be answered with yes/no and is not a good decision. Typically, this is a cross-check when doing an assessment and not a real central decision the person is trying to make. On the other hand, 'How will the enemy courses of action affect friendly objectives?' requires more than a yes/no answer and is a decision that is pertinent to the person's goals. Further, if a question's only purpose is to discern a single piece of information, it is not a decision, rather it is an information requirement and belongs in the SA requirements portion of the hierarchy. You should ask 'why do you want to know that?' to determine which goal and decision the information need belongs to.

5.4.3 SA requirements

To determine the SA requirements, take each decision in turn and identify all the information the operator needs to make that decision. The information requirements should be listed without reference to technology or the manner in which the information is obtained. When delineating the SA requirements, be sure to fully identify the item for clarity, e.g., instead of 'assets,' identify as 'friendly assets' or 'enemy assets.' Although numerous resources can be utilized to develop an initial list of SA requirements (e.g., interview notes or job manuals), once a preliminary list is created, the list needs to be verified by domain experts.

Typically, the experts will express many of their information needs at a data level (e.g., altitude, airspeed, pitch). You will need to probe them to find out why they need to know this information. This will prompt them to describe the higher-level SA requirements—how this information is used. For example, altitude may be used (along with other information) to assess deviations from assigned altitude (Level 2 SA) and deviations from terrain (Level 2 SA). You should probe the experts carefully to learn how the information is used—what the expert really wants to know from the data he or she lists. Generally when you find a higher-level assessment that is being made, you will need to probe further to find all of the lower-level information that goes into that assessment. 'What else do you need to know to determine enemy strengths?'

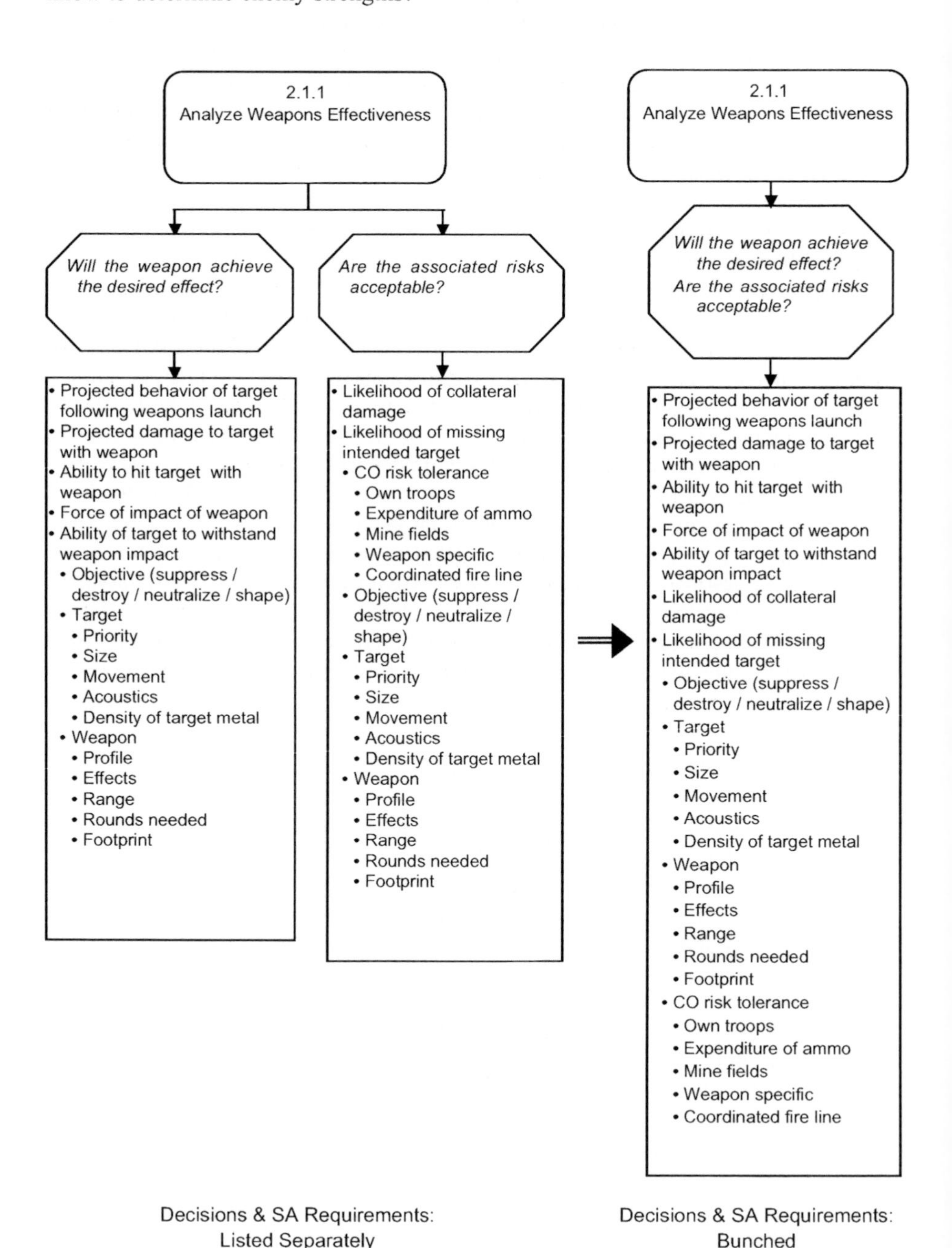

Figure 5.9 Separated and bunched decision requirements.

As you go through the interview you should also listen closely for predictions or projections the person may be making when reaching a decision. Try to distinguish between what they are assessing about the current situation, 'where is everyone at?' (i.e., location of troops) and what they are projecting to make a decision, 'where will the enemy strike next?' (i.e., projected location of enemy attack). These items form the Level 3 SA requirements in the analysis. Generally, a number of Level 1 and Level 2 SA items will go into each Level 3 projection. At times, however, you will also find that Level 3 projections can go into Level 2 SA assessments.

The SA requirements can be listed in the GDTA according to the level of SA by using an indented stacked format (refer to Figure 5.2). This format helps to ensure the SA requirements at each level are considered and generally helps with the readability of the document. The organization provides useful structure. However, the three levels of SA are general descriptions that aid in thinking about SA. At times, definitively categorizing an SA requirement into a particular level will not be possible. For example, in air traffic control the amount of separation between two aircraft is both a Level 2 item (distance now) and a Level 3 item (distance in future along their trajectories). Not all decisions will include elements at all three levels—in some cases, a Level 2 requirement may not have a corresponding Level 3 item, particularly if the related decision is addressing current operations.

SA requirement blocks that are frequently referenced also can be turned into callout blocks. For example, weather issues at all levels of SA may frequently be of concern across many subgoals in the hierarchy. For ease and to reduce redundancy, a 'weather conditions' block can be created. This block can be called within the SA requirements section of multiple subgoals as 'Weather' and fully defined at the end of the GDTA (see Figure 5.10).

Weather Conditions

- Temperature
- Precipitation
- Wind
  - Direction
  - Magnitude
  - Surface winds
  - Aloft winds
- Visibility
- Day/Night
- Ambient Noise
- Tides
- Cloud Ceiling
- Moon Phases

Figure 5.10 SA requirements block: Weather.

Consistency is important for the SA requirements, and SA requirement callout blocks can be useful to ensure that each time a frequently used item is referenced all of its relevant information requirements (at all levels of SA) are included. In general, the different ways this weather information is used (i.e., the Level 2 assessment and Level 3 projections) will be listed in each subgoal block where the item is referenced. This is because the Level 2 and 3 assessments made from a set of information will be different, depending on the decisions being made and the relevant goals. So in one area they may want to know the impact of weather on troop movement, and in another area they may want to know the impact of weather on radio communications ranges.

5.4.4. General

After completing the process of organizing the interview notes into the relational hierarchy, existing manuals and documentation can be referenced to help fill in the holes. Exercise caution, however, since these sorts of documents tend to be procedural and task specific in nature; the information in the GDTA is concerned with goals and information requirements, not current methods and procedures for obtaining the information or performing the task. Although operators often don't explicitly follow written procedures found in the related documentation, evaluating the use of this information in the hierarchy can help to insure the GDTA is complete and can spark good discussions of how the operator actually performs a job and what the operator's true information needs are.

5.5 FUTURE INTERVIEWS

Once a draft GDTA is created, it can serve as a tool during future interview sessions. A good way to begin a session is to show the primary goal hierarchy to the interviewee and talk about whether the goal hierarchy captures all the relevant goals. After this discussion, one section of the GDTA can be selected for further review, and each component of that section (i.e., goals, subgoals, decisions, and SA requirements) can be discussed at length. Use the draft to probe the interviewee for completeness, 'what else do you need to know to assess this?', and to determine higher-level SA requirements, 'how does this piece of data help you answer this question?'. Showing the entire GDTA to the participant is generally not a good idea; the apparent complexity of the GDTA can be overwhelming to the participant and consequently counter-productive to data collection. If time permits after discussing one section of the GDTA, another subset can be selected for discussion.

After each round of interviews is complete, revise the GDTA to reflect new information. New information gleaned from iterative interviews allows for reorganizing and condensing of the goals within the structure to create better logical consistency. For example, when several participants bring up a low-level goal as an important facet of the job, the goal may need to be moved to a higher place in the hierarchy. Furthermore, as more information is uncovered concerning the breadth of information encompassed by the various goals and subgoals, the goals and subgoals can often be rewritten to create a more descriptive and accurate category. As the GDTA is reviewed further, commonalities become apparent that can be used to assist refining the charts. Often goals will seem to fit more than one category. In these cases, the goals should be positioned where they seem to belong most and can then be called out to other areas within the hierarchy as needed.

5.6 INTERVIEW ISSUES

5.6.1 Maximizing data collection

Obtaining time with domain experts is often a difficult proposition, so maximizing the time when it is available is essential. Several suggestions can be offered to this end. First, conducting the interview with two interviewers present minimizes downtime; one interviewer can continue with questions while the other interviewer quickly reviews and organizes notes or questions. The presence of two interviewers will also have a benefit beyond the interview; constructing the goal hierarchy is easier when two people with a common frame of reference (developed in part from the interview session) work on the task.

Second, the number of interviews performed on a given day should be limited. Time is needed to organize information gleaned from each interview and to update the charts to insure that subsequent sessions are as productive as possible. Interviewing too many people without taking time to update notes and charts can induce a 'brain fog' and negatively impact not only the interview sessions, but also the resulting data quality. Further, interviewer fatigue can be a considerable factor if too many interviews are scheduled without a break or if the interviews last too long. Two hours per session is generally the maximum recommended.

Finally, after discussing a set of GDTA charts with one participant, the charts should be updated to reflect the changes before being shown to another participant. Showing a chart to another participant before it has been edited may not be the best use of time; interviewer notations and crossed out items on the charts can be confusing to the participant and thus counterproductive.

5.6.2 Participant personality factors

Not all interviews will go smoothly and result in optimal data collection. Even the most experienced interviewer encounters sessions that wander from the direction the interviewer intends to pursue. Several things can contribute to this suboptimal experience and should not be taken as indicative of the interviewer's skill or ability to collect appropriate data. Although the majority of interview participants are willing to help and desire to give relevant information, participant personality factors can sometimes negatively influence the interview. Four interviewee personality types that can hamper or derail an interview will be considered, along with suggestions to mitigate their effect.

Chatty Cathy. Some participants have a propensity to wander in their conversation and to provide very few pauses during which an interviewer can inject questions. This situation can occur with people who are trying very hard to be helpful (and thus try to convey everything they know as quickly as possible), as well as with people who are naturally inclined to be talkative. Try to direct the conversation and maintain focus on the issues at hand, even if you have to interrupt. Say 'please wait for a minute. That was very interesting. Can I ask more about it before we continue?' This should not offend the interviewee.

Quiet Ken. On the opposite end of the spectrum from the chatty participant is the quiet participant. These participants are less outspoken and tend to provide very direct, brief answers. Try to draw out the participant and foster a continuing conversation in which the participant is comfortable providing lengthier, more descriptive answers. 'Tell me more about why you want to know that. Tell me about how you have used that information in the past.'

Skeptical Sam. Skepticism can be rooted in any number of factors (e.g., negative experiences with similar efforts or doubt regarding the interviewer's qualifications), many of which have nothing to do with the interview at hand. In any case, try to develop a rapport with the participant that will move past the participant's skepticism and allow for quality data collection. It helps if the participant understands how the data will be used. Stress that the quality of systems they will use in the future depends on the quality of the analysis to determine real user needs.

Agenda Jill. Participants may come to the interview with preconceived notions of what should be done in the domain the researcher is investigating. In these cases, the challenge is to help the participant move past their agenda in order for a valuable information exchange to occur. Often in these situations, simply allowing the participants to express their opinions about the real problems and issues in their job is necessary. After having their say, participants are more willing to answer questions, especially if the questions can leverage some aspect about which the participant feels so strongly.

Remember not all experts are equally good at expressing their knowledge, even if they are very good at doing their jobs. You may need to speak to several before finding one who is very good at expressing this type of information.

5.6.3 Audio/video recording

Audio or video recording of interview sessions is not generally recommended. The presence of the recording device may make participants more hesitant to speak their minds and less likely to relate examples of when SA was poor or mistakes were made. Fear of reprisal for admitting mistakes or for criticizing current systems/procedures can sometimes hamper the investigative process. This fear is typically less apparent and the information exchange more open when no recording devices are present. Recording the sessions also creates the additional burden of requiring extra work to transcribe or analyze the tapes afterwards. Participant anonymity should always be insured, in accordance with approved Human Subject Use Guidelines.

5.7 ORGANIZATIONAL TIPS

Creating a GDTA can be a seemingly overwhelming task at times. Listed next are some tips to simplify data management.

1. Once a draft of the GDTA is created, number the pages to minimize confusion and facilitate the discussion. During a collaborative session, locating page 7 is easier than locating 'Subgoal number 1.2.3.1'.
2. When reorganizing the GDTA, manually notate on a clean copy of the GDTA all the changes. The working copy of the GDTA will quickly become messy as various options are considered for organizing the goals and SA requirements. Once the next round of changes have been decided, make all the notations on a clean copy before trying to change the charts in the computer. Charting software or an outlining program make the job easier.
3. As soon as the reorganization process allows, make a new initial structure page to reduce confusion and provide an overall structure for future changes.
4. Do not get too hung up on delineating the different SA levels; these are not concrete, black and white items, and it is conceivable that a particular item can change SA levels over time. For example, something may be a comprehension item because a person must integrate information, but as technology improves, the information may be given directly to a person and thus become a Level 1 (perception) issue instead of a comprehension issue. Furthermore, the nature of the SA requirements may mandate nested levels (i.e., it can have nested Level 2s or 1s, and possibly even Level 3s). Thinking about the different SA levels is mainly an aid to make sure you consider how information is used.
5. As the GDTA hierarchy is being constructed, make notes of any questions that arise concerning how or where something fits. This list can be brought to the next interview session and clarification requested.
6. Begin organizing notes into a preliminary GDTA as soon as possible after the interview while the conversation is still fresh in your memory.
7. If the same thing is being considered at various places to achieve essentially the same goal, combine them. For example, the questions 'How does the enemy COA affect ours?' and 'How does the enemy COA affect battle outcome?' are essentially the same and can be combined.
8. When considering whether an item qualifies as a goal, ask the question 'Why?'. The answer to the question will help position the item within the hierarchy. For example, in the current example, asking the question 'Why?' for the possible goal 'Evaluate/analyze the situation' brings an answer such as 'To support friendly COA development'. Thus,

'Evaluate/analyze the situation' is not a separate major goal, rather it should be encompassed within the hierarchy of 'Support friendly COA development.'

5.8 GDTA VALIDATION

Creating a comprehensive GDTA for a particular job is hard work. It will take many interviews with subject matter experts; it is not uncommon for it to take anywhere from 3 to 10 sessions, depending on the complexity of the position. Even then, there can be a fair degree of subjectivity involved on the part of the analyst and the experts. These problems are common to all task analysis methods, and GDTA is no exception.

To help insure that the final GDTA is as complete and accurate as possible, it is best to try to validate it with a larger group of subject matter experts. Printouts of the final GDTA can be distributed to 10–20 experts with instructions on how to interpret it and asking them to identify missing information or errors. Needed corrections can then be made.

Bear in mind that in some cases a particular expert will report that he or she does not consider certain information or do particular subsets of the GDTA. This should not mean that you should eliminate those items, however, if other experts report using that data or performing those subgoals. We are looking for ideal SA—what people should ideally consider—and some people may not be as expert in considering the full range of factors as others. In addition, some people will have had slightly different experiences than others and simply may have never been in a position to execute on all possible subgoals. As the GDTA will form the basis for future design, however, we want to make sure that the full breadth of possible operations and information requirements is considered.

An additional way to validate the GDTA is through observation of actual or simulated operations by experienced personnel in the position. While it can be difficult to always know exactly what the operators are thinking (unless they are instructed to 'think aloud,' performing a verbal protocol), these observations can be used to check the completeness of the GDTA. It should be possible to trace observed actions, statements, and activities back to sections of the GDTA. If people are observed to be performing tasks that there is no apparent reason for in the GDTA, or looking for information that is not identified in the GDTA, follow up should be conducted to determine why. Any additions to the GDTA that are needed should be made based on these validation efforts.

Once the GDTA has been completed, it forms the basis for designing systems that will support operators in achieving a high level of SA across these requirements, as needed in meeting each goal. Chapters 6 through 11 provide a detailed discussion of design guidelines that can help the designer develop system designs that will be effective at transmitting required SA to operators in a wide variety of challenging conditions.

CHAPTER SIX

Principles of Designing for SA

6.1 FROM THEORY TO DESIGN

The way in which information is presented to the operator through the interface greatly influences SA by determining how much information can be acquired in the limited time available, how accurately it can be acquired, and the degree to which that information is compatible with the operator's SA needs. In general, our goal is to create system interface designs that transmit needed information to the operator as quickly as possible and without undue cognitive effort. Reducing the mental workload associated with system interfaces has been a consideration in design efforts for many years. At the same time, the level of SA provided (the outcome of that mental effort) needs to be considered. Several general principles for creating systems that enhance the operator's SA are presented here. These principles were developed based on an understanding of the factors that affect SA in complex systems (Chapters 2 and 3). Following the presentation of these design principles, we will show a case study that illustrates the application of some of these design principles to a real-world system design problem.

Principle 1: Organize information around goals

Information should be organized in terms of the operator's major goals, rather than presenting it in a way that is technology-oriented—displayed based on the sensors or systems which created the information (e.g., oil pressure or temperature). It should be organized so that the information needed for a particular goal is co-located and directly answers the major decisions associated with the goal. The SA requirements analysis provides the input needed to determine which information is required for addressing each goal. This analysis should guide the determination of how to group information across multiple displays and to ensure that all the needed information is provided, to the extent possible.

Principle 2: Present Level 2 information directly—support comprehension

As attention and working memory are limited, the degree to which displays provide information that is processed and integrated in terms of Level 2 SA requirements will positively impact SA. For instance, directly portraying the deviation between a current value and its expected (or required) value is better than requiring the operator to calculate this information based on lower level data. Level 2 SA requirements vary between systems, but typically include considerations of prioritization, congruence between data values and requirements, and the effect of various factors on other systems or on operator goals. In some cases, a system calculation will be required to provide this information. In other cases, the presentation of the information in juxtaposition with other information (relevant comparators) will make comprehension easy for the operator. For example, showing the current altitude of an aircraft portrayed next to a line designating its assigned altitude makes it easy for operators to visually detect deviations from assigned levels. In some cases, the Level 2 SA information is what is really needed and low level data will not be required. In other cases, both the Level 1 and 2 data are important for building complete SA.

An example of how a technology-centered approach, which simply presents data generated by system sensors, can be replaced by a human-centered approach, which presents the information the operator really needs in terms of integrated Level 2 data, is illustrated by the Oz project (Still & Temme, 2001). Traditionally pilots have been required to mentally calculate the flight envelope and where their aircraft parameters fall within that envelope (e.g., the deviation between the current aircraft state and its limitations). Available thrust, as affected by drag, airspeed, and configuration, were mentally calculated based on aircraft dials and gauges.

The Oz concept (Figure 6.1) presents a graphical image to the pilot that directly portrays this complex relationship. In this representation, the pilot only needs to keep the horizontal bar, or strut, between the left and right edges of the bent 'wings' shown on the screen. The length of the strut shows how much thrust is available, and how much is in use (the inner portion of the bar). The edges of these wings show the maximum safe structural speed and the minimum flying speed the pilot must stay between, as well as the optimal minimum drag speed shown by the V in the wing. This drag curve is automatically adjusted for aircraft configuration (weight, landing gear, flaps, etc.) and changes in density altitudes, saving the pilot from mental calculations. What he really needs to know—where his current aircraft performance falls relative to aircraft limits and optimal flight control values—is always directly displayed. A number of other factors are also integrated into the display, including information on turn rate, pitch, and roll, which are graphically depicted and readily determined with a quick glance. Research shows that not only can people learn to fly much more quickly with the Oz display than with traditional cockpit displays, but they are better able to retain desired flight control under heavy turbulence and have more mental resources available for attending to other tasks while flying (Still & Temme, 2001).

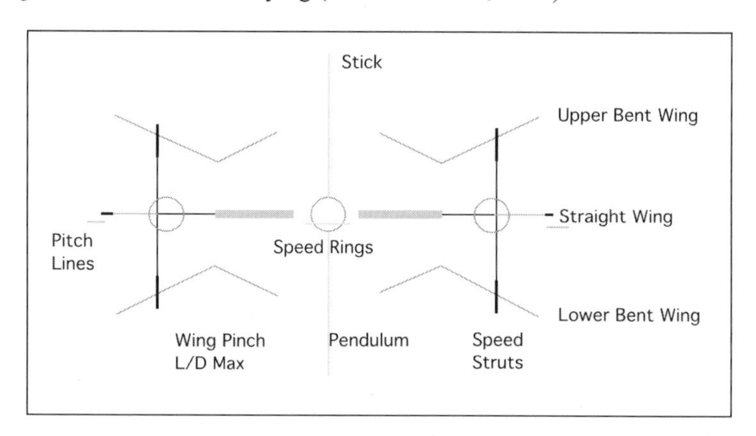

Figure 6.1 Oz Cockpit Display (from Still & Temme, 2001).

Principle 3: Provide assistance for Level 3 SA projections

One of the most difficult and taxing parts of SA is the projection of future states of the system. Projection requires a fairly well-developed mental model. System-generated support for projecting future events and states of the system should directly benefit Level 3 SA, particularly for less experienced operators. While creating system projections for some information may be difficult, it can be accomplished for many operator tasks. For instance, a trend display, graphing changes in a parameter over time, can be very useful for helping operators project future changes in that parameter (see Figure 6.2).

Displays that allow operators to anticipate possible occurrences (such as a display of aircraft detection envelopes for enemy aircraft) have been shown to support the operator's own ability to create accurate projections (Endsley & Selcon, 1997) (see Figure 6.3).

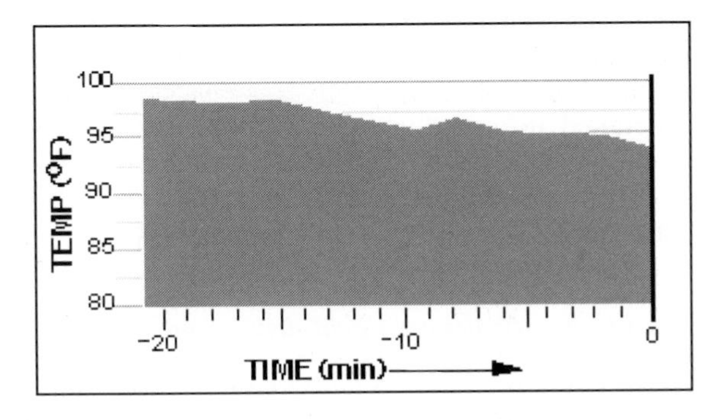

Figure 6.2 Making projections from trend graphs.

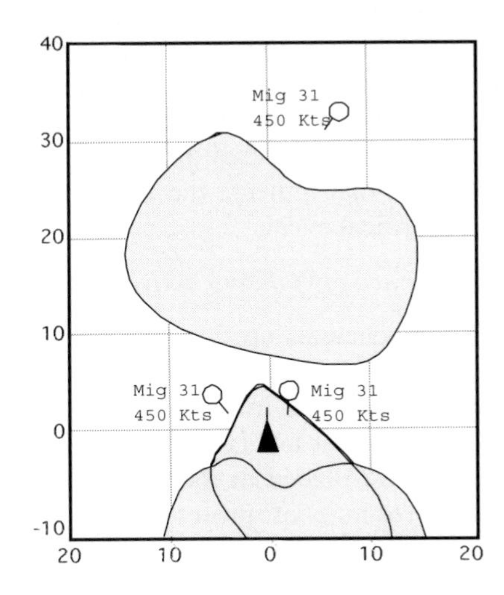

Figure 6.3 Aircraft detection envelopes enable higher levels of projection–Level 3 SA (Endsley & Selcon, 1997).

In this example, missile launch zones were calculated, not statically, but dynamically in relation to the pilot's changes in altitude, attitude, and airspeed. These zones then provided the dynamics needed to allow the pilots to generate good predictions of how their path and flight control actions would affect their vulnerability to the other aircraft. In other cases, information on component dynamics can be used to create displays of projected component behavior over a designated period of time.

Principle 4: Support global SA

A frequent problem for SA occurs when attention is directed to a subset of information, and other important elements are not attended to, either intentionally or unintentionally (Jones & Endsley, 1996). Designs that restrict access to information only contribute to this attentional narrowing. Excessive menuing and windowing, for example, often obscures information on other windows that should signal operators to attend to other, more important information.

Attentional narrowing can be discouraged through the use of displays that provide the operator with the 'big picture' or *global SA*. Global SA—a high level overview of the situation across operator goals—should always be provided. Simultaneously, detailed information related to the operator's current goals of interest should be provided as required. In many systems, a global SA display that is visible at all times may be needed. Global SA is critical for accurately determining which goals should have the highest priority and for enabling projection of future events.

Principle 5: Support trade-offs between goal-driven and data-driven processing

Designs need to take into consideration both top-down and bottom-up processing. The design of the system around operator goals (Principle 1) will support goal-driven processing. The big picture display that supports global SA (Principle 4) will support data-driven processing by directing the operator as to where to focus attention to achieve high priority goals. The key is to ensure that these two approaches complement each other. As environmental cues with highly salient features (e.g., flashing lights, bright colors, loud sounds) will tend to capture attention away from current goal-directed processing, these characteristics should be reserved for critical cues that indicate the need for activating other goals, and should be avoided for noncritical events.

Principle 6: Make critical cues for schema activation salient

In that mental models and schemata are hypothesized to be key features used for achieving the higher levels of SA in complex systems, the critical cues used for activating these mechanisms need to be determined and made salient in the interface design. If particular fuel levels, for instance, trigger different classes of situations, markers should be placed at these 'breakpoints.' In particular, those cues that indicate the presence of prototypical situations will be of prime importance. Kaplan and Simon (1990) found decision making is facilitated if the critical attributes are perceptually salient.

Determining just what these critical cues are is often quite difficult however. Operators often cannot verbalize the subtle cues that they may attend to. Those that can be verbalized and are known to be important, should be coded on the display with salient cueing that supports operator decision making.

Principle 7: Take advantage of parallel processing capabilities

The ability to share attention between multiple tasks and sources of information is important in any complex system. System designs that support parallel processing of information by the operator should directly benefit SA. While people can only visually take in so much information at one time, they are more able to process visual information and auditory information simultaneously. Similarly, they can process tactile information (felt through the skin) simultaneously with auditory or visual information, as these different modalities draw on different cognitive resources (Wickens, 1992).

Tactile displays placed on the arm, head, or other body part, for example, provide information by electrically or mechanically stimulating the skin. This alternate form of information transmission is being examined as a way of increasing SA in visually dense cockpits and space applications (Gilliland & Schlegel, 1994; Rochlis & Newman, 2000; Zlotnik, 1988). Similarly, sophisticated auditory displays that provide cues that make the sounds appear to be spatially distributed are being explored as a means of enhancing SA (Begault & Wenzel, 1992; Calhoun, Janson, & Valencia, 1988; Calhoun, Valencia, & Furness, 1987; Endsley & Rosiles, 1995; Forbes, 1946; McKinley, Erickson, & D'Angelo, 1994; Wenzel, Wightman, & Foster, 1988). These displays convey spatial information in new ways that don't load a person's visual channel. The degree to which multiple modalities can be used to increase information throughput is an active research area that promises to aid SA in many domains that suffer from information overload.

Principle 8: Use information filtering carefully

The problem of information overload in many systems must still be considered. The filtering of extraneous information (not related to SA needs) and reduction of data (by processing and integrating low level data to arrive at SA requirements) should be beneficial to SA. In many systems, information filtering has been proposed as a means of aiding SA. In these cases, the computer presents just that information it determines the operator needs at any one time. It is thought that such an approach will reduce data overload and thus improve SA. Unfortunately, this type of information filtering approach can actually seriously degrade SA (Endsley & Bolstad, 1993).

First, information filtering deprives people of the global SA they need to be predictive and proactive. Operators do not immediately have SA simply by looking at instantaneously presented information. A certain amount of time is required to orient oneself to a situation and to ascertain the players and their critical features. Furthermore, SA is developed across a period of time by observation of system dynamics, and information filtering is detrimental to this process. The development of SA on system dynamics and trends built up over time will be lacking.

Second, the operator needs to be able to respond to not only immediate crises, but to look ahead to what is coming up in order to identify developing situations (Level 3 SA). This allows them to plan ahead to avoid unwanted situations, to develop a tactical strategy for dealing with possibilities, or to prime themselves for possible actions thus minimizing reaction time. These actions are only possible if operators can look ahead to develop this higher level of SA.

Third, individual differences must be considered with respect to the formation of information filtering schemes. Individual operators may need and use different types of information to form their SA and doing so would not be possible with computer-driven information filtering.

Presenting information in a clear and easy to process manner, with the operator in charge of determining what they will look at when, is far better than computer-driven strategies for providing only subsets of information. The operator will be able to maintain a higher level of system awareness (understanding of what the system is doing) and a better level of predictive awareness that is needed for high levels of SA and proactive behavior. Only information that is truly not needed should be eliminated.

6.2 CASE STUDY: SA-ORIENTED DESIGN

To more clearly illustrate these design principles, a real-world case study is provided based on maintenance control centers in the U.S. Federal Aviation Administration (Endsley & Kiris, 1994c). The air traffic control system is in a period of transition. The replacement of antiquated equipment for controlling air traffic affects not only controllers, but also those responsible for overseeing and maintaining the vast network of systems that make air traffic control possible. These systems include beacons and communications equipment that span the country, as well as a wide variety of systems that support pilots in landing at airports in different weather and visibility conditions.

In the past, technicians had to drive long distances to find out how a particular system was functioning and try to diagnose the problem if something was wrong. If a problem did exist, it would take considerable time to drive back to headquarters to get parts. Yesterday's manual system for maintaining these airway facilities is being replaced with a newer system featuring centralized monitoring of remote information at maintenance control centers (MCC). The MCCs are responsible for ensuring the ongoing integrity of this complex, geographically dispersed system. Many challenges must be met to achieve this transition. The success of the new system in realizing critical aviation goals of reliability and safety depends on the degree to which the system supports maintenance personnel in their tasks, including monitoring system status for operational problems, diagnosing problems and determining solutions, and conducting preventive maintenance activities.

The case study's objective was to evaluate existing MCC systems in terms of the degree to which they support the SA needs of the MCC specialist, and recommend a redesign of the system as needed to improve SA. This objective was addressed through three main efforts: (1) determining the major tasks and SA requirements of an MCC specialist, (2) evaluating the degree to which the existing systems support those needs in accordance with the SA design principles provided here, and (3) development of a new system interface for improving the SA of MCC specialists.

6.2.1 MCC SA requirements analysis

A detailed analysis was undertaken to determine the SA requirements of the MCC specialist (using the methodology described in Chapter 5). The analysis was conducted through a review of available documentation, and through observation and interviews with MCC specialists to determine goals, decisions, and information requirements. The analysis was conducted at the Austin, Texas MCC which, at the time of the study, had been operating for several years and was considered one of the best in the country. The analysis incorporated input from three MCC specialists. In addition, a visit was paid to a second MCC located at the Dallas-Fort Worth Airport to provide a comparative basis for assessing between-MCC variations in procedures and responsibilities. The list of SA requirements developed was reviewed by specialists at the Austin MCC for accuracy and completeness, and modified accordingly.

A list of SA requirements was developed, including (1) data requirements (Level 1 SA), (2) higher level information regarding the significance of that data when integrated (Level 2 SA), and (3) projection of future events and trends regarding facilities in the sector (Level 3 SA) that allows specialists to be effective and timely decision makers.

This assessment of SA requirements incorporates the information needed to perform the major goals and tasks of the MCC specialist. These were defined as: (1) monitoring of sector status, (2) control (real-time and proactive), (3) communication and coordination of activities within the sector, and (4) reporting of sector status and activities. The SA requirements for supporting each of these goals is provided in Figure 6.4. The figure shows for each major goal a delineation of the major subgoals, and Levels 3, 2, and 1 SA requirements associated with each subgoal.

6.2.2 MCC system evaluation

Following the detailed SA analysis, an evaluation was made of the existing systems available for supporting the MCC specialist in performing his or her functions. The SA requirements analysis was used as the key starting point in the analysis. It points out exactly what information is required by operators when involved in pursuing each of their various goals. The analysis was used to determine the degree to which the required information was made readily available through the system. If certain data was needed but was not present, or was difficult to obtain (in various reference manuals, for instance), this fact was highlighted. The SA analysis was used to point out the higher level integrated and goal-directed information that is required by operators, but not supplied directly by the system design. Which information must be used together was also considered so that appropriate grouping and arrangement of information could be determined.

In addition, the degree to which the interface enhanced or compromised situation awareness was assessed based on the SA-oriented Design principles presented in this chapter. This assessment addressed more than whether information was present,

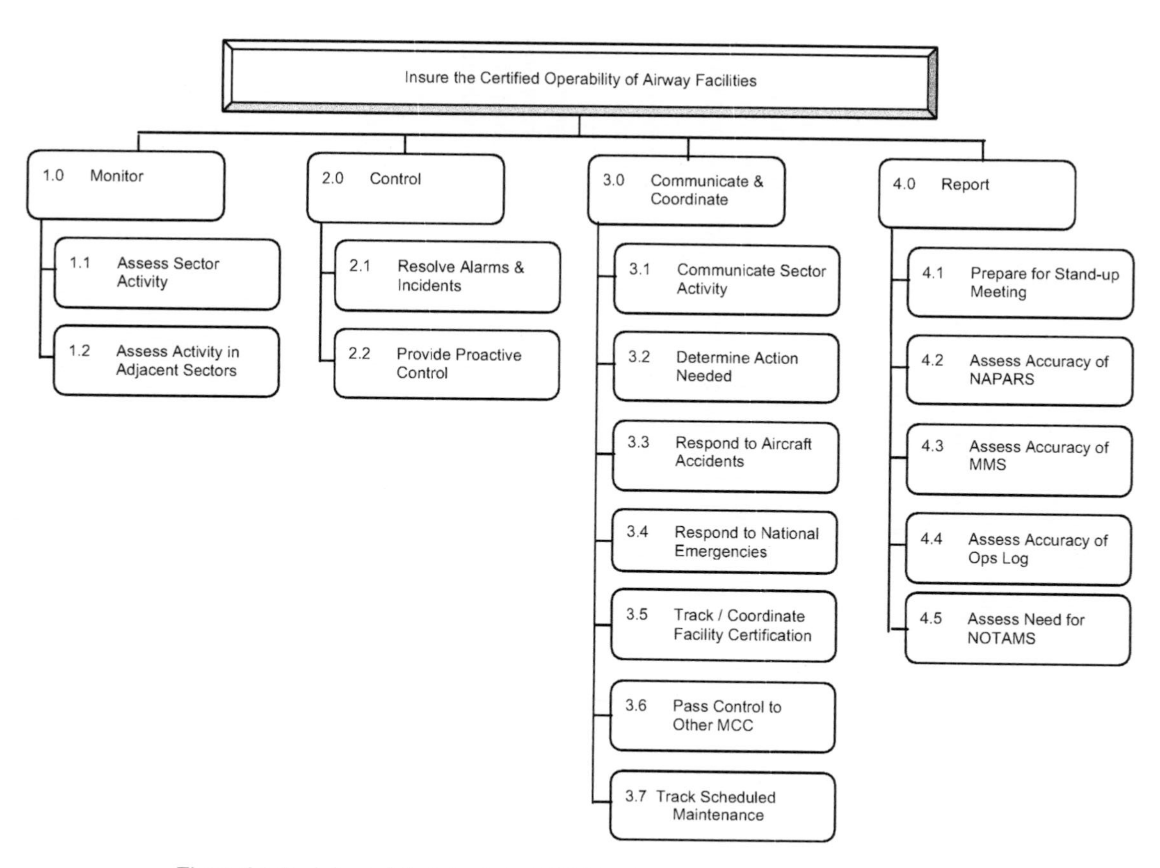

Figure 6.4 Goal-directed Task Analysis: MCC Situation Awareness Requirements

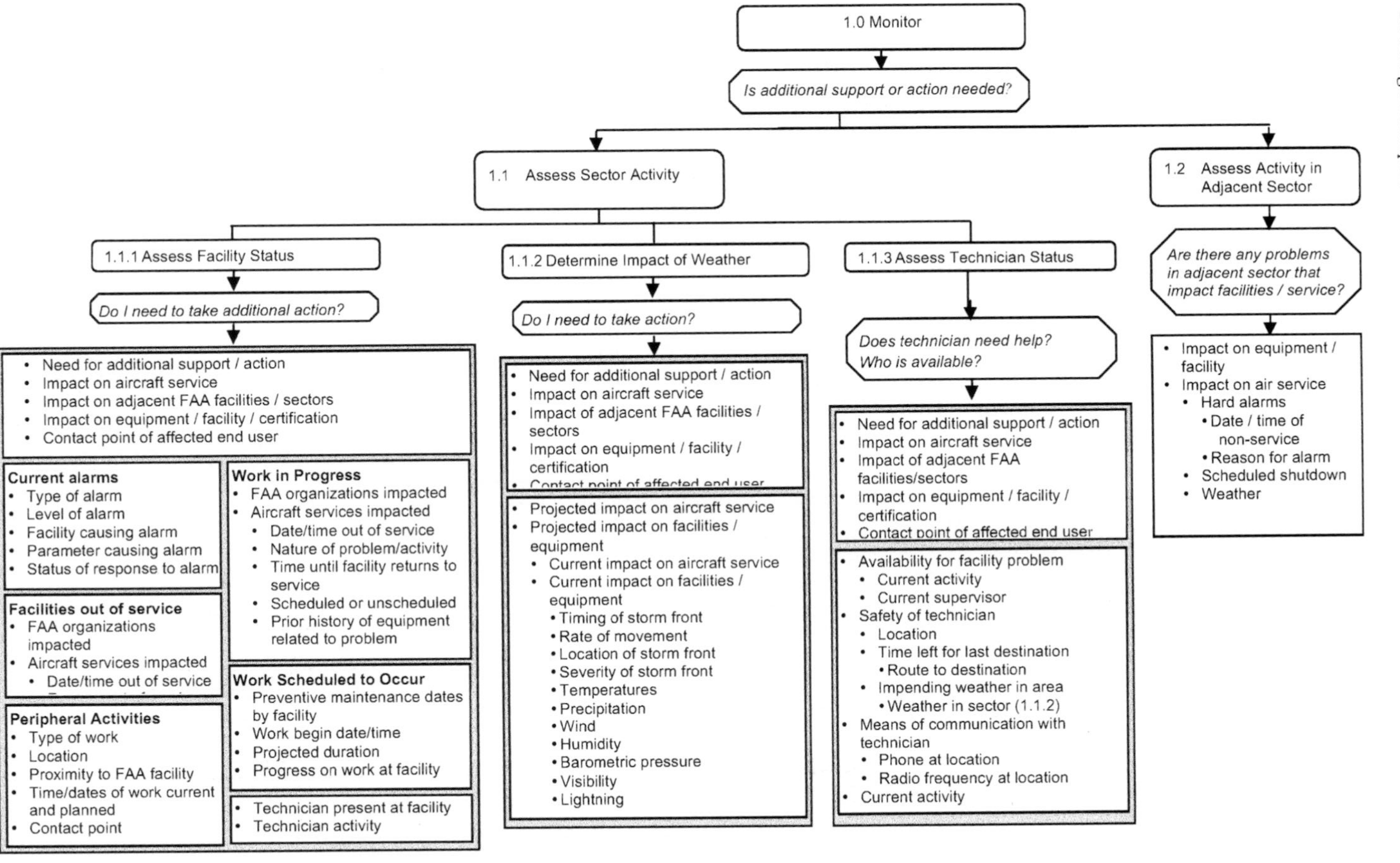

Figure 6.4 Goal-directed Task Analysis: MCC Situation Awareness Requirements (Continued)

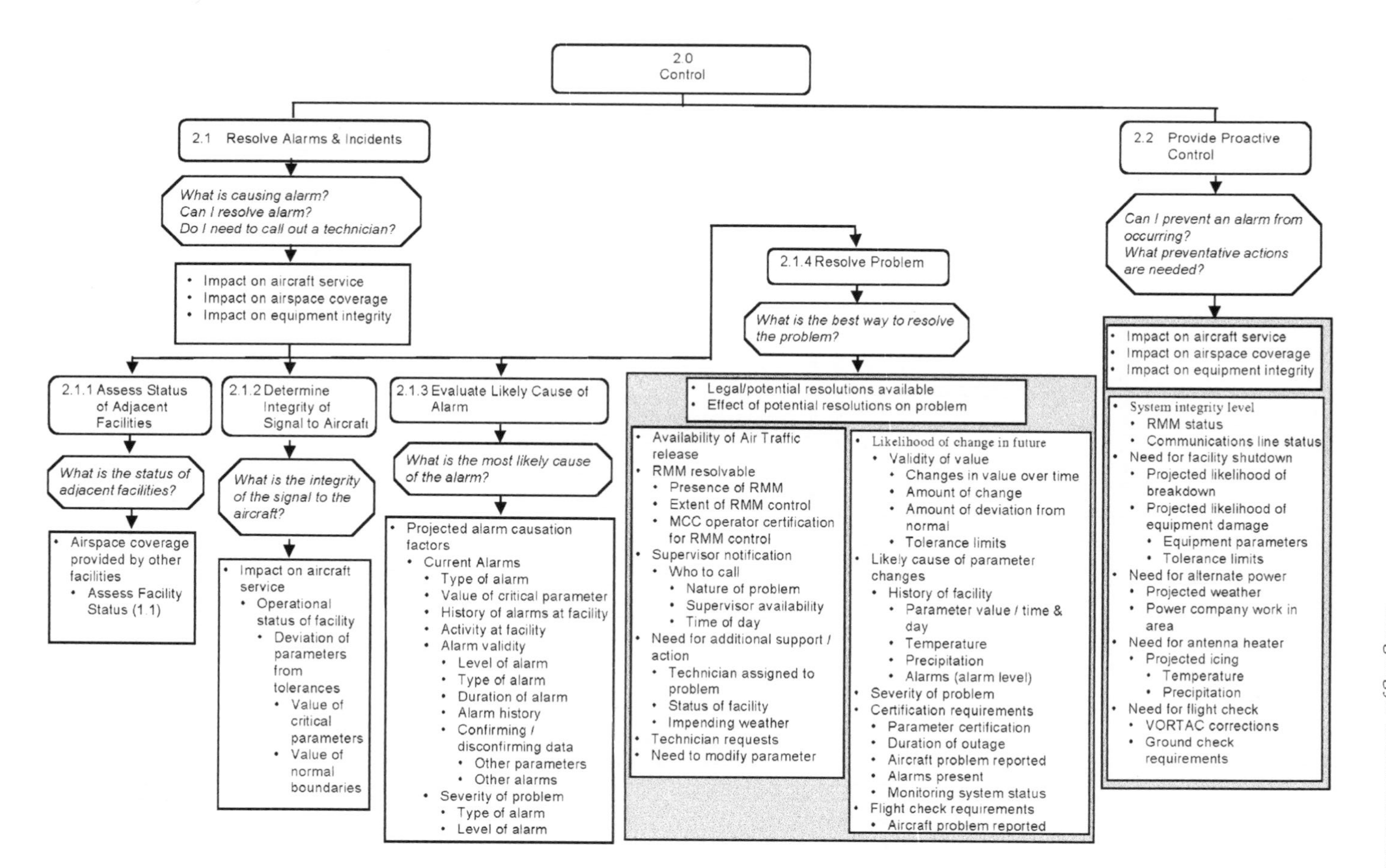

Figure 6.4 Goal-directed Task Analysis: MCC Situation Awareness Requirements (Continued)

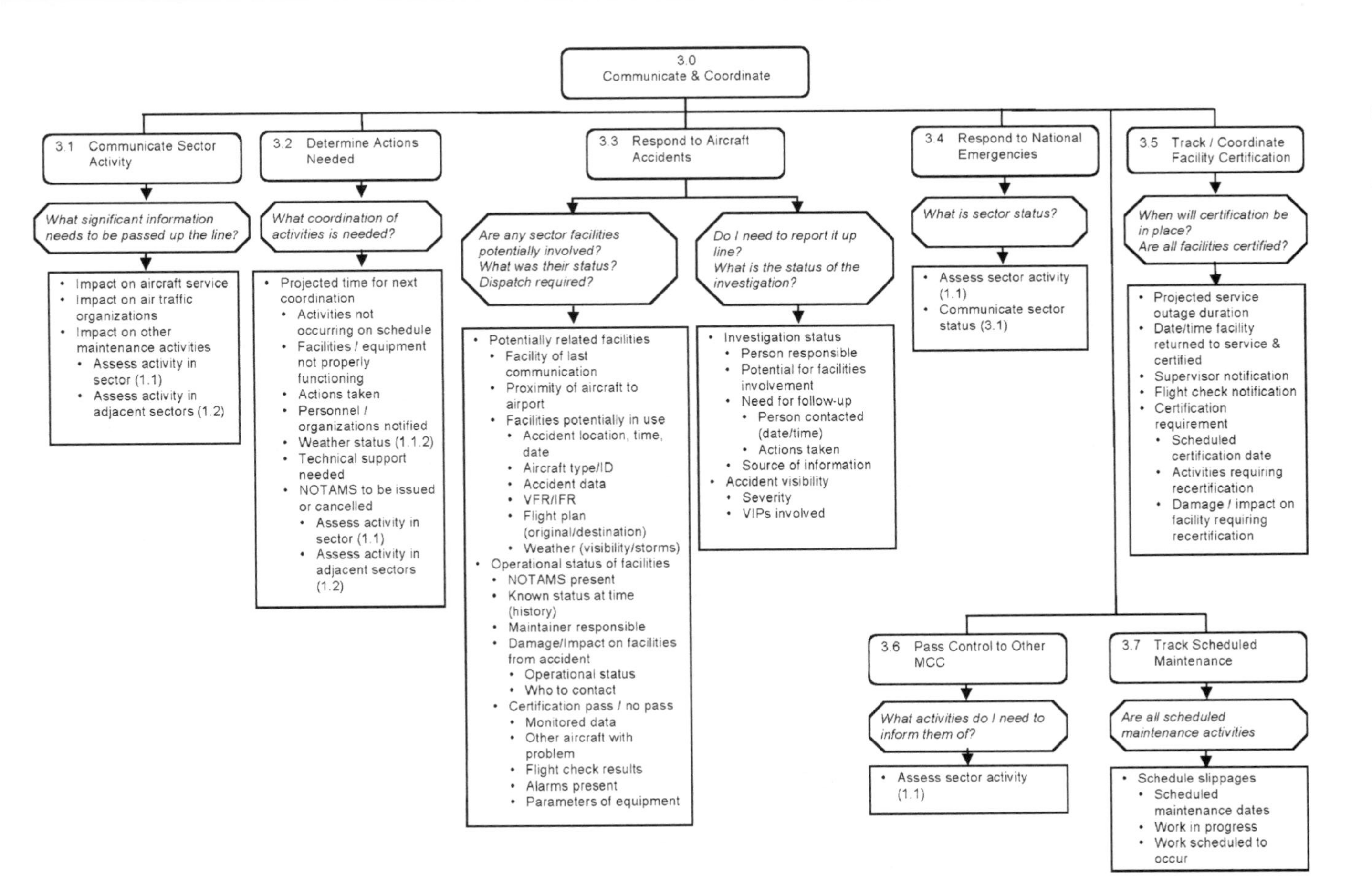

Figure 6.4 Goal-directed Task Analysis: MCC Situation Awareness Requirements (Continued)

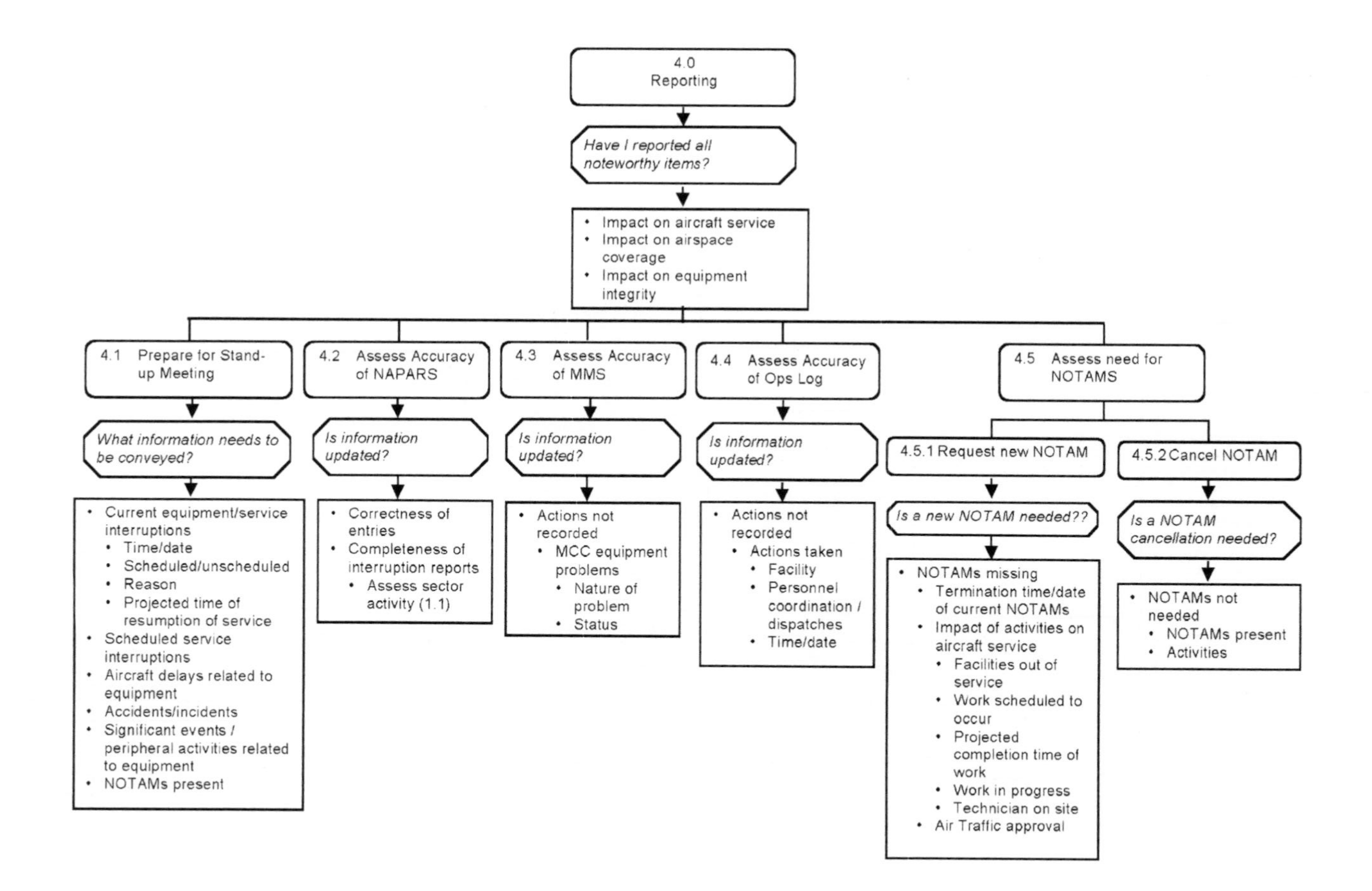

Figure 6.4 Goal-directed Task Analysis: MCC Situation Awareness Requirements (Continued)

Table 6.1 Comparison of existing system to SA requirements.

1.0 MONITORING	NAPERS	NOTAMS	Scheduled Maint. Log	MMS	Outage Scheduler	Card File	Notepad Activity Log	MCS Site Status	MCS Site Directory	MCS Site Info	MCS Active Alarms	MCS Reports	MCS Utility	FS Monitor	Weather Map
1.1 ASSESS SECTOR ACTIVITY															
Contact point of affected end user															
Impact on aircraft service															
Impact on adjacent FAA facilities/sectors		F													
Impact on equipment/facility/certification		I													
Need for additional support/action		I													
1.1.1 FACILITY STATUS	I													F	
Current alarms											F			F	
Type of alarm								F			F	F		F	
Level of Alarm											F	F		F	
Facility causing alarm	F							F			F	F		F	
Parameter causing alarm								F			F	F			
Status of response to alarm	F						P				P			P	
Facility out of service	F										I				
FAA organizations impacted	I	F									I				
Aircraft services impacted	I	F	P		P						I				
Date/time of non-service	F	F	P		P		P	F							
Reason out of service	F	F	P		P		P	I			I				
Work in progress	F	P													
FAA organizations, services impacted	I	P	P		P										
Scheduled/Unscheduled	F	P	F		F										
Time facility out of service	F	P			P		P								
Time until facility returns to service		P					P								
Nature of problem/activity	F	P	P		P		P								
Prior history of equipment				P								F			
Work scheduled to occur															
PM dates/facility			F		F										
Work begin date/time-projected duration	P	P	F		F										
Progress on work at facility	F			P											
Peripheral activities (const., utility, telco work)							P								
Type of work		P					P								
Location		P					P								
Proximity to FAA facility		P					P								
Time/dates of work (current/planned)		P					P								
Contact point							P								
Technician present at facility (who)				P			P							P	
Technician activity				P			P							P	
1.1.2 WEATHER IN SECTOR							P								
Impact on aircraft service (potential/actual)															
Impact on facilities/equipment (potential/actual)															
Impact on travel/road conditions															
Timing of storm front															
Rate of movement														P	
Location														F	
Severity of storm front															
Temperatures															
Precipitation														F	
Wind															
Humidity															
Barometric pressure															
Visibility															
Lightning															
1.1.3 TECHNICIANS															
Availability for facility problem															
Current activity			P	P	P		P								
Supervisor						F									
Safety															
Location															
Time left at last location/destination							P								
Route to destination							P								
Impending weather situation in area															P
Weather in sector															P
Means of communication with technician															
Phone at location						F									
Radio operating at location (frequency)						F									
Current activity				P			P								
1.2 ASSESS ACTIVITY IN ADJACENT SECTOR															
Impact on equipment/facility		F													
Impact on air service		F													
Hard alarms															
Date/time of non-service		F													
Reason		F													
Scheduled shutdowns		F	F		F										
Weather															

FULL (F) PARTIAL (P) INFERRED (I)

Table 6.1 Comparison of existing system to SA requirements (continued).

2.0 CONTROL	NAPERS	NOTAMS	Scheduled Maint. Log	MMS	Outage Scheduler	Card File	Notepad Activity Log	MCS Site Status	MCS Site Directory	MCS Site Info	MCS Active Alarms	MCS Reports	MCS Utility	FS Monitor	Weather Map
2.1 RESOLVE ALARMS/INCIDENTS															
Impact on aircraft service															
Impact on airspace coverage															
Impact on equipment integrity															
2.1.1 STATUS OF ADJACENT FACILITIES															
Airspace covered by other facilities															
Assess facility status (1.1.1)															
2.1.2 INTEGRITY OF SIGNAL AVAILABLE TO A/C								Inferred							
Impact on aircraft service								Inferred							
Facility operational status		Partial					Full	Inferred			Inferred			Full	
Deviation of values from tolerances								Full			Full				
Value of critical parameters								Full			Full				
Value of normal boundaries															
2.1.3 LIKELY CAUSE OF ALARM															
Projected alarm causation factors															
Type of alarm								Full			Full			Full	
Value of critical parameters								Full			Full				
History of alarms at facility				Partial								Full			
Activity at facility			Partial		Partial		Partial								
Alarm validity															
Level of alarm (hard/soft)											Full	Full		Full	
Type of alarm								Full			Full	Full		Full	
Duration of alarm								Inferred			Full	Inferred			
Alarm history				Partial								Full			
Confirming/Non-confirming data	Inferred	Inferred	Inferred	Partial	Inferred			Inferred			Inferred	Inferred		Inferred	Inferred
Other parameters								Full			Full	Full			
Other alarms								Full			Full	Full			
Severity of problem												Full		Full	
Type of alarm								Full			Full	Full		Full	
Level of alarm											Full	Full			
2.1.4 RESOLVE PROBLEM															
Legal/potential resolutions available															
Effect of potential resolutionon problem															
Air Traffic release availability		Partial													
RMM resolvable															
Presence of RMM								Full	Full		Full				
Extent of RMM control								Inferred							
MCC operators certified on facility															
Technician/supervisor notification															
Who to call	Full	Full	Inferred		Inferred	Full			Inferred	Full	Full				
Nature of problem	Full		Inferred		Inferred			Full				Inferred			
Availability of supervisor															
Time of day								Full			Full				
Need for additional support/action	Full		Inferred		Inferred										
Technician assigned to problem	Full		Inferred		Full										
Status of facility	Full						Partial								
Impending weather															Inferred
Need to modify facility parameters															
Likelihood of change in future												Inferred			
Validity of value															
Changes in parameter over time											Inferred	Inferred			
Amount of change												Inferred			
Amount of deviation from normal								Partial			Partial	Partial			
Tolerance limits								Full			Full	Full			
Likely cause of parameter change															
History of facility															
Parameter value/time & day												Full			
Temperature															
Precipitation															
Alarms (alarm level)												Full			
Severity of problem															
Certification requirements															
Parameter certification															
Duration of outage								Partial				Full			
Reported A/C problem present															
Alarms present								Full			Full				
Monitoring systems in service								Full			Inferred				
Flight check requirements															
A/C reports of problems															
Technician Requests						Partial	Partial								
2.2 PROVIDE PROACTIVE CONTROL															
Impact on aircraft service															
Impact on airspace coverage															
Impact on equipment integrity															
System integrity level															
RMM working								Full							
Communication line status															
Need for facility shutdown															
Likelihood of facility breakdown															
Likelihood of equipment damage															
Need for alternate power															
Projected weather															Full
Power company work															
Need for antenna heater															
Impending icing															Full
Temperature															Full
Precipitation															Full
Need for flight check															
VORTAC corrections															
Ground check requirements															

FULL ■ PARTIAL ■ INFERRED ■

Table 6.1 Comparison of existing system to SA requirements (continued).

3.0 COMMUNICATE AND COORDINATE	NAPERS	NOTAMS	Scheduled Maint. Log	MMS	Outage Scheduler	Card File	Notepad Activity Log	MCS Site Status	MCS Site Directory	MCS Site Info	MCS Active Alarms	MCS Reports	MCS Utility	FS Monitor	Weather Map
3.1 COMMUNICATE SECTOR ACTIVITY															
Impact on aircraft service		Full													
Impact on air traffic organizations															
Impact on other maintenance activities															
Asses activity in sector															
Assess acitivity in adjacent sectors															
3.2 DETERMINE ACTIONS NEEDED															
Projected time for next coordination	Inferred						Partial								
Activities not occuring on schedule	Full	Inferred	Inferred		Inferred										
Facilities/equipment not properly functioning	Full	Full					Partial	Full			Full	Full			
Actions taken	Full		Inferred	Full	Inferred		Full								
Personnel/organizations notified	Full	Partial	Full		Full		Full								
Weather situation															Full
Technical support needed															
NOTAMS to be issued/cancelled	Inferred	Partial	Inferred		Inferred										
Assess sector activity															
Assess activity in adjecent sectors															
3.3 RESPOND TO A/C ACCIDENTS															
Potentially related facilites															
Facility of last communication															
Proximity to airport															
Facilities potentially in use															
Accident location, time, date															
A/C type, ID															
Accident data															
VFR/IFR															
Flight plan (origin/destination)															
Weather (visibility, storms)															Partial
Operational status of facilities															
NOTAM's present		Full													
Known status at time (history)	Partial	Partial	Partial	Partial	Partial							Full			
Maintainer responsible						Full			Full						
Damage/impact on facilities from accident															
Need for dispatch															
Operational status								Full			Full	Full			
Who to contact						Full				Inferred					
Facilities pass/no pass certification															
Monitored data								Inferred							
Other aircraft w/problem															
Flight check results															
Alarms present								Full			Full	Full			
Parameters of equipment								Full			Partial	Partial			
Investigation status							Full								
Person responsible						Full	Full								
Potential for facilities involvement							Full								
Need for follow-up							Full								
Person contacted/date, time							Full								
Actions taken							Full								
Source of information							Full								
Accident visibility															
Severity															
VIPs on board/involved															
3.4 RESPOND TO NATIONAL EMERGENCIES															
Assess sector status															
Communicate sector status															
3.5 TRACK/COORDINATE FACILITY CERTIFICATION															
Supervisor notification						Full									
Flight check notification						Full									
Certification requirement															
Scheduled certification date			Full		Full										
Activities requiring recertification															
Damage/impact on facility from accident															
Projected out of service duration															
Date/time facility returned to service/certified			Full		Full										
3.6 PASS CONTROL TO OTHER MCC															
Assess sector activity															
3.7 TRACK SCHEDULED MAINTENANCE															
Schedule slippages		Partial	Inferred	Partial	Inferred										
Scheduled maintenance dates		Partial	Full		Full										
Work in progress	Full	Partial		Partial											
Work scheduled to occur	Partial	Partial	Full		Full										

FULL ■ PARTIAL (grey shading) INFERRED (hatched)

Table 6.1 Comparison of existing system to SA requirements (continued).

	NAPERS	NOTAMS	Scheduled Maint. Log	MMS	Outage Scheduler	Card File	Notepad Activity Log	MCS Site Status	MCS Site Directory	MCS Site Info	MCS Active Alarms	MCS Reports	MCS Utility	FS Monitor	Weather Map
4.0 REPORTING															
Impact on aircraft service															
Impact on air traffic organizations															
Impact on other maintenance activities															
4.1 STAND-UP MEETING															
Current equipment/service interruptions	Full	Full					Full	Inferred			Inferred	Inferred			
Time/date	Full	Full	Full		Full		Full	Full				Full			
Scheduled/unscheduled	Full		Full		Full		Full								
Reason	Full						Full	Inferred			Inferred	Inferred			
Projected time of resumption of service	Partial	Full	Full		Full		Full								
Scheduled service interruptions		Partial	Full		Full										
A/C delays related to equipment		Full													
Accidents/incidents							Full								
Significant events/peripheral activities related to equipment							Full								
NOTAMS present		Full													
4.2 NAPERS															
Correctness of entries	Full	Partial	Partial		Partial										
Completeness of interruption reports	Full	Partial	Partial		Partial										
Assess sector activity															
4.3 MMS															
Actions not recorded															
MCC equipment problems															
Nature of problem															
Status															
4.4 OPS LOG															
Actions not recorded															
Actions taken															
Facility				Partial											
Personnel coordinations/dispatches				Partial											
Time/date				Partial											
4.5 NOTAMS															
4.5.1 REQUEST NEW NOTAM															
NOTAMs missing															
Termination time/date of current NOTAMS		Full													
Impact of activities on A/C service			Inferred		Inferred										
Facility out of service	Full							Inferred			Inferred	Inferred			
Work scheduled to occur	Partial		Full		Full										
Projected completion time of work	Partial		Full		Full										
Work in progress	Partial	Partial		Partial											
Technician on site	Partial			Partial	Full										
Air traffic approval															
4.5.2 CANCEL NOTAM															
NOTAMs not needed															
NOTAMs present		Full													
Activities completed							Full								

FULL ■ PARTIAL (grey) INFERRED (hatched)

but also whether it was present in a format that contributes to achieving and maintaining situation awareness in a complex, dynamic, and distributed system.

Table 6.1 presents the results of this evaluation. A particular SA requirement may be met either fully (black square) or partially (grey square) by a given system or source, or it may provide enough information that the requirement can be inferred by a knowledgeable specialist (indicated by diagonal slashes). Blanks in the table indicate that the information is not determinable directly from a given system. This information must be either (1) acquired from another system or source, (2) acquired through verbal communications with technicians or other personnel, or (3) ascertained by the specialist by integrating information across the various sources of information available. Many of the SA requirements were being met by these latter two methods.

In the existing MCC, a large portion of specialists' time was spent in acquiring SA. With a limited number of facilities to monitor and control, and limited duties, the MCC specialists appeared able to meet this challenge, however they will likely be severely taxed in the future as the number of systems being monitored and reported in to the MCC and their responsibilities grow. It was projected that the number of facilities monitored and controlled through the MCC would continue to increase, as will the requirement for the MCC to remotely affect modifications to facility parameters, respond to problems, and assume a proactive role in avoiding alarm conditions. In addition to increasing the scope of the MCC specialists' responsibilities, this will also increase the number of specialists working in an MCC, which imposes certain additional communication and coordination requirements.

The following summarizes the major issues present in the existing MCC system that create situation awareness difficulties and limit the performance of specialists in fulfilling their role.

Lack of integration of information

First, while a fair amount of the information that is needed by the MCC specialists was found to be present in one form or another, there was a severe lack of integration of this information. They were required to go to several different displays or reports, or page through several windows to find the information needed for a single problem or goal. Furthermore, much of the information present in any one source was only partial information. For example, the specialists spent a great deal of time ensuring that the current status of *all* of the facilities and equipment on *all* of the parameters was being represented in a display. Finding correct historical information that has all of the details they needed was difficult. Specialists were required to consult external manuals to find what the value of a particular parameter *should* be. This lack of integration imposed unneeded workload and could lead to situation awareness errors.

Lack of required information

Much of the needed information was not present at all. Alarm information was found to be inadequate for real-time control. Information such as the allowable ranges for parameters, normal values of parameters, and real-time trend information are needed. Furthermore, this information needs to be integrated with work-in-progress and scheduled maintenance information so that alarm validity and impact can be quickly assessed. This task was found to be very cumbersome and difficult for the MCC specialists.

The ability of the MCC specialists to assume real-time control of facilities was also severely limited by a lack of remote maintenance monitoring (RMM) sensors. They were receiving remote information on only a few facilities and systems in the sector. For the systems they did receive RMM on, they were limited in their ability to perform tasks remotely (either technically or organizationally). This places the specialist in a position of making only a very limited diagnosis of a problem (if any) before notifying field personnel, thereby making this notification task their main function. Furthermore, there was little online help available or system support provided for making control actions remotely. For the same reasons, specialists were almost completely unable to assume the proactive control needed to avert problems before they developed.

Poorly presented or lacking information to support diagnosis

The ability of the specialists to make accurate diagnoses of problems was found to be severely limited by the poor availability and presentation of historical information and the integration of that information with parameters of interest. The only historical information obtainable was not available in real time (a considerable delay was involved in getting it sent over the network). It was in a difficult to read format, which required a great deal of processing to remove extraneous and bogus readings, and from which ascertaining meaningful trends was very difficult. There was no direct relationship between the data and other parameters of interest, which must be inferred or determined separately. This entire process was so cumbersome and difficult that very limited use was being made of the historical information that was available.

Weather information was also found to be very limited. The Austin MCC had a very nice color map that showed precipitation across the sector. This information was derived from National Weather Service information and was more than most MCCs had; however, this still met only a fraction of their needs with regard to current and projected weather information. Acquiring more detailed information was very difficult and the information that was available was often not directly related to the facilities of interest.

Major functions not supported

Communication and coordination among technicians, sector management, air traffic organizations, and outside contractors and service personnel were found to be one of the MCC's most important functions. This coordination activity created a considerable memory load because specialists were required to keep track mentally of who is where, who knows what, who will be getting back to them with information when, when changes will occur, etc. With only one specialist and a small range of control, this task was somewhat achievable. However, with multiple personnel and many different activities going on across a sector, information and events can easily fall through the cracks. The probability of missing something important, of falling behind in knowing what is going on in the sector, or of not communicating critical information across specialists will become much greater. Better support was needed for this function beyond the unstructured notepad in use.

Integration time consuming

Formulating up-to-date reports of sector status for transmitting up the management chain is also an important function of the MCC. In the existing system this task was far more time consuming than it should be, as specialists must integrate information from numerous reports and sources. It was very difficult for the specialists to be sure they had covered everything.

Information not made explicit

Information about the facilities and equipment in the sector—their history, status and unique characteristics, local procedures and personnel—was conveyed only informally in the existing system. This sector-specific information is as critical as well-developed technical knowledge of the equipment to the specialists' ability to evaluate alarms and determine the impact of situations on overall goals. MCC specialists develop this knowledge through extensive experience in the sector and in the MCC. This makes it very difficult for specialists from other sectors to rotate in to meet manpower or emergency demands or even for technicians from the same sector to train in the MCC or take over functions when needed. While the MCC specialist position will always require a high level of technical knowledge across a wide variety of equipment, the need also exists for sector-specific information to be made more explicit. This would allow greater transferability of specialists across MCCs when needed and make it easier to transition duties between specialists within an MCC on a daily basis.

Needed communications sporadic

Locating and communicating with technicians in the field posed a challenge. MCC specialists had only a limited ability to find individuals and to project where technicians would be. This made it difficult to optimize manpower utilization when they were required to decide whether to notify a supervisor or call out a technician for a problem, or whether technicians working at remote sites might need assistance.

Information on adjacent sector activities not provided

Almost no current information on activities in adjacent sectors available to the MCC. Notice to Airmen (NOTAMs) were the only source. NOTAMs are generated to inform pilots of equipment problems that might affect their flight. As such, NOTAMs rarely contain the type of information the MCC specialist wants to know and they are often not current in terms of the status of present activities as they are issued to cover large blocks of time. While not much information is needed about adjacent sector activities, in cases where there are facilities that affect operations across sectors, a limited amount of current information would be useful to the decision-making process of the MCC specialists.

Global SA poor

Obtaining complete and accurate status information on sector facilities and activities at a glance (global SA) was not possible. While the status display at the Austin MCC goes a long way toward meeting this goal, certain information is still missing. For instance, when arriving in the morning, specialists cannot tell from the display if certain alarms have been deactivated. They therefore could easily be led into a false sense of well-being if they did not check additional logs to find out what RMM and alarms were functioning. Specialists must seek other sources to find out what significant events had occurred and what activities were in progress or scheduled for the day.

Data overload

As with many complex systems, the large volume of data being presented was difficult to wade through. The specialists could not get the information they wanted quickly. The information was not optimally organized and represented to meet their needs. Codes for representing operationally relevant differences were not always present (e.g., for distinguishing alarms incurred by shutting down a facility to perform other maintenance work from a true alarm for a given parameter). Sorting through these factors challenged the specialists' ability to ascertain what was going on.

Poor user interfaces

The existing systems needed improvements in their user interface. The different system interfaces with which the specialists interacted completely lacked standardization. Not only does this impose an unnecessary training burden, but it also greatly increases the probability of making errors through negative transfer (mistakenly operating one system in the way another familiar system works). Reliance of the systems on command-based control structures further impinged on limited working memory and negative transfer problems. In addition, the systems had very limited capabilities to do multiple windowing. This increases the difficulty imposed by having needed information spread out across many different displays and systems.

Due to these many shortcomings, the principles described in this chapter were applied to design a system interface that supports much higher levels of SA.

6.2.3 Situation awareness-oriented interface design

A prototype of a new interface was developed to address the shortcomings identified with the existing system. In addition to providing MCC specialists with the needed situation awareness requirements, this interface is an illustration of how to implement the concepts inherent in an SA-Oriented Design.

First, information is oriented around specialist goals. It supports the maintenance of a global picture of the entire system, while making perception of specific problems salient, and facilitating diagnosis and information gathering relevant to these problems. Projection of future events is supported by providing specialists with information on current and past trends of system parameters. The new system prototype includes several features designed to promote SA and specifically deal with some of the problems noted with the existing system.

1. The system is designed to provide MCC specialists with a broad overview of the state of the airway facilities in their area at all times, while allowing them to get detailed information on specific facilities easily. A Status Map (Plate I) is displayed at all times on one of two monitors. It is never covered by other display windows.
2. The windowing environment and each display are organized around goal attainment and are designed to allow easy switching between goals. The goals and subgoals derived from the goal-directed task analysis were used to provide primary and secondary groupings of information.

3. The comprehension of alarms and projection of future trends is supported by the addition of graphically presented historical data on the prior behavior of system parameters and easy comparison to other sector activities. These graphs can be easily modified by the user to meet a wide range of diagnostic needs.

4. Information support is provided for conducting activities that required information that was either implicit or buried in various documents (e.g., alarm values, contact information, and facility information). This will allow for improved SA, which is needed for multiple MCC specialists to be able to effectively work in the same facility.

5. The system is designed to be easy to use, requiring minimal training and low memory load.

6. The system specifically supports specialists in their efforts to keep up with many activities by grouping information by location, and by aiding communication between multiple specialists in an MCC.

7. The interface is designed to allow frequent and important tasks to be conducted easily (e.g., performing a manual check on the status of facility parameters and RMM each morning).

Hardware

The system consists of two large-screen color monitors, one of which displays the Sector Status Map, and one that displays detailed facility information in moveable, layerable windows. A third large-screen color monitor is provided to display sector weather information. Supporting PC monitors will continue to exist to provide access to the maintenance record system provided by technicians and to support peripheral activities of the MCC specialist. A printer and telephone support are also provided. Hardware and software links to existing information sources are provided so that needed information is automatically accessed and integrated into the new SA-oriented system interface.

Status Map

The Status Map, shown in Plate I, is displayed at all times. It shows the status of the facilities in the sector(s) that the MCC specialist is responsible for. Color symbology is used to convey a high-level understanding of facility status at a glance. All symbol shapes and colors are redundant to guard against any human color perception deficiencies. A green, filled circle designates all facilities are okay at a location; an unfilled blue circle designates technician work in progress at a location; a black diamond indicates that a facility has gone down (not in service); and an unfilled square indicates that a facility is unmonitored. Additionally, a yellow square indicates a soft alarm and a red triangle indicates a hard alarm. It is very easy to tell not only what is known, but also what is not known.

The SA-oriented system interface supports goal-directed activities by providing access to numerous types of information directly from the Sector Status Map. The software is designed to work in a Windows-type environment. When any button is selected on the Status Map, the corresponding display window will be shown on the second monitor at the workstation. Multiple windows can be shown, moved around at will on the display, or closed independently by the MCC specialist. The functions on the Status map are described below.

- *Status Button.* Provides an overview of the status of all displayed facilities (see Plate II). In addition, double-clicking the symbol associated with the facility (either on the Status Map or the Current Status window previously accessed) brings up the Status Overview window associated with that facility (see Plate III).
- *Facilities Button.* Provides additional information on a particular facility of interest (see Plate V).
- *Activity Button.* Provides an overview of all current activities (scheduled maintenance, unscheduled outages and maintenance, and certifications) (see Plate VI).
- *History Button.* Provides information on the prior behavior and performance of parameters of interest for any monitored facility (see Plate VIII).
- *Telephone Button.* Provides communications and supporting information for each facility (see Plate X).
- *Set-up Button.* Provides specialists with functions for customizing the system to their MCC's particular needs (see Plate XI).
- *Print Button.* Allows any window to be printed to obtain a hard copy for the specialist's use.
- *Notes Button.* Provides an area for general notes to be recorded.

Most of the buttons available on the Status Map can also be accessed from other windows, thereby maximizing the versatility and the ease of use of the displays.

Facility Status

Clicking the Status button on the Status Map (Plate I) brings up the Current Status Overview window, shown in Plate II, providing the status of all facility locations in the displayed sector(s). This window displays a short description of the status of each facility location. The description field for each facility location may be scrolled using the scroll bar. Clicking the name of any facility location brings up the status window for that facility location (see Plate III). Two other buttons are present on the Current Status Overview window. The Update button causes the RMM software to update the readings on all monitored facilities. The Show All button causes the Status windows for all the facility locations to be displayed in layered windows for easy review by the MCC specialists.

The Facility Status window provides detailed information on the status of facilities at each location in one integrated display. The status window for each facility location contains a number of fields of information, as shown in Plate III. In the top field, a report on all current alarms for facilities at that location is shown, with the most recent at the top. In the second field, status information for the facility location is shown. This information is entered by the MCC specialist, and might include activities going on at a facility, actions taken, etc. This field provides a log of information pertinent to a particular facility or location that allows transfer of information across specialists and shifts, and a reminder to a specialist of actions or activities of interest. Information entered in this field is the same information that is shown in the status field of the Current Status Overview window (Plate II) for all facility locations. The information, which can be entered in either window, is automatically updated in the other window. The third field in this display shows all NOTAMs that are currently in effect at the facility location. Each of these fields is scrollable using the scroll bar at the right.

The Activity, History, Telephone, and Last functions are also available from the Status window. In addition, several actions may be taken from the Status window. Update causes the RMM to update the reading of monitored parameters for facilities at that location. Acknowledge Alarm allows specialists to acknowledge a new alarm. Control brings up the Control window for the facility that is showing a current alarm, as discussed next.

Control

Plate IV shows the Control window, which allows the MCC specialist to take several control actions for the indicated facility. Pressing Reset reinitializes a facility, usually in response to an alarm. Shutdown turns off a facility and Start turns on a facility that is not operating. Disconnect disconnects a facility from RMM, and Connect reinstates RMM. Diagnostics allows diagnostics available through RMM to be performed. The Activity, History, Status, Facilities, Telephone, and Last functions are also available from the Status window.

Facility

Clicking the Facilities button on the Status Map, or on any other display with this button, brings up the Facilities window, shown in Plate V. This window allows MCC specialists to access information about any facility at any location within the sectors listed. The MCC's sector of responsibility, any sectors it may occasionally have responsibility for (at night, for instance), and all surrounding sectors are included in the sector list.

To obtain information on a facility, the appropriate sector is selected in the sector list. This causes the locations in that sector to be listed in the location list, as shown in Plate V. The location of interest may then be selected and the facilities associated with that location are listed in the facility list. In the facility list, the identifiers for all facilities at the selected location are shown. Those facilities with RMM capability are shown in bold text. Clicking the facility identifier selects the facility that the specialist is interested in. The Show All button selects all facilities at a given location. All three fields—Sector, Location, and Facility—are scrollable using the scroll bar at the right of the respective field. Once a facility has been selected, the buttons at the bottom of the window can be used to obtain the type of information desired: Activity, History, Telephone, Control, or Last.

This method of facility selection allows easy access to information by MCC specialists who may not be familiar with every facility in a sector, and serves as a memory prompt in everyday use. In addition, if the MCC specialist knows the identifier code of the facility of interest, he or she may enter that code directly in the Location and Facility fields below the selection lists. This allows more rapid access for common facilities.

Activity

Selecting Activity on the Status Map displays the Activity Overview window, which contains a list of all activities in the sector(s) displayed on the map, as shown in Plate VI. Selecting Activity from any other window reduces this to a list of all activities in the indicated location or facility. The window lists activities currently ongoing or scheduled for the indicated sector or location. Unscheduled activities are listed in red, ongoing scheduled activities are listed in blue, and activities that are scheduled, but not yet begun, are shown in white. This color coding provides an important classification of activities for the specialist. For each activity, the facility, type of activity, out-of-service (OTS) time and date, and estimated return to service (EST) time and date are listed, along with an indication of whether the necessary NOTAM has been issued and necessary approvals received. Activities are listed in descending order by OTS date and time.

Several actions are possible from this window. Clicking the New button allows entry of new activities. History provides information on prior behavior of the selected facility. Certifications causes the Activity Overview window to list only those activities listed as having certification in the Type field. The Last button causes the last displayed window to be reopened.

Clicking the Show button provides detailed information and coordination logs for the activity that has been selected with the mouse (see Plate VII). The Activity window lists all pertinent information regarding a specific ongoing activity: the type of activity, OTS date and time, technician, estimated return-to-service (EST RTS) date and time, and actual return-to-service (ACT RTS) date and time are listed, along with whether NOTAMs have been issued, all approvals received, and certifications performed, if needed. (This information, entered by the MCC specialist in this window, is automatically updated to the Activity Overview window.)

The Activity window provides a log of all pertinent coordination actions taken relative to a given activity. Each of the possible coordination activities for an activity is listed on the left. The table allows specialists to designate which organizations are affected and which have been notified of the activity. Approvals required for taking the facility out of service (time, date, and initials) and returning it to service (time, date and initials) are recorded. Thus, this window provides a central organized repository for coordination requirements that allows any MCC specialist to rapidly assess the status of coordination efforts, even if accomplished by another person.

The current status of the activity is listed in the upper right-hand box. Space is also provided for recording pertinent information about a given activity in the Remarks field. A stack of Activity windows may be rapidly viewed by clicking the edge of each Activity window in turn. In addition to logging information and coordination relative to an activity, several other functions may be accessed from the Activity window: Last, History, and Telephone.

History

The History window (Plate VIII) will be presented for the indicated facility when the History button is selected from any window in the system associated with a particular facility. Clicking the History button on the Status Map causes the Facilities list to be displayed (Plate V). Once a facility has been selected and the History button designated, the prior history of monitored parameters for the indicated facility will be displayed, as shown in Plate VIII.

This window shows a list of prior alarms for the facility and a graphic depiction of the prior behavior of monitored parameters for the facility. In addition, the current value of the parameter is displayed digitally in the upper right-hand corner. The soft alarm limit is displayed in yellow and the hard alarm limit is prominently displayed in red (so that specialists do not need to look up these values for each facility), allowing specialists to rapidly assess the significance of current and history values.

This graph, which is automatically displayed according to default values established by the specialists in the Set-up window (Plate XI), can be easily reconfigured to best display the information of interest. The Scale button (shown at the top of Plate VIII) presents a window allowing the minimum and maximum values for the Y scale (the parameter values) to be specified. The Time button presents a window allowing the time period displayed to be specified by clicking the box next to several commonly used time periods (from the present time). The specialist may also wish to specify a beginning and ending time and date for the display, which can be achieved by clicking Other in the Time Selection window. As logged data collected by RMM typically have many erroneous values, a Filter may be applied to screen out values which are obviously too high or too low to be valid. To view history information on another parameter, the New button may be selected, causing a list of parameters to be presented for selection. Once the information of interest is presented as desired, a printout may be obtained by selecting Print. Other functions available from this window include facility Status, Activity, Telephone, and Last.

Some parameters of interest provide only discrete data (functioning/ nonfunctioning). For these parameters, a History window showing outages is presented, as shown in Plate IX. The timing and duration of scheduled and unscheduled outages of each code type is displayed. Exact starting time and date, ending time and date, and ID number for any outage can be obtained by clicking the line representing the outage. The Time, New, and Print functions are also provided.

Communications

A major function of the MCC specialist is communicating with technicians, supervisors, and FAA personnel in various facilities. These functions are supported in a central location for each facility by selecting the Telephone icon on any window. This window lists the phone numbers for the facility, internal phone numbers, and radio contact information (see Plate X). Phone numbers are also listed for the duty station, supervisor, and technicians certified on the facility in order of call out (as specified in the Set-up window, Plate XI). This is an example of how sector specific information can be made explicit, allowing easier control by other specialists who may not be as familiar with a particular facility. Phone numbers for all relevant organizations are listed, as well as information on travel to the site, which technicians often need help with.

Phone numbers may be manually dialed or auto-dialed by selecting the number with the mouse and clicking the Telephone button in the upper left-hand corner. The Notes button displays a list of notes relevant to that facility, providing communication across shifts and specialists, and for those not familiar with the site. Activity, History, Status, and Last functions are also available.

Set-up

The Set-up window, shown in Plate XI, may be accessed by selecting Set-up in other windows in the interface. This window allows customization of the system interface for a particular MCC. The Show Map field allows the selection of the map to be shown on the Status Map. Any sector or group of sectors may be selected for viewing at any time. This allows for a quick look at an adjacent sector or for the transfer of control from one MCC to another as needed.

The list of facilities provided in the Facilities window may need to be modified as new facilities are added, taken out of service, or RMM added. In addition, the personnel section of the Set-up window allows a current list of personnel, their availability and current phone numbers to be maintained in a centralized database, which will automatically update the communications information for each facility. Information on any particular individual can be viewed by entering the person's initials and clicking Find. Information can be added manually by clicking New, or updated automatically from the MMS by clicking Update.

The Personnel Information window (not shown) lists the pertinent information for each technician associated with the sector. For each person, initials, full name, and phone number(s) are listed. The facilities for each location in a sector for which the individual is to be listed on the call-out list are designated. The list indicates that the individual should be listed first, second, etc. If the person is scheduled to be unavailable due to vacation, training classes, disability, etc., this can be indicated, taking them off the communications lists for that time period.

Weather Map

The current weather in the sector and surrounding sectors is shown on the third color monitor provided for the MCC technicians (Plate XII). The map shows the sector boundaries and major facility locations and cities. Superimposed on this is a color map showing precipitation in the area. Clicking any facility location brings up a list of weather parameters for that site. This data is automatically updated hourly by the system through sensors at the sites and/or the National Weather Service. In addition, the specialist may request a more recent update of this data by clicking the Telephone button on the display to call the local facility.

6.2.4 Summary of interface design case study

The new interface design illustrates several of the design principles discussed in this chapter. First, the data that the MCC specialists need is organized around their major goals and subgoals (Principle 1). Separate displays are provided for monitoring facility status, diagnosing alarms, communicating, coordinating, and control. By providing the information in an integrated fashion, SA of relevant information for pursuing each goal is at the specialists' finger tips, and the likelihood of important information being overlooked is significantly reduced.

A global overview of the situation is always present (Principle 4), which allows the specialist to rapidly determine what the highest priority goals are (i.e., 'are there other events present that should take precedence over what I am currently doing?'). Information that is known, as well as what is not known (e.g., facilities that have gone unmonitored) is made explicit and salient on the display. (See Chapter 7 for a discussion on the display of areas of uncertainty.) The buttons allow rapid switching between multiple goals associated with a facility and between the global overview and any particular goal (Principle 5). Critical information that indicates a need to switch goals (e.g., a new alarm), is made highly salient through the use of both color on the display and an audio alarm (Principle 6).

The History window supports SA by directly presenting comprehension (comparison of current parameter value to alarm limit values) and projection information (trends over time allowing factors affecting parameter changes to be ascertained) (Principles 2 and 3). These displays allow operators to rapidly determine what they really want to know. 'Is the alarm real? What does it relate to? Will it go away?' Although the specialists initially believed some sort of automation or information filtering would be required to manage the data overload in the MCC, with the new system design this proved to be unnecessary (Principle 8). Good organization and integration of data was more than sufficient to overcome the problems that were present.

Limited use of multiple modalities for supporting parallel processing was illustrated by this example (Principle 7). Audio indicators were provided for each alarm type (red and yellow) that helped to insure any new alarms were attended to promptly. In this environment, MCC specialists spend a considerable amount of time communicating over the telephone or radio, so they are fairly well loaded in terms of both visual and auditory information.

The explicit representation of information that is sector specific, and usually only known by an individual who has worked in that sector for a long time, makes good SA possible for even new specialists. Overall, less chance exists that information will fall through the cracks, that specialists will miss information, or that information will not be communicated between personnel. The overall ease of obtaining needed information greatly reduces workload and enhances SA. With the new system design, the addition of more remotely monitored facilities and increased responsibilities for the MCC specialist could be easily accommodated.

These eight principles form a core foundation for SA-oriented Design. We will next address several key issues that affect SA in many systems. The first of these issues is the management of information certainty (or its correlate, uncertainty), which is critical to operators' confidence levels in information—a key component of SA.

STATUS
FACILITIES
ACTIVITY
SET-UP
HISTORY
PRINT
NOTES
Ft Worth
Dallas
Big Spring
Abilene
Midland
Brownwood
Waco
San Angelo
Lampasas
Temple
College St.
Llano
Austin
Houston
San Antonio

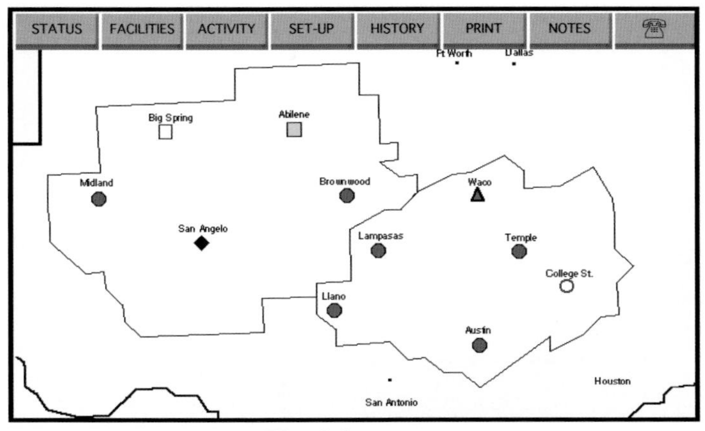

Plate I Status map.

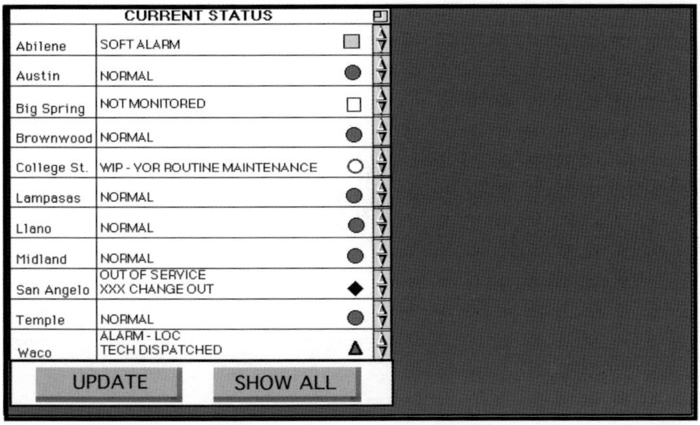

CURRENT STATUS
Abilene
SOFT ALARM
Austin
NORMAL
Big Spring
NOT MONITORED
Brownwood
NORMAL
College St.
WIP - VOR ROUTINE MAINTENANCE
Lampasas
NORMAL
Llano
NORMAL
Midland
NORMAL
San Angelo
OUT OF SERVICE
XXX CHANGE OUT
Temple
NORMAL
Waco
ALARM - LOC
TECH DISPATCHED
UPDATE
SHOW ALL

Plate II Current Status overview.

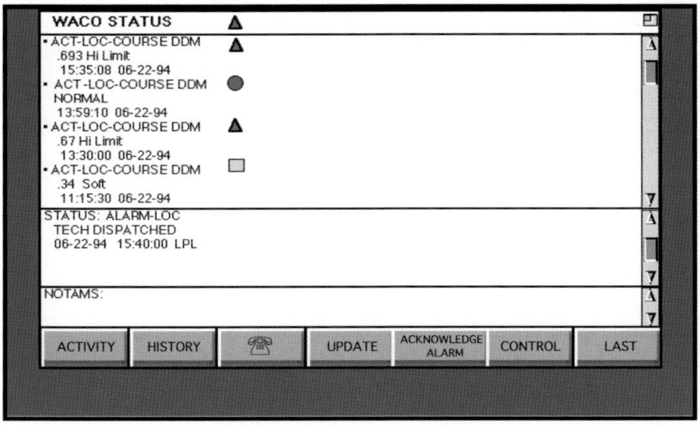

WACO STATUS
• ACT-LOC-COURSE DDM
.693 Hi Limit
15:35:08 06-22-94
• ACT-LOC-COURSE DDM
NORMAL
13:59:10 06-22-94
• ACT-LOC-COURSE DDM
.67 Hi Limit
13:30:00 06-22-94
• ACT-LOC-COURSE DDM
.34 Soft
11:15:30 06-22-94
STATUS: ALARM-LOC
TECH DISPATCHED
06-22-94 15:40:00 LPL
NOTAMS:
ACTIVITY
HISTORY
UPDATE
ACKNOWLEDGE ALARM
CONTROL
LAST

Plate III Facility Status display.

CLL - VOR
.693 Hi Limit
15:30 06-20-94
RESET
DISCONNECT
SHUTDOWN
CONNECT
START
DIAGNOSTICS
ACTIVITY
HISTORY
STATUS
FACILITIES
LAST

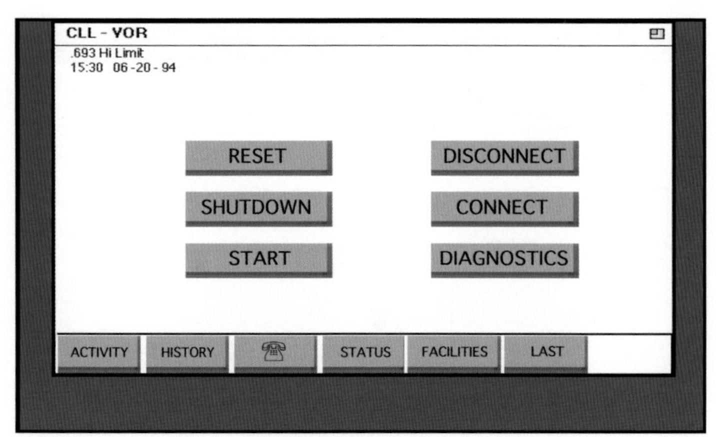

Plate IV Control display.

FACILITIES
SECTOR
AUSTIN
DALLAS
HOUSTON
LOCATION
ABILENE
AUSTIN
BIG SPRING
BROWNWOOD
COLLEGE ST.
LAMPASAS
LLANO
MIDLAND
SAN ANGELO
TEMPLE
WACO
FACILITY
ACT-ARTS
ACT-ASI
ACT-DME
ACT-DMUX
ACT-FDIORB
ACT-GS
ACT-LOC
ACT-LOM
ACT-MCR
ACT-MM
ACT-NDB
ACT-OM
ACT-PCS
ACT-RMSC
ACT-RTR
ACT-RVR
ACT-VOR
ACT-TACR
SHOW ALL
FACILITY:
LOC
FAC
SHOW
ACTIVITY
HISTORY
STATUS
CONTROL
LAST

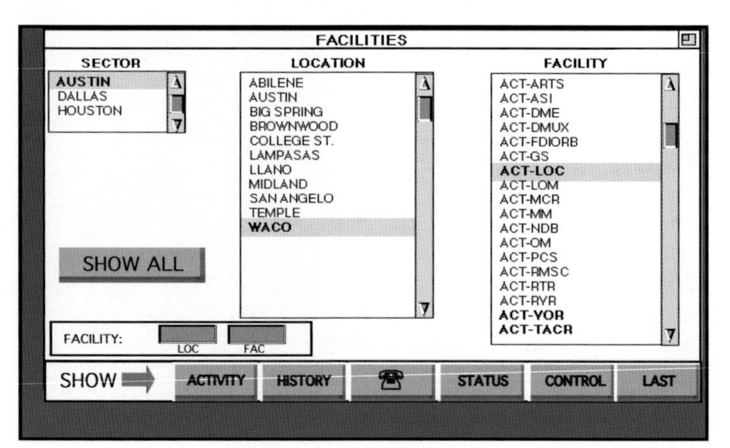

Plate V Facilities display.

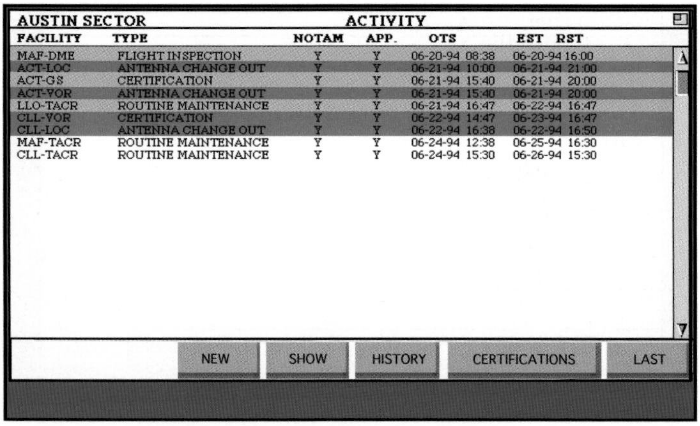

AUSTIN SECTOR
ACTIVITY
FACILITY TYPE NOTAM APP. OTS EST RST
MAF-DME FLIGHT INSPECTION Y Y 06-20-94 08:38 06-20-94 16:00
ACT-LOC ANTENNA CHANGE OUT Y Y 06-21-94 10:00 06-21-94 21:00
ACT-GS CERTIFICATION Y Y 06-21-94 15:40 06-21-94 20:00
ACT-VOR ANTENNA CHANGE OUT Y Y 06-21-94 15:40 06-21-94 20:00
LLO-TACR ROUTINE MAINTENANCE Y Y 06-21-94 16:47 06-22-94 16:47
CLL-VOR CERTIFICATION Y Y 06-22-94 14:47 06-23-94 16:47
CLL-LOC ANTENNA CHANGE OUT Y Y 06-22-94 16:38 06-22-94 16:50
MAF-TACR ROUTINE MAINTENANCE Y Y 06-24-94 12:38 06-25-94 16:30
CLL-TACR ROUTINE MAINTENANCE Y Y 06-24-94 15:30 06-26-94 15:30
NEW
SHOW
HISTORY
CERTIFICATIONS
LAST

Plate VI Activity Overview display.

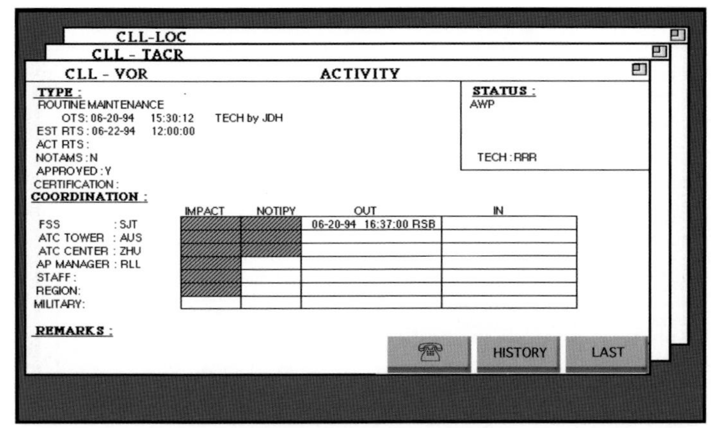

Plate VII Activity display.

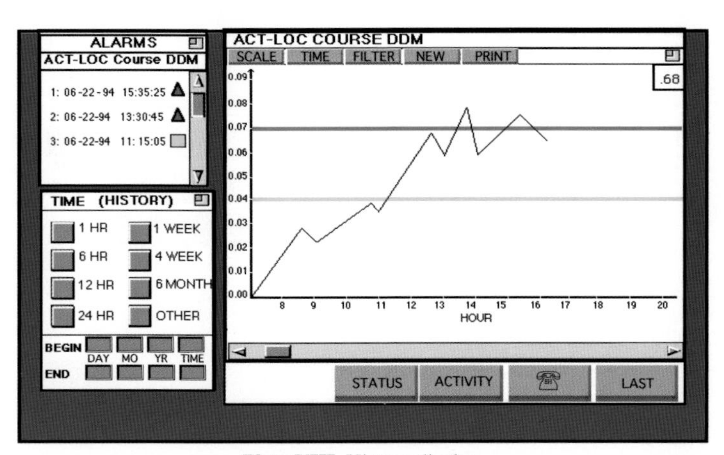

Plate VIII History display.

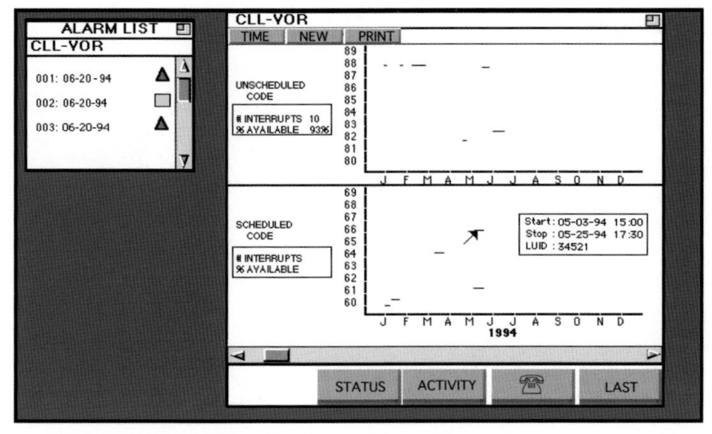

Plate IX History Outage display.

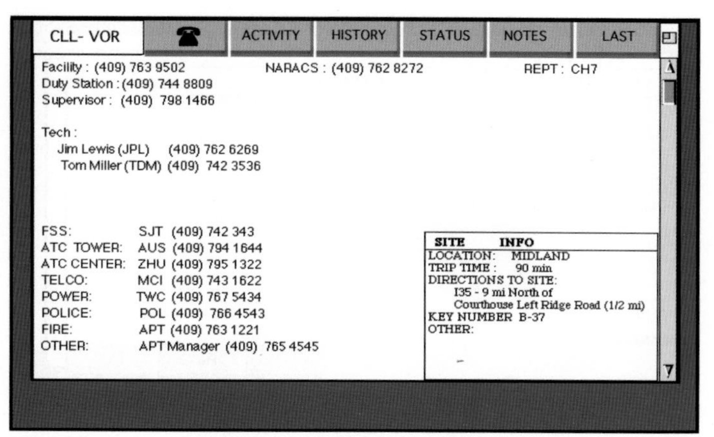

Plate X Communications display.

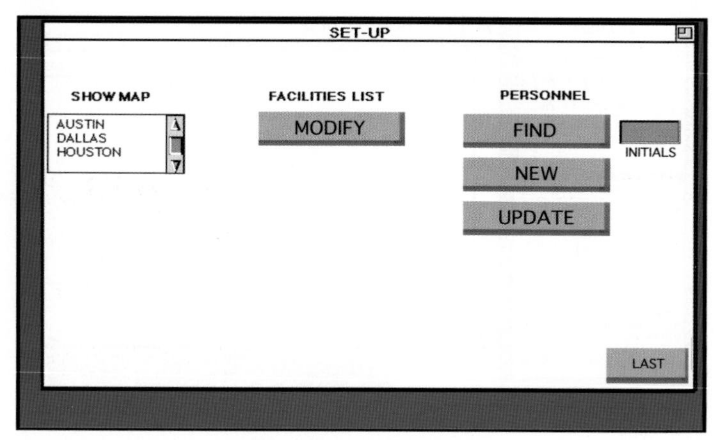

Plate XI Set-Up display.

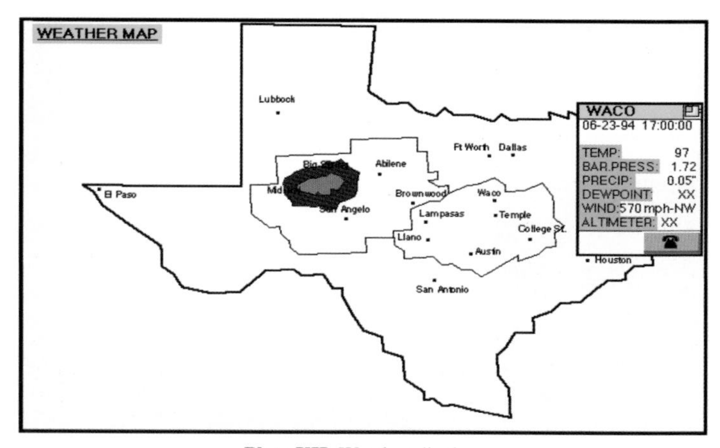

Plate XII Weather display.

CHAPTER SEVEN

Confidence and Uncertainty in SA and Decision Making

7.1 UNCERTAINTY

Uncertainty is a part of life. This is no less true for those system operators for whom we seek to design system interfaces. Uncertainty is not simply an inconvenience that can be wished or designed away, but is a significant factor in shaping human decision making and action. As such, it is very important that system designs support the operator's need to understand the degree and nature of the uncertainties in the situation. Confidence level in information, a correlate of uncertainty, is an important component of SA in most domains. Because of its unique nature, we will describe different types and sources of uncertainty, the role that uncertainty plays in situation awareness, and ways in which people actively manage uncertainty in complex domains. We will then describe design recommendations for the display of information reliability or certainty in order to support SA and decision making.

7.2 TYPES AND SOURCES OF UNCERTAINTY

Uncertainty plays a role throughout the decision process (Figure 7.1). There is a level of uncertainty associated with the basic data perceived by the individuals (u_1), the degree to which they gather a correct comprehension of that data (u_2), their ability to project what will happen in the future (u_3), and the degree to which their decisions will produce desired outcomes (u_d).

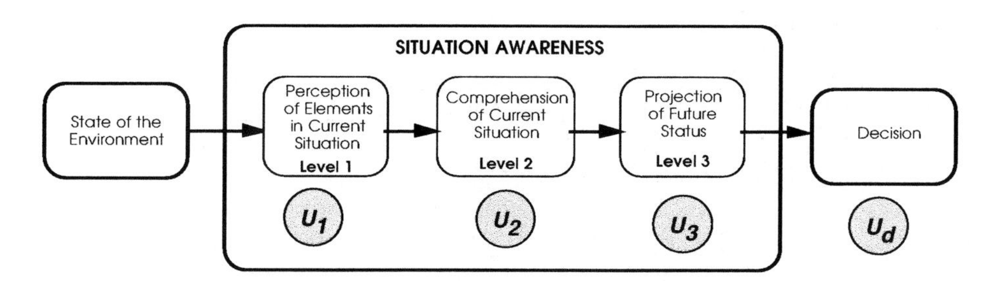

Figure 7.1 Types of uncertainty in the decision process.

7.2.1 Level 1—Data uncertainty

The first type of uncertainty relates to the basic data available to the operator regarding the state of the environment. In general, sensors measure or detect events and objects in the environment or their properties, and will have some degree of error or uncertainty associated with those measurements. This might include a mechanical sensor that detects the location, speed, and heading of an aircraft, or another person who reports the presence of enemy tanks in a given sector. In an operating room, sensors are used to measure patient heart rate and the concentration of oxygen in the blood. All of these sensors' measurements are known to have some error associated with them. A number of factors contribute to the degree of uncertainty associated with the data produced by sensors.

(1) *Missing Information.* One of the most frequent sources of uncertainty is that of missing data (McCloskey, 1996). In military scenarios, the enemy actively works to deny or hide relevant information, making missing data an almost constant problem. Pilots or commanders must actively work to find enemy aircraft or troops and determine their relevant attributes (e.g., numbers, formation, heading, type), but they will almost always have to operate with only partial information. In the medical arena, uncertainty may exist due to missing information on the nature of an underlying injury as x-rays or other tests can only detect part of the needed data, or missing information on a patient's past medical history, previous inoculations or allergies, as the patient may be unconscious or unable to remember (Xiao & MacKenzie, 1997). In almost every domain, missing information is a significant source of uncertainty.

(2) *Reliability/Credibility of the Data.* The information available to the operator has a certain degree of uncertainty that is generated based on the degree of unreliability of the sensors or sources detecting and measuring the underlying phenomenon. No sensor or person is 100% accurate. A verbal report of enemy activity from an inexperienced soldier will be viewed as less reliable (more uncertain) than one from a seasoned veteran. A target detection on radar may be viewed as more reliable (less uncertain) than one from an infrared sensor. Different medical tests are known to have different rates of reliability for detecting various medical conditions. Over time, operators learn which sensors are more reliable than others, just as they learn which people are more reliable and trustworthy than others.

More importantly, people also learn *when* these systems are reliable, and the factors that lead them to be unreliable. For example, a particular sensor may be very reliable most of the time, but function poorly with cloud cover or high humidity levels. A given individual may be very reliable most of the time, but not on Friday nights when they are known to drink too much. This type of gauging of the reliability of information is far more useful than a simple number (e.g., 75% reliable), because people are better able to calibrate the degree of confidence they can place in a particular piece of data provided by a system or a person *at any given time.*

Both the likelihood that the sensor or measure will correctly detect the underlying event (its veracity) and the granularity of the reported data (degree of specificity) are relevant to the source's reliability. In other words, the more detailed the sensor is about what it is reporting, the higher the person's confidence can be in its accuracy. For instance, air traffic controllers get reports of aircraft altitude that are accurate to within 300 feet of the actual altitude. Global positioning system (GPS) signals can report an object's location (depending on the number of satellites in the triangulation) to within 50 feet. There is an inherent degree of uncertainty associated with these accuracy levels as there is room for variance.

(3) *Incongruent/Conflicting Data*. Often, multiple sensors provide information on the same underlying phenomenon. When those different data sources are in agreement, even if each is of lower reliability, they can act to increase subjective confidence in the accuracy of the data (Gilson, Mouloua, Graft, & McDonald, 2001). Many times, however, multiple sources may report conflicting data. This incongruence can occur due to different sensors having different reliability levels associated with them, or because they are actually reporting on a different underlying phenomenon. While people may use different strategies to resolve such conflicts (discussed later in this chapter), it is safe to say that incongruent data can lead to lower levels of confidence or certainty, and can significantly slow down decision making and degrade performance (Bolstad & Endsley, 1998). Determining which sensor is correct can be a difficult and time-consuming activity for the operator.

(4) *Timeliness of Data*. While in some environments data are detected and reported to the operator on an almost continuous basis, in many others data collection occurs only sporadically or periodically. Aircraft position reports are updated every 6 seconds on an air traffic controller's display (per the sweep of the radar), thus introducing a degree of uncertainty associated with changes in the aircraft's location, heading, and altitude since the last update. Intelligence information in a battlefield may be hours or days old, thus introducing uncertainty as to what has changed in the interim, even if the data are completely accurate at the time they were originally collected.

(5) *Ambiguous or Noisy Data*. In addition, many data are embedded in contexts which are inherently noisy or ambiguous, leading to uncertainty. An ECG signal may pick up numerous movement artifacts that can obscure the signal the doctor is looking for. Vehicles and tanks may be camouflaged, making their detection against a background difficult and introducing uncertainty as to whether an object is present or not.

Each of these five factors—missing data, reliability of sensors, incongruent data, time lags, and noisy data—can introduce uncertainty and affect a person's confidence in their SA, not only in a quantitative, but also a qualitative fashion. Knowing that an object was definitely at a certain location 5 minutes ago is quite different from a partially reliable report of its location at present. The first case provides the individual with certain information (e.g., an enemy tank is definitely in the area), and a range of limited possibilities from which to extrapolate its current status (e.g., it must be within X miles of this location). The second case allows for the possibility that no tank was there at all, but could be the result of an unreliable sensor or reporter. The implications for further data collection and decision making are significant. The uncertainty that exists for one sensor which has 80% reliability is quite different than that produced by two sensors with a combined reliability rate that is the same, or by two sensors that report different information. Thus, while uncertainty can result from any of these different sources, the qualitative nature of that uncertainty is, in itself, information that is used to feed decision making and action and should not be lost or obscured from the operator.

7.2.2 Level 2—Comprehension uncertainty

Internally, people form a degree of confidence associated with their understanding of what is happening. Given certain cues and data input, a commander may be 90% sure that the enemy is massing in a particular sector. Based on several tests and physiological measures, a doctor may be 80% confident in a particular diagnosis.

Similarly, many systems aggregate data or apply different types of algorithms to classify or categorize underlying data. Automatic target recognition systems or sensor fusion algorithms, for example, may combine the output of multiple sensors to produce a target location or identification. This output will have a certain degree of reliability based on both the quality of the underlying data and the quality of the software algorithm used to determine the output.

Categorization mapping (the degree of fit of the signal characteristics to known classes of a specific parameter) falls into this area. Even if a system is 100% accurate in detecting an aircraft in a given location, uncertainty may exist associated with mapping its features to known types of aircraft (e.g., 80% probability that it is friendly and 20% probability that it is an enemy). Finding ways to meaningfully present confidence level information associated with classification/categorization and data aggregation is very important for SA if these systems are to be used successfully.

7.2.3 Level 3—Projection uncertainty

Projecting what will happen in the future is inherently uncertain in most systems. While good knowledge of the situation and good mental models of how objects behave can act to reduce that uncertainty, to some extent a degree of uncertainty regarding what will happen in the future will always exist. A fighter pilot can try to anticipate in which direction an enemy plane will veer when his weapons lock on to it, but he cannot know for certain. A surgeon can believe that the artery she's about to clean plaque out of will function normally afterwards, but she can't know for certain that it can be repaired without further problems. Even though predictions are often far from perfect, people generally strive to be predictive and thus proactive. The difficulty of projection is compounded by not only the uncertainty of the underlying data, but also the ability of the person (or any system) to accurately predict future behavior and events based on that information.

7.2.4 Decision uncertainty

Finally, uncertainty often exists in a person's degree of confidence that a particular selected course of action will result in the desired outcome. Considerable research has been conducted on decision making under uncertainty (Kahneman, Slovic, & Tversky, 1982), and on decision making in naturalistic settings (Klein, Orasanu, Calderwood and Zsambock, 1993) which acknowledges this type of uncertainty. For the most part, this book focuses on the display of the first three types of uncertainty (pertaining to SA).

7.3 ROLE OF CONFIDENCE IN LINKING SA AND DECISION MAKING

While the actual accuracy of an individual's SA significantly affects the quality of their decisions, their confidence in that SA (degree of certainty) also plays a significant role. Such subjective assessments of confidence in one's SA may provide a critical link between SA and performance. That is, peoples' subjective assessment of the quality of their SA may be important in determining how they will choose to act on that SA (conservatively if it is low or boldly if it is high), independent of the actual quality of their SA.

Christ, McKeever, and Huff (1994) and Endsley and Jones (1997) have discussed how SA and confidence work together for a person to make a decision. As shown in Figure 7.2, if SA is good and confidence in that SA is high, a person is more likely to achieve a good outcome, as it will have been possible to make good decisions and plans based on that SA. A military officer who has a high degree of confidence in his knowledge of enemy forces will formulate and execute a plan of action for meeting and overcoming those forces.

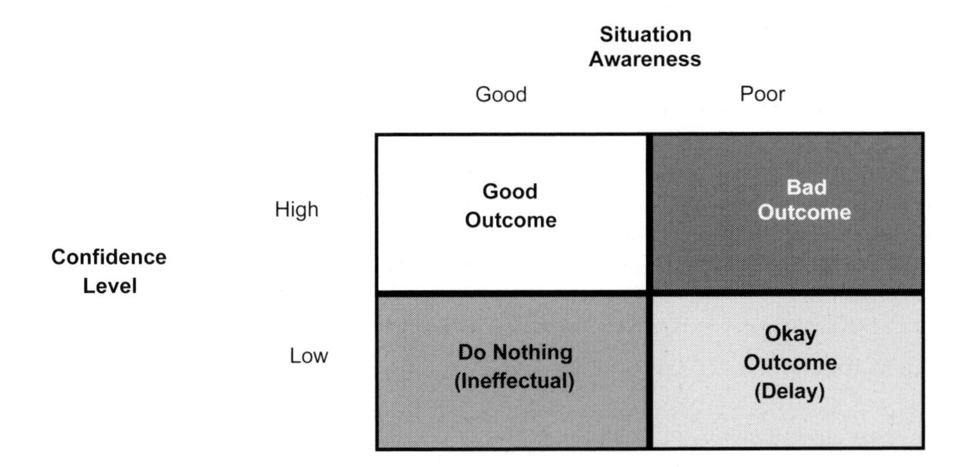

Figure 7.2. Relationship between situation awareness and confidence (from Endsley & Jones, 1997).

If, with equally good SA, the person has a low level of confidence in that SA, however, they most likely will not act on it, choosing to gather more information or behave protectively, and thus be ineffective. In this case, the same military commander may delay action and continue to try to collect more information on the enemy, even though his actual SA may be the same as in the previous case. This lack of confidence (if not properly calibrated with the actual quality of the SA) can contribute to indecisiveness and delays that produce poor outcomes. This illustrates the importance of communicating information reliability so that people can properly calibrate their confidence level

If a person with poor SA recognizes that it is poor (low confidence level), he or she will correctly choose not to act or act protectively and will continue to gather more information to improve SA, thus averting what could be a very bad outcome. In this case, the military commander would be correct in delaying action in order to gather more information.

The worst situation is that of the person who has poor SA, but has a high level of confidence in that erroneous picture. Not only will this person be likely to act boldly and incorrectly, but often will draw in others who will be fooled by the misplaced confidence. Of the four possible situations, this one is the most dangerous. The military commander would be apt to boldly lead his troops into disastrous situations where many lives would be lost.

A critical issue, therefore, is ensuring not only that operators have as good a picture of the situation as possible, but also that they are able to attribute the correct amount of confidence or certainty to that picture.

The amount of confidence or certainty that an individual requires before acting can vary. Rules-of-engagement (ROEs) exist in most military environments that require a certain degree of confidence prior to action. For example, in a non-wartime scenario, a fighter pilot may not be allowed to fire unless fired upon. Under other ROEs, he may be able to fire only if he has a visual sighting of an enemy target (providing a high degree of confidence in SA for this element), but not an electronic identification (lower confidence). Under more relaxed ROEs, an electronic identification may be considered adequate for action. The ROE thus acts as a confidence threshold for decision making. Confidence in one's knowledge of the certainty of information must exceed certain thresholds for action.

In other environments, similar although often unspoken thresholds for decision making exist. A nurse might monitor a patient's progress for some period of time until enough symptoms manifest themselves to warrant calling the physician at home. Once a person has collected enough information to be reasonably sure of the situation, he or she will decide to act.

7.4 MANAGEMENT OF UNCERTAINTY

In most environments, information certainty is not a given. People actively work to manage or reduce uncertainty to the greatest degree possible. Several strategies for reducing uncertainty play a significant role in the decision-making process and have an impact on SA.

7.4.1 Search for more information

One of the most obvious strategies for reducing uncertainty is searching for more information. This search may involve looking through displays that are present, but it usually also involves taking active steps to collect the needed data. Many times, the operator determines which areas of the environment are searched by a particular sensor. A pilot may point the radar in a particular direction. A doctor may request additional tests. A commander may request an unmanned aerial vehicle (UAV) to fly over a particular area with a camera to survey it for enemy activity, or he may position his soldiers at a particular vantage point where they can report enemy movements. Frequently, this search is proactive. That is, people set things up in advance to be able to collect data they may need in the future (e.g., sending out scouts, or implementing reporting procedures).

People will also use a strategy of searching for more information to deal with conflicting or unreliable data. 'If sensors A and B do not agree, can another source of information be found to resolve the conflict? If system X is known to be unreliable, is there any other information that can confirm its report?'

People are active participants in generating the data needed for building their SA. This search for information may continue until enough information is acquired for a satisfactory solution (at an acceptable level of confidence), or until the time available for making a decision is exceeded. Often this is the case, and people must act even with more uncertainty than they desire.

7.4.2 Reliance on defaults

Frequently, needed information will simply not be available, or there will be insufficient time to obtain it. In this case, people rely on default values in their mental models to fill in missing data (Endsley, 1988; 1995c). For example, a pilot may be able to observe that a particular aircraft is traveling at 450 knots, and from this infer what type of aircraft it most likely is, because she knows at about what speed different aircraft typically fly. A nurse may assume, in the absence of other information, that the 'usual' procedure will be followed unless the doctor explicitly specifies an alternate procedure. These default values act as substitutes in the absence of actual data from the current situation, allowing experienced individuals to be fairly effective, even with limited information. A certain degree of uncertainty will continue to exist, however, until data confirming the accuracy of these defaults, or replacing them, is obtained.

7.4.3 Conflict resolution

The most difficult uncertainty to resolve may be data that is in conflict. Rather than simply taking some mathematical average, people often work to resolve conflicting data by determining which of the data sources is incorrect. They may turn off and on a piece of equipment, check connections, or take a new reading to see if a piece of data is erroneous. They may seek an additional reading (i.e., send someone to check) to resolve which data source is incorrect. An active process to find some factor that would lead to an erroneous reading from one of the conflicting data sources will be pursued and should be supported through the user interface.

If no means are available to resolve conflicts, people may use a composite technique whereby both the number of sensors reporting a particular event and the reliability of those sensors are used as data input (Bolstad & Endsley, 1998, 1999a; Montgomery & Sorkin, 1996). People appear to be sensitive to the reported reliability of sensors and the number of sensors reporting confirming or conflicting data in making their assessments. An operator responsible for detecting incoming aircraft will consider both how many sensors are reporting the presence of an aircraft, and the reliability level of those sensors in determining whether an aircraft is actually present.

7.4.4. Thresholding

Often people may not act to reduce uncertainty past a certain point. As long as they are reasonably confident of some piece of information, they will proceed without increasing that certainty level. For example, police officers may only need to produce enough information to show a reasonable probability that a crime has been committed in order to obtain a search warrant from a judge. They do not need to continue to collect enough data to be 60% or 80% confident, but can act as soon as 'enough' data has been collected. Similarly, decision makers often will not act to ascertain whether condition A or B is present, because they will take the same action regardless. A doctor may not work to determine exactly which organism is responsible for an infection, but will prescribe the same antibiotic regardless. Mechanics will often replace a part they suspect is causing problems rather than work to determine with any degree of certainty if it is actually faulty or what aspect of the part is faulty. It simply is not necessary to reduce uncertainty in many cases.

7.4.5 Bet-hedging and contingency planning

Because it can be almost impossible to predict what may happen in the future, a frequent strategy used by experts is to plan for the worst case of possible future events. By actively playing 'what if' to anticipate possible occurrences and preparing for what they would do under those contingencies, they simply ignore the problem of uncertainty. Even if such events are of a low probability of occurrence, this strategy leaves people prepared to deal with possible negative events in the future. The projected consequences of events are more important to this strategy than worrying about their projected likelihood. People may prepare to quickly detect these negative events (e.g., by setting up observers in a particular location, or by actively monitoring for certain leading indicators) or to respond to the contingencies (e.g., by leaving a company of soldiers in a particular area, just in case), as well as pursue more likely future scenarios.

7.4.6 Narrowing options

While it may be very difficult to predict what may happen in the future (e.g., which direction the enemy will attack from), another strategy for managing uncertainty is to actively reduce the possible options for what may occur. For example, an army commander may act to eliminate certain directions of attack by blocking passes so that the enemy only has one or two options available, for which the commander can prepare in advance. An air traffic controller will work to position aircraft so that even if aircraft make errors in following their clearances or make unanticipated control actions, collisions are not likely. Rather than trying to ascertain what may happen, this strategy seeks to reduce the uncertainty by reducing the number of things that could occur. Thus uncertainty is effectively reduced.

7.5 DESIGN PRINCIPLES FOR REPRESENTING UNCERTAINTY

Given that all uncertainty will never be eliminated and that a person's degree of confidence in information can be as important to decision making as the information itself, helping people to determine how much confidence to place in information is critical to SA. Several design principles can be put forth to help guide this process.

Principle 9: Explicitly identify missing information

People will often mistakenly treat information that is missing as if it would be in agreement with other presented information, i.e., no news is good news. For example, Bolstad and Endsley (1998) found that if the data from one out of three sensors is missing, people will treat it as if its reading would be the same as the two sensors that are present, when in fact it could be conflicting. If the missing data had in fact been in conflict, their decisions would have been quite different.

Banbury *et al.* (1998) showed a similar phenomenon. When a target identification algorithm showed only the most likely identification (e.g., 93% probability that it is an enemy aircraft), pilots responded one way—they were very likely to attack the enemy aircraft. When the remaining possibility was explicitly shown (e.g., 7% friendly aircraft or 7% other enemy aircraft), responses were quite different—they were less likely to attack the aircraft if the possibility that it was friendly was shown. When the missing possibility was not shown, they treated it as if it would agree with the most likely possibility (e.g., it would be another enemy). This problem can directly lead to fratricide.

Missing information on spatial displays can be particularly insidious. For example, it can be quite difficult to tell whether no information presented in an area means there are no targets or objects of interest in that area, or whether that area has not been searched, or sensors are not working. The first case represents 'no hazards' and the second case represents 'hazards unknown.' These are very different situations for the decision maker and great care needs to be taken to help prevent a 'false world' of no hazards from being presented.

While many times the operator may 'know' through training or experience that certain information is not being shown, people are so visually dominant that in times of stress or when task load is high they can easily forget or not attend to this nonpresent information. For example, in 1996, a Boeing 757 aircraft loaded with passengers crashed in mountainous terrain near Cali, Colombia. When the pilots got off course through a problem with the flight management system, they continued their descent while bringing the plane back to the correct course, apparently completely unaware of the mountainous terrain below them. The graphical display in front of them, which they normally use when flying, did not show any terrain information. This 'false world' they constructed may have contributed to their decision to continue descending (Endsley & Strauch, 1997). The missing information was simply not salient to them.

When that missing data is highly relevant (e.g., as in the case of terrain information on the navigation display of the commercial airliner), design changes should be made to insure that the needed information is presented to the operator. In many cases, technological limitations will exist such that missing data is inevitable. In these cases, it is important for the operator to recognize that information is missing from the display.

An indication of which information is missing can be shown in a variety of ways. In the case study in Chapter 6, we used symbology to explicitly identify to the operator which sensors were not reporting data so that a false sense of 'no problems there' would not be generated. In a military display where only certain geographical areas have been searched, shading may be used to show areas where the presence or absence of enemy troops is unknown. Figure 7.3 shows a shaded area where no information is available on enemy presence, encouraging the commander to understand that no enemy reported in this area does not mean that none are present.

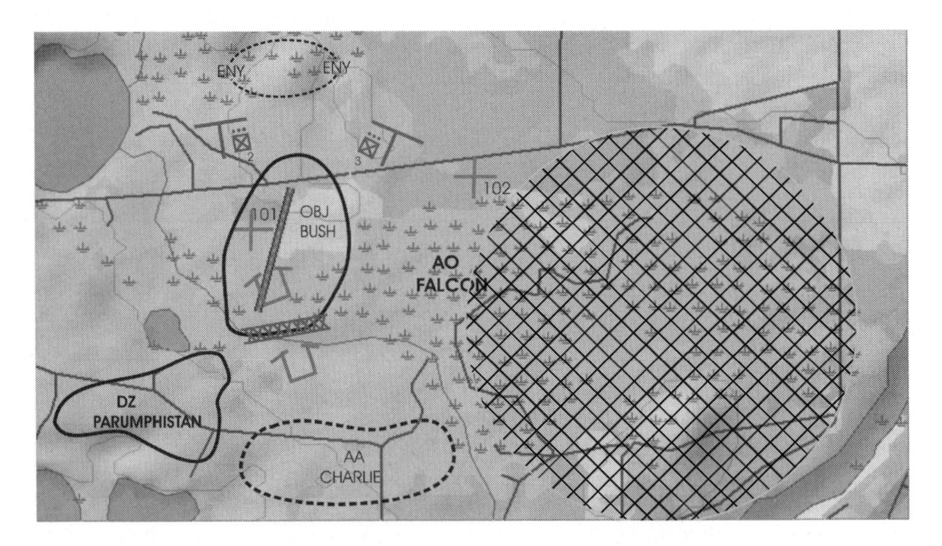

Figure 7.3 Representation of areas of uncertainty.

In the case of sensor fusion or categorization algorithms, it appears that explicitly identifying the lower probability matches, in addition to the highest probability match, is warranted in certain cases (Figure 7.4). When consequences are high, such as a possibility of fratricide or system failure, it may be worthwhile to explicitly represent other options to encourage people to consider lower probability possibilities. As and example, the left display shows the most likely type of aircraft detected. The right display shows both the most probable match, the Su-77 and also another aircraft, the F16, for which the system is only 12% confident of the match. The display on the right would be more likely to encourage people to consider reliability of the display's representation of the actual situation.

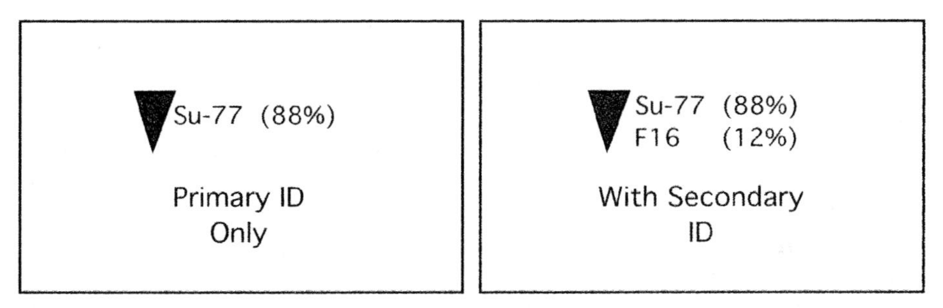

Figure 7.4 Display of system certainty for multiple matches.

Principle 10: Support sensor reliability assessment

People are able to use information in their judgments on the reliability of sensors in order to weight the data presented, although they may not do so optimally (Berg, 1990; Bolstad & Endsley, 1998; Montgomery & Sorkin, 1996). They also seem to weight information in the center of their visual field more highly under time stress (Sorkin, Mabry, Weldon, & Elvers, 1991) and more highly on the left hand side when presented in reading order (Bolstad & Endsley, 1999a). Therefore the most reliable sensors should be allocated to more prominent real estate in the display.

While sensor reliability information can be displayed numerically (e.g., 80%, 65%), Montgomery and Sorkin (1996) showed that luminance levels could be used to convey the relative reliability of sensors, such that more reliable sensors are brighter than less reliable sensors. Other presentation techniques that draw on information salience and analog processing should also be considered (see Principles 11 and 13).

Most importantly, people can benefit from information that tells them when certain information may not be reliable or which provides them with information to check the reliability of a given reading. That is, they need to know not just reliability in general, but factors that allow them to assess a sensor's reliability at any particular time. This situational context information, when relevant, should be made readily available within the area of the displayed data. See Principle 14 for some ways to accomplish this support.

Principle 11: Use data salience in support of certainty

Some researchers have sought to directly display the accuracy of information on a display. For example, Kirschenbaum and Arruda (1994) showed rings of uncertainty (95% confidence intervals) around the displayed position of ships to submarine officers. Similarly, Andre and Cutler (1998) showed uncertainty rings around aircraft position symbols whose displayed locations were imprecise (Figure 7.5).

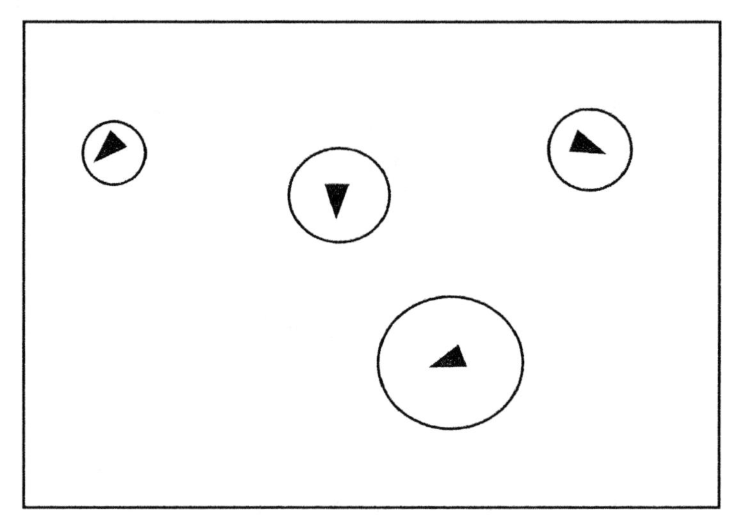

Figure 7.5 Positional uncertainty conveyed via rings.

While this approach may be successful in displaying positional uncertainty, it has an unfortunate side effect. The information about which the most uncertainty exists becomes the most prominent on the display, drawing the operator's attention to it. When multiple symbols are presented, this misplaced salience will result in more attention to the more uncertain objects than to others, which may be counter to where the operator's attention should be directed.

An alternative approach is to use display salience to convey certainty information. For example, shading can be used to show positional uncertainty without misdirecting the operator's attention The shaded area around the triangular symbols in Figure 7.6 shows the certainty range for the position, with the outer areas decreasing in both certainty and amount of shading. In another example, the thickness of the lines and fill color are used to show target certainty information in a way that makes the targets about which the most is known the most salient to the operator. In Figure 7.7 the filled triangle depicts a target which is more certain (i.e., based on more reliable sensors), than that with a thick black outline, which in turn is based on more reliable sensors than those with a thinner outline.

Uncertainty regarding the range of future position projections should be very carefully presented so as not to overwhelm the operator. For example, projection envelopes or fans, such as those shown in Figure 7.8, have been proposed as a way of displaying the possible future position of an aircraft. While not a problem for a single aircraft, such a display becomes overwhelming when multiple aircraft are presented, with the fans becoming more salient than the actual current positions of the aircraft. Again shading and luminance levels can be used to help better direct information salience in the display (Figure 7.9).

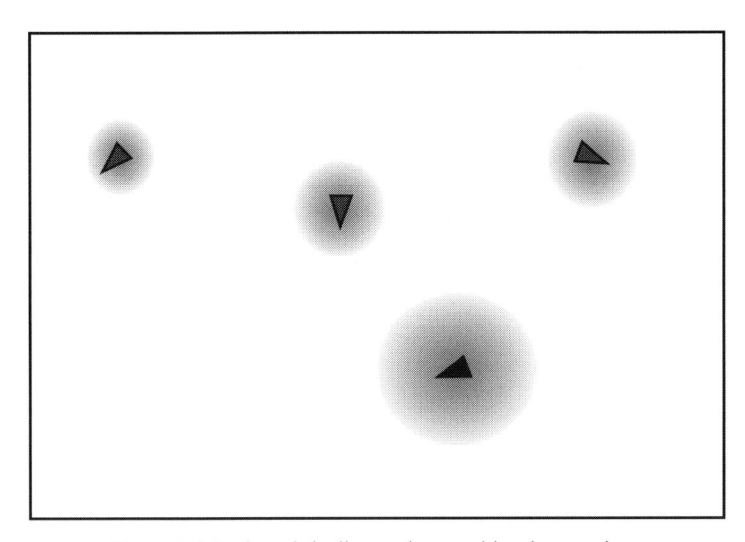

Figure 7.6 Graduated shading to show positional uncertainty.

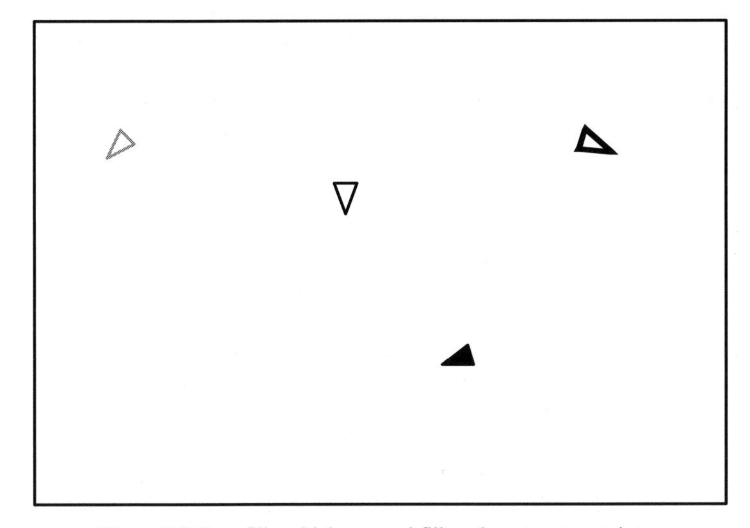

Figure 7.7 Use of line thickness and fill to show target certainty.

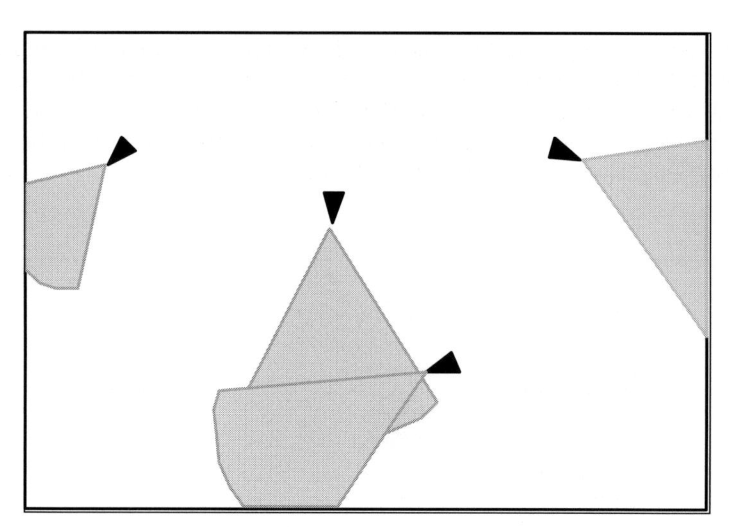

Figure 7.8 Use of fans for uncertainty in location projection.

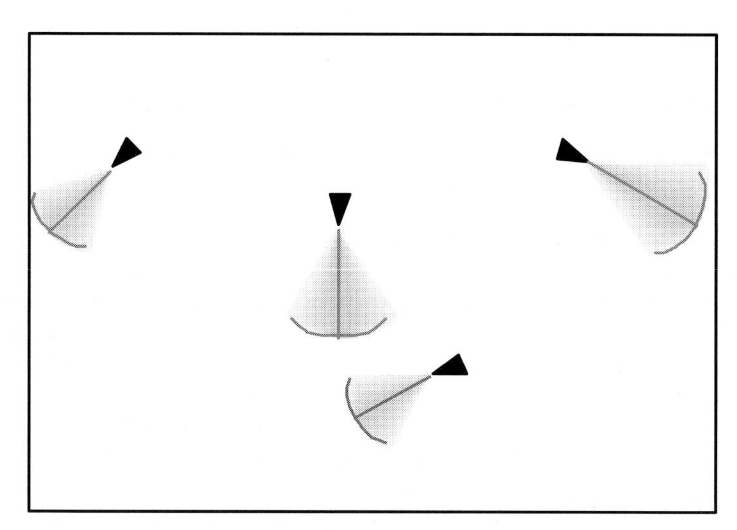

Figure 7.9 Use of shading and whiskers for uncertainty in location projection.

Principle 12: Represent information timeliness

Just as information may be uncertain because it is unreliable, it may also be uncertain because a period of time has elapsed since the data was collected. If this period is very short and predictable, such a lag may not pose a significant issue. If the period is long or if it is highly variable (such as is the case with verbal reports in a battlefield scenario), it will be necessary to explicitly represent the timeliness of such data so that it can be considered appropriately in decision making. Shading or luminance levels can be used to show information timeliness in a way that will make the most timely, and thus the most accurate information the most salient (Figure 7.10). These techniques can be used even for information lags that are shorter. For example, the luminance of a displayed aircraft symbol may gradually decline until the next 'sweep' of the radar refreshes it to full intensity. Such cues provide clear but subtle information to the operator on the accuracy of displayed information.

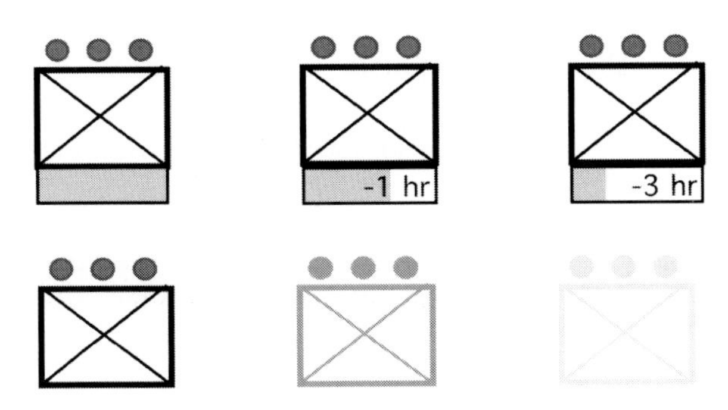

Figure 7.10 Display of information timeliness.

Principle 13: Support assessment of confidence in composite data

As more systems employ classification algorithms, sensor fusion and decision support systems, the need will arise to apprise operators of the reliability or confidence levels of these systems' outputs. The reliability can be affected by the quality of the underlying data (missing data, confirming or incongruent data, and accuracy of the data), and the algorithm itself (e.g., the degree to which the data fit prescribed categories). More research is needed on how to accomplish this task effectively. By their very nature, such systems generally obscure cues that operators currently use to infer information reliability (e.g., which sources produced the underlying data, or contextual information on its collection). It remains an open question as to whether numerical probability estimates of system confidence in its recommendations can be used as effectively by operators.

A few studies have examined this question. In general, when system confidence levels associated with recommendations or classifications are close (e.g., 45% for A and 53% for B), decision time can be significantly slowed (Selcon, 1990). Research indicates that people are relatively poor at handling probability information (Kahneman *et al.*, 1982).

Endsley and Kiris (1994a) showed that a categorical (*high*, *medium*, or *low*) indication of system confidence in its recommendations led to slightly faster decision times when compared to no information (i.e., operators had to make their own decision as to which alternative was right). Numerical percentages, analog bars, or rankings of alternatives led to much slower decision making (Figure 7.11). Finger and Bisantz (2000) used a computer blending technique that degraded symbols to show their degree of fit with categories on either end of a continuum (e.g., friendly or hostile) (Figure 7.12). They found that people performed equally well in identifying targets with this symbology as with numerical probability presentations on the goodness of fit to the categories.

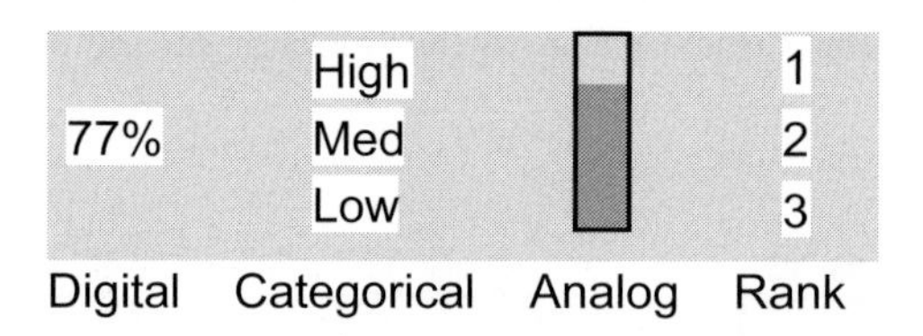

Figure 7.11 Display of system confidence level in alternatives.

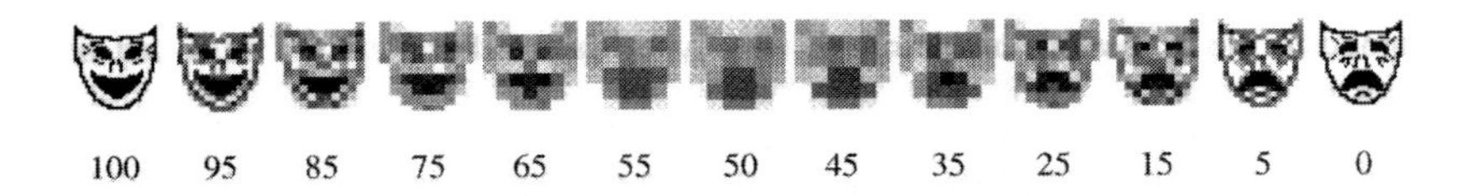

Figure 7.12 Display of degree of category membership (from Finger & Bisantz, 2000). Reprinted with permission from *Proceedings of the XIVth Triennial Congress of the International Ergonomics Association and 44th Annual Meeting of the Human Factors and Ergonomics Society*, 2000. Copyright 2000 by the Human Factors and Ergonomics Society. All rights reserved.

It is not known how much access people may need to the underlying data that goes into system generated composite data, or if a means can be developed to better inform people of the qualitative as well as quantitative nature of the reliability of such estimates.

Principle 14: Support uncertainty management activities

Uncertainty management is often sorely neglected in display design. As operators need to assess the quality of particular data readings, means should be provided to make such assessments rapid and easy to accomplish. Confirming or negating information and relevant contextual information should be clearly presented proximally to the display information. For example, if the accuracy of a particular signal is prone to interference, a readout of signal strength could be displayed directly below the data reading. If its reliability is affected by humidity or some other value, that information should be displayed proximally, so that a rapid assessment of a particular reading's reliability can be made. This situation-specific information will be far more useful than a generic presentation of overall reliability (e.g., 67%), as it allows the operator to know how much to trust a particular piece of data in the current instance, or how to adjust the system to obtain more reliable data.

If the system infers information (e.g., likely target identification) based on other data, this needs to be displayed in a way that allows a rapid determination of which data is known and which is inferred, for example through the use of color coding, line thickness, or shading.

Finally, the amount of precision or certainty that operators actually need to have in different situations should be determined. Systems that display too much precision (97.6%) may slow down both computer and human information processing. Systems that work to determine and display the highest probability events or matches to data may neglect that what the operator really needs is support in anticipating and preparing for lower probability but higher consequence events, for which they need to prepare (e.g., the enemy could attack through the sewers). The ways in which individuals manage uncertainty in a particular domain should be carefully studied and provided for in the display design to support their need for understanding how much confidence to place in information and to support their need to reduce uncertainty appropriately.

CHAPTER EIGHT

Dealing with Complexity

8.1 A SIMPLIFIED VIEW OF COMPLEXITY

Complexity has become a ubiquitous part of modern life. Complexity is at the heart of many systems, from those that send astronauts into space to office desktop computers. The technologies that surround us embody complexity in both their form and their function. While, to a certain degree, this complexity cannot be avoided, it can be better managed, and must be if we are to develop systems that will allow users to have high levels of situation awareness when working with these systems. Left unchecked, complexity directly works against SA. It reduces people's ability to perceive needed information among the plethora of that available. It can also significantly affect the ability of users to understand and predict what will happen when working with a system (Level 2 and 3 SA) because complexity makes it much more difficult to form a good mental model of the system or task domain.

To more fully understand what factors make something complex and how to design systems to deal with that complexity, we will first describe several different components that add up to determine the level of complexity that the user must cope with. This can be thought of as multiple layers of complexity (Figure 8.1), including: (1) the complexity of the system, (2) the operational complexity that the user must deal with, and (3) the apparent complexity that is brought about by the system interface features (cognitive complexity, display complexity, and task complexity). Finally, the user's mental model of the system greatly moderates the level of complexity that will be perceived by a given individual.

8.1.1 System complexity

First, there is the level of complexity of the overall system the user is dealing with. A system can be very complex (e.g., an automobile) or very simple (e.g., a bicycle). Complexity at the system level can be thought of as a function of:

- the *number of items* (e.g., components, objects, features) incorporated in the system,
- the *degree of interaction* or interdependence of those items in creating functionality,
- the *system dynamics*, indicating how fast the status of items and interactions will change within the system, and

- the *predictability* of such changes (e.g., the degree to which the dynamics are driven by simple rules versus a dependency on multiple factors or component states).

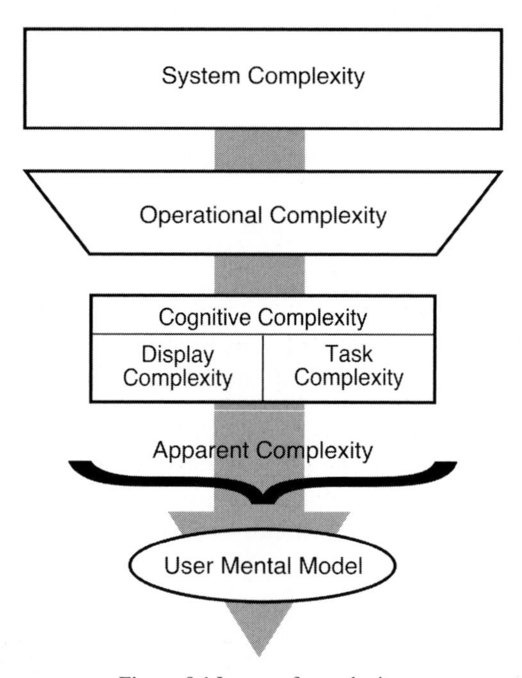

Figure 8.1 Layers of complexity.

A system with fewer items, less interaction between items, and a slow rate of change of items will clearly be less complex than one in which these variables increase. System complexity is frequently driven by feature creep (Figure 8.2). A sundial was once sufficient for telling the approximate time of day. Watches and clocks were then developed that involved many internal mechanisms to provide greater precision in time keeping, including the provision of a hand for designating minutes and even seconds. They also provided the ability to tell time in poor weather and with a greater degree of portability. Today's watches and clocks no longer simply provide the time of day, they also include alarms, timers, calendars, snooze functions, a variety of alarm sounds and timing functions, and a radio, all in the same device! These are no doubt useful functions, but are supplied at a price.

Such increased functionality requires more components (mechanical, circuitry, or software) and more connectivity among those components. The behavior of the system requires more rules and more combinations of factors and system states. For example, due to the large number of features available, setting the time on a multifunction watch requires multiple button presses performed in a specific sequence using a particular subset of the available control buttons; any other sequence of button presses will activate a different function on the watch. Conversely, simple systems are void of the complexity that comes with multiple features and a variety of possible system states—setting the time on a wind-up watch that only tells the time requires simply pulling out the only control knob and turning it until the watch hands show the desired time. Resetting the time on a watch and setting the alarm time on clock-radio are no longer as simple as they used to be.

8.1.2 Operational complexity

System complexity does not necessarily translate into complexity for the user, however. For example, the automobiles of today are far more complex than they were in the 1920s. They have more components and each of those components is far more complex. Automatic transmissions, antilock breaks, and fuel injection systems are all fairly complex subsystems that did not exist in early automobiles. Yet automobiles are not more complex to operate today. In fact, they are actually less complex for the driver. In the early days, drivers needed to know quite a bit about the internal workings of the automobile. They had to properly prime or choke the engine prior to starting it and may have needed to operate a handheld crank. Today, a simple turn of the key is all that is needed, and few drivers understand the workings underneath the hood.

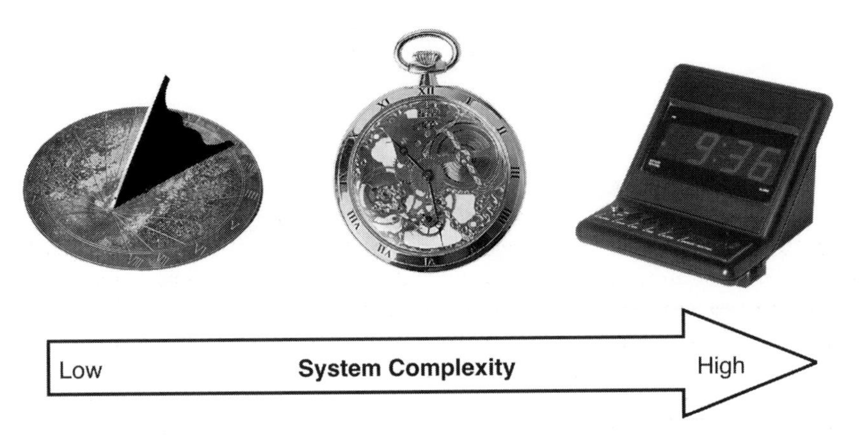

Figure 8.2 Features, components, interactions, and predictability drive system complexity.

For the automobile mechanic, however, complexity has significantly increased. The use of integrated circuitry and internal sensors and diagnostic systems greatly increases the complexity of what the mechanic must deal with. Most people no longer perform their own repairs, but must take their cars to specially trained mechanics. In the early days of driving, most people could and did perform their own repairs.

Similarly, while elevators have gotten more complex (in terms of the number of mechanisms involved), they have not gotten more complex to operate. In fact, rather than requiring specially trained operators, people now use elevators quite easily with no special training at all.

These examples illustrate that system complexity is moderated by the *operational complexity* required for a particular operation (Figure 8.3). While the automobile may be very complex, the automobile driver is required to understand very little of that complexity in order to drive it. The operational complexity for the driver is fairly low. The operational complexity for the mechanic is quite high, however. In order to repair or maintain the car, the mechanic must have a much greater understanding of how it functions. Operational complexity can be thought of as that portion of system complexity relevant to the achievement of a set of operational goals (e.g., the goals of the mechanic or the driver).

In many cases, automation has been designed with the objective of reducing operational complexity. By doing things automatically, the cognitive requirements on the user should be less; for example, automatic transmissions reduce the cognitive requirements of the driver by removing the need for the driver to maintain awareness of the current gear, to remember the rpm's associated with each shift point, and to operate the clutch.

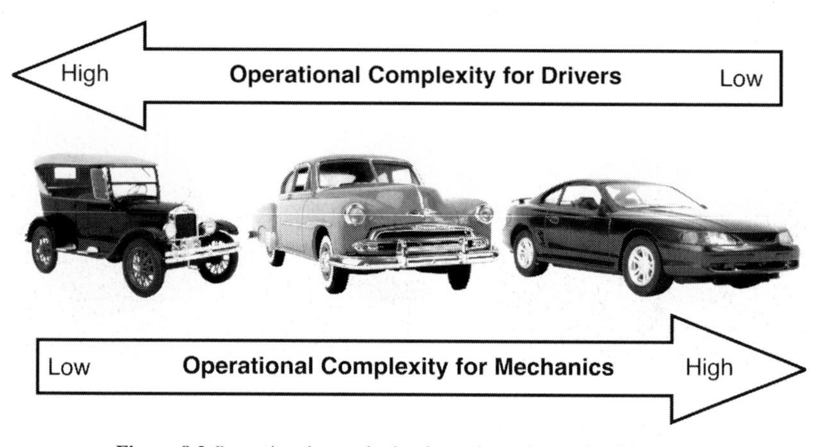

Figure 8.3 Operational complexity depends on the goals of the user.

Oftentimes, however, this type of reduction in complexity has not occurred. In fact, the introduction of automation has frequently increased the operational complexity of the system. This is because, unlike the example of the elevator where users never need to know how the system works, users of most systems where automation has been employed do need to know how the automation works. They are typically required to oversee the automation (i.e., act as a supervisory controller), take over when the system is performing improperly, and interact with it to perform joint tasks. In these cases, the addition of automation effectively increases the operational complexity of the system. Such automation can be seen in the cockpit of modern aircraft—automated flight systems require pilots to be aware of multiple, often non-intuitive modes of operation, thereby increasing cognitive load and operational complexity. Unless automation is so reliable that the human never needs to take over for it or understand how it works in order to interact with it, automation will increase the operational complexity of the system. It's not just the complexity of the automation that matters from an operational complexity standpoint, it's the need to interact with it to achieve operational goals. Dealing with automation will be addressed in Chapter 10.

8.1.3. Apparent complexity

The *apparent complexity* of a system has to do with the complexity of the representation of the system shown to the operator. A system with a given level of operational and system complexity can be displayed in a simple or in a complex way. For example, a personal computer can be operated via Microsoft® DOS or Windows or the Apple Macintosh™ operating system. A microwave oven can have a fairly simple interface, requiring only the turning of a knob, or a complex one, requiring multiple button presses and an understanding of which buttons in which order will accomplish desired functions (Figure 8.4). The complexity imposed on the user by these very different representations for interacting with these systems is significantly different in a number of ways. The apparent complexity, therefore, is a function of its cognitive, display, and task complexity.

Figure 8.4 Apparent complexity results from the interface that provides translation for the operation of the underlying system.

8.1.3.1 Cognitive complexity

Cognitive complexity refers to the complexity of the logic used by the system in its operations. Cognitive complexity has also been described as the ease or difficulty of acquiring the meaning that is potentially derivable from a particular representation (Eggleston, Chechile, Fleischman, & Sasseville, 1986). It is a direct result of the ease or difficulty a person has in mapping system functioning into an accurate and meaningful mental representation of the system. To some degree this is a reflection of the underlying complexity of the system, filtered by the operational needs of the user. A number of people have tried to model the cognitive complexity of systems including Chechile and his colleagues (Chechile, 1992; Chechile, Eggleston, Fleischman, & Sasseville, 1989; Chechile, Peretz, & O'Hearn, 1989), Kieras and Polson (1985), McCabe (1976), and Rouse (1986). Based on these efforts, it appears that factors contributing to cognitive complexity include:

- the number of objects, states, symbols, and properties of the system components,
- the number of linkages or relationships between them,
- the breadth and depth of hierarchical layering present in the linkages (hierarchical clustering), and
- the degree of branching present in the linkages (logic propositions) leading to conditional operations or modes of operations.

The home thermostat in Figure 8.5, for example, involves a number of components and linkages for controlling temperature. In this example, temperature may be controlled in three different ways:

- directly by pressing the up and down arrows, which will set a temporary target temperature for an unspecified period of time,
- through a programming feature that allows the user to specify the target temperature for four different periods of time, each of which must have their beginning times entered, for both a cold and heat setting (i.e., eight different settings and their time periods must be entered),
- by pressing Hold Temp twice followed by entering the number of days using the up and down arrow keys, which will set a temporary target temperature for a longer period of time.

Considerable logic regarding how each function works (linkages between components) must be learned by the user, as well as how each cluster or layer of the system interacts with the others. For instance, from this example, it is not clear how these three modes override the others, yet there is underlying logic that dictates this. Two switches are provided for controlling the system to be activated: System (Heat, Cool, or Off) and Fan (On or Auto). Additional logic and branching is present that must dictate how these two functions interact. Does Fan On override System Off? If System is set to Cool and the Fan is On, will it run continuously or will it stop when the temperature is reached? The answers to these questions are determined by a complex set of control logic that has a large effect on the cognitive complexity the operator must deal with in order to understand the displays and determine an appropriate course of action to operate the system.

Sarter and Woods (1994b) and Vakil and Hansman (1997) make the point that inconsistencies in the underlying structure also drive up the cognitive complexity of a system, as has been the case in aircraft automation. In many cases new functions or modules are added to the system over time that work in different ways

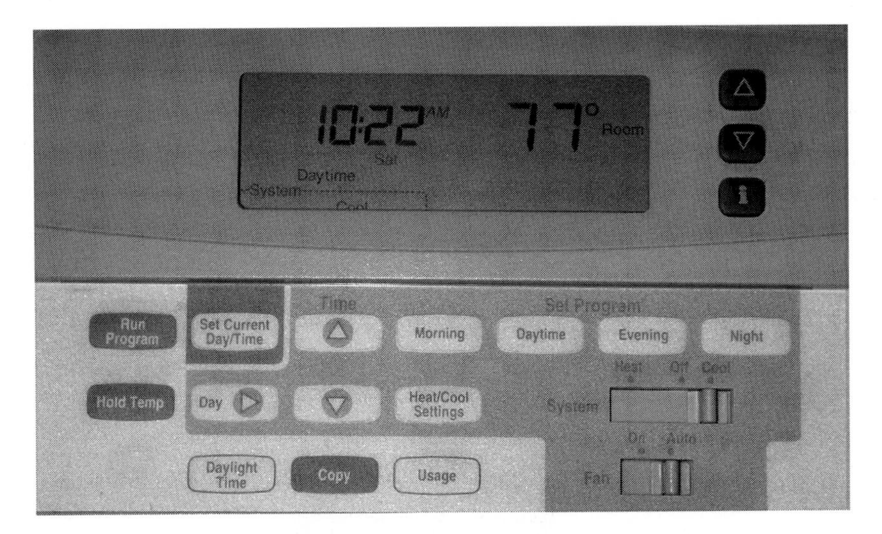

Figure 8.5 Cognitive complexity in a standard home thermostat.

than existing modules. Exceptions and special cases in the logic increase the overall 'entropy' or complexity of the system (Vakil & Hansman, 1997). These factors all act to significantly strain the ability of the user to develop a good mental model of how the system works, which leads to shortcomings in SA (Endsley, 1995b; Jones & Endsley, 1996). In the aviation domain, which has seen a large increase in the use of automation, the inability of pilots to develop robust mental models of the automation, even after prolonged use, suggests an inordinately high level of cognitive complexity within these automated systems; consequently, these systems have received considerable attention (McClumpha & James, 1994; Sarter & Woods, 1994a; Wiener, 1993; Wiener & Curry, 1980).

In addition to these factors, which to a large degree reflect basic system properties, cognitive complexity, from a user standpoint, is also a function of its *compatibility* with the users' goals and tasks and mental models. Users need to be able to make a clear mapping between the system's functionality and their goals (e.g., in order to increase the amount of light in the room, activate the light by moving the light switch from the 'off' position to the 'on' position). The thermostat example in Figure 8.5 is more poorly aligned with user goals.

Related to this is the issue of the *transparency* or observability of the system, which greatly influences its cognitive complexity. Many systems provide operators with very little information on which to judge or understand the behavior of the system. They are forced to infer functionality and causation from observed behaviors in order to figure out what the system is doing (as in the above example of the thermostat). System outputs may be displayed (e.g., the aircraft is now in a particular mode or the system detects a traffic conflict), but the user is forced to determine what logic may have been at work to generate such outputs. In systems that contain a lot of branching or inconsistencies in system logic, this inferencing can be quite error prone (Sarter & Woods, 1994b). In addition, systems with low transparency often provide little feedback to the user. Users need to be able to determine what effect their actions are having on the functioning of the system. Without such feedback, users not only have trouble understanding system state, but also in developing mental models of how the system works.

In cases with low transparency, developing a mental model to direct system interactions is very difficult. For example, the entertainment system remote control shown in Figure 8.6 provides a great deal of functionality for controlling a variety of devices. Unless the user has a mental model of how these devices are physically connected to one another, however, it is almost impossible to determine the right sequence of keys to successfully operate the desired devices.

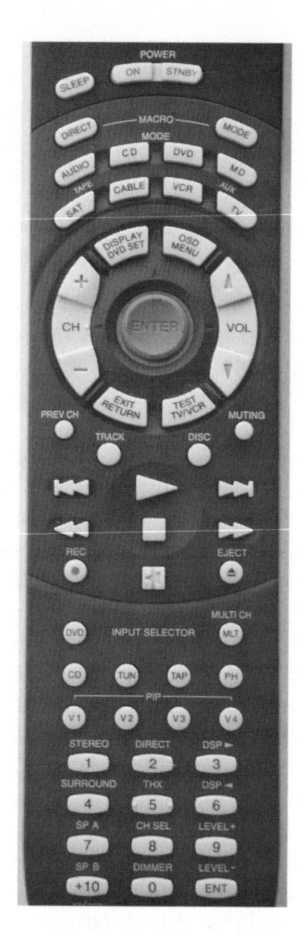

Figure 8.6 A typical universal remote control with low transparency.

For example, in order to successfully play a DVD and turn up the volume (assuming the desired DVD is in the system), requires 13 button presses in a particular order. Determining what that order might be, and how the various components or buttons affect each other is not readily apparent from the device.

(Hint: the correct order is Audio, On, DVD (input selector), DVD (mode), On, TV, On, 9, 1, DVD (mode), Play, Audio, Vol (up arrow).) In the family of one of the authors, of five people only one person (the person who set it up) can successfully operate this device after three years of use and numerous training sessions with the other family members. Memorizing long series of keystrokes is very difficult to do, and generating the correct sequence requires an understanding of how the system works. Developing this understanding from the interface provided (and remembering it) is almost impossible as no clues are provided.

8.1.3.2 Display complexity

While cognitive complexity is largely driven by the characteristics of the system, how it functions, and which information is portrayed to the user, the *display or perceptual complexity* refers to *how* the system information is presented to the user. Tullis (1983) describes four factors that determine the complexity of alphanumeric displays which can be extended to a broader range of displays that may be present in complex systems:

- overall density,
- local density,
- grouping, and
- layout complexity.

Overall density of the display is defined as a ratio of the number of items being displayed per the display space available. Very dense displays with many items (whether text on a printout, icons on a radar display, or objects in a visual scene) have a greater degree of perceptual complexity.

Local density of the display is determined by the number of other items near each item on the display. A display with ten objects clustered together provides more display complexity than the same size display with the ten objects distributed within the display.

Grouping of those items, used to provide perceptually salient chunks of information items, acts to reduce display complexity. Grouping can occur on the basis of functional similarity or some other criteria (e.g., source of information or time of occurrence). To support SA, grouping on the basis of the use of the information for addressing a given goal, decision, or higher level SA requirement is recommended. (See Principles 1 and 21.)

Layout complexity refers to the extent to which the arrangement of information follows a predictable scheme. Predictability of information displayed can be provided by something as simple as a left to right or top to bottom ordering of information or use of alphabetical ordering. In some systems, predictability is provided by additional structures that greatly reduce the display complexity. For example, considerable structure is provided to air traffic controllers by the use of a limited number of routes or airways on which aircraft will transition across the displayed air sector. These routes add a great deal of predictability regarding where aircraft can be found on the display and how they will move across the display over time.

While Tullis' work largely focused on single displays, in many systems of interest multiple displays of rapidly changing information will be present. A number of additional factors need to be considered with regard to display density in these cases.

Display complexity can be affected by the *degree of consistency* between different displays used by the same individual. In many systems, multiple computer display windows may be viewed at different times, or at the same time on different monitors. Displays that follow a consistent scheme for the presentation of similar information will be less complex (e.g., if the control buttons are always arranged in the same place in the same order, or if similar information order schemes are used on different displays). Consistency in terminology and labeling similarly reduces complexity. Consistency (or standardization) essentially acts to improve the predictability of the information location and meaning.

Certain characteristics of the displayed data also affect complexity, including the *speed with which information changes* on the display, the *reliability* of the data displayed, and the *difficulty of acquiring* needed information. Displays where information changes more slowly (due to slower dynamics of the underlying system, or slower rate of travel for a moving visual scene, for example) will be less complex than those in which information changes rapidly. Uncertainty associated with the information displayed (often a reflection of imperfect sensors, system algorithms, or information sources) induces a higher degree of complexity. Similarly, conflicting or incongruent data will be viewed as significantly increasing display complexity. Surprisingly, missing data does not seem to provide much of a problem and appears to be handled as if the missing data would merely confirm the information that is present (e.g., the possibility that the missing data might be incongruent appears to be discounted (Banbury *et al.*, 1998; Bolstad & Endsley, 1998)).

The number of steps that need to be taken to acquire needed information for a given decision also increases display complexity. If operators must search across or call up multiple displays to acquire needed information, a higher degree of complexity will be present.

In relation to SA, while operational and cognitive complexity act to make it more difficult for people to achieve high levels of comprehension and projection (Level 2 and 3 SA), high levels of display complexity will make it more difficult for them to achieve good Level 1 SA—perception of needed information. Higher display complexity can drive up the amount of time it takes to search for and acquire needed information, and can increase errors in that process (Sanders & McCormick, 1992; Wickens, 1992). Interestingly, research by Chechile, Peretz, and O'Hearn (1989) found that cognitive complexity factors explained four times the variance in performance scores as perceptual complexity factors, although both were needed to fully explain complex performance.

8.1.3.3. Task complexity

Task or *response complexity* is a function of what the operator has to do to accomplish goals. This is a function of a number of factors, including how many simultaneous or alternating goals the operator has and how many steps or actions have to be taken in fulfilling these goals. In some jobs, users must juggle many competing or concurrent goals. This alone adds complexity to the tasks they must perform. In addition, they need to be able to map actions to those goals that need to be performed with the system. The directness of that mapping (a function of cognitive complexity), and the number of steps needed to achieve a given goal directly effects the task complexity. The thermostat and remote control devices in Figures 8.5 and 8.6, while accommodating a number of user goals, each require a number of steps to accomplish those goals, leading to a moderate degree of complexity, even though the underlying systems are not particularly complex themselves. These are two simple household devices. In very complex systems, such as a power plant, task complexity would be magnified a thousand-fold across many different subsystems. A system in which users have more goals to juggle and must perform more steps in order to achieve their goals will be more complex.

In addition, the difficulty of mapping actions or steps to particular situations will increase task complexity. Environments with heavy proceduralization and very clear steps to perform in a given situation will be less complex than those with nebulous novel situations where deciding what to do is more difficult. Campbell (1988) states that task complexity is a function of :

- multiple paths to a goal or end state,
- multiple desired end states,
- conflicting interdependence among the paths to those multiple desired end states, and
- uncertainty or probabilistic linkages among paths and outcomes.

Task complexity acts to increase overall mental workload (e.g., by increasing the amount of information the operator must remember in order to perform a task or by increasing the amount of information through which an operator must sort in order to determine an appropriate action) and can make other tasks associated with SA, such as prospective remembering (i.e., remembering to do something in the future), more error prone (Wichman & Oyasato, 1983).

8.1.4 Role of the user's mental model

Cognitive, display, and task complexity will all combine to contribute to the degree of complexity perceived by the operators, moderated by the ability of the operators to deal with that complexity through their mental model of the system. The internal capability of the operator to deal with the apparent complexity, which is dependent on the completeness and clarity of the mental model, can diminish the effects of the apparent complexity. Hence, experienced air traffic controllers, pilots, and drivers can deal with more complexity than novices in these fields.

Designers often strive to create systems that will be consistent with existing mental models of users. The cognitive complexity of how the system works may not be less, but it will be perceived as significantly less complex by users who already have a mental model of how the system works (usually based on experience with a similar system). The Apple Macintosh computer, for example, was designed around the model of a physical desktop. Early word processing systems were designed around the model of a typewriter.

Many times designing to fit a pre-existing mental model is not possible, however, and users need to develop a mental model for the new system through training or experience. Of course, developing a good mental model is significantly more difficult when the underlying system is complex. Not only are there potentially more components, component interactions, branching, layering, and exceptions to develop an understanding of, but also there is frequently less opportunity to gain experience with the system in all possible modes or states (Endsley, 1996a). Complexity is the enemy of the mental model.

8.2 DESIGN PRINCIPLES FOR TAMING COMPLEXITY

From this discussion of factors contributing to complexity and their ill effects on SA, we can generate several design principles for managing complexity in system design that should help to support operator SA. Many of these are quite simple but are often overlooked in system design.

Principle 15: Just say no to feature creep—buck the trend

The simplest and most direct way to deal with complexity is reduce it. In the process of system design, this often means fighting an uphill battle. Engineers and marketers press to add more and more features to systems in the belief that people want or need them. And if one simply asks users, 'Do you think this feature would be useful?', many will say yes. 'Who knows? I might need that someday!' Users rarely understand the complexity such choices generate, however, or what it costs them in terms of time and errors in performing their main tasks.

A useful process during system design is to seriously question the addition of multiple features that may be only infrequently used, or used by only small subsets of the user population. If necessary, use rapid prototyping to assess the costs of these additions to the system in terms of poorer user performance on frequent tasks. These data can be used to demonstrate to others involved in the design the costs associated with adding features as well as the perceived benefits. Only those features that are really needed should be included in the system design in order to best reduce complexity.

Principle 16: Manage rampant featurism through prioritization and flexibility

As often it will be virtually impossible to limit the growth of system features as much as might be needed to simplify a particular system, a second option is to try to manage the features that are present to minimally affect the user. Features that may be infrequently used (e.g., customizing alarm criteria or setting the date) can be made slightly harder to access than more commonly used functions (although not hard to figure out how to find!). On non-software oriented systems (e.g., the remote control discussed earlier), buttons associated with less commonly used features can be minimized (e.g., made smaller, put in less prime locations, or have smaller labels).

Many software systems allow users to customize the toolbars with the features they use most often. This allows a self-prioritization that helps manage the proliferation of features in these software packages. While not eliminating the problems of system complexity, approaches such as these can help reduce the cognitive complexity imposed on the user.

Principle 17: Insure logical consistency across modes and features

Inconsistencies in the logical functioning of the system dramatically increase complexity. Differences in operational logic, display of information, and different sequences of inputs that are not directly necessary for the operation of that mode or feature should be reduced or eliminated. For example, if a telephone menu requires the user to press 2 for 'no' in one part of the system, it should maintain this convention throughout. The more consistency in functioning, display, and control that is provided between modes, modules, or features of the system, the easier it will be to deal with any underlying complexity.

Principle 18: Minimize logic branches

Similar to minimizing complexity through fewer features, this principle seeks to minimize complexity by reducing the linkages and conditional operations associated with systems. 'If-x-then-y' logic is hard enough for people to learn. 'If-x-then-y-unless-z' or 'if-x-and-z-then-y' logic is even more challenging for people striving to develop mental models. The use of modes (discussed further in Chapter 10), creates a significant increase in complexity. Even if people develop a mental model of all the modes of a system (which can be difficult due to branching), people have a hard time maintaining awareness of the current mode in order to properly interpret system information and perform actions. The best remedy is to avoid modes as much as possible.

Principle 19: Map system functions to the goals and mental models of users

A clear mapping between user goals and system functions should be present. In the remote control device example discussed earlier (Figure 8.6), user goals are usually limited (e.g., watch a movie, watch TV, play a CD). Yet the interface offered does not map directly to those goals. Rather it requires the user to understand how the systems work and how they are connected in order to achieve goals. A less complex device would provide single-step control for major goals. One button would be all that is required to configure the audio system, DVD player, and TV for playing a DVD.

Alternately, direct support can be provided to help users create a mental model of the system. For example, some Web sites provide a branching map on the side that allows users to see where they are in the nested structure of the system. The interface of the remote control or thermostat can show how components are linked and which override others by drawing lines on the face plate and through organization of the function buttons on the device.

Principle 20: Provide system transparency and observability

Most importantly, a high degree of transparency and observability of system behavior and functioning is needed for those components that the user is expected to oversee and operate. This can be directly provided through system displays that show users not just what the system is doing but why it is doing it and what it will do next. Hansman and his colleagues (Kuchar & Hansman, 1993; Vakil, Midkiff, & Hansman, 1996) demonstrated a significant improvement in pilots' understanding of aircraft flight management systems when a vertical situation display was provided that showed the projected altitude transitions of the aircraft in its current automation mode. In a fairly clever application, Miller (Miller, 2000; Miller, Hannen, & Guerlain, 1999) provided pilots with displays that showed what the aircraft inferred the pilot was trying to do (goals) that drove automation behavior. Rather than being forced to guess what the system was doing and why, pilots were able to easily monitor this goal information and correct it as needed. These are but two simple examples of how better system transparency can be achieved. Surprisingly, very little information is usually provided to users to support understanding about how the system functions.

Principle 21: Group information based on Level 2 and 3 SA requirements and goals

Grouping information in a display to reduce complexity and facilitate search and readability is certainly not a new design principle. Designing to support SA provides a foundation for grouping that is not technology-centered (based on the device producing the information), but user-centered (based on how people need to use the information). User goals form a high-level grouping criteria (see Principle 1). Information that is used to support the same goal needs to be centralized on a common display so that users are not required to search across multiple systems to derive the SA needed for goal-driven behavior. Within those displays, the higher level SA requirements derived from the cognitive task analysis (Chapter 5) direct which information needs to be grouped together within each display.

Principle 22: Reduce display density, but don't sacrifice coherence

Excessive display density (too much data to wade through) can slow down search and retrieval of needed information. While extraneous data should obviously be eliminated, great care should be taken not to reduce display density at the cost of decreasing display coherence. That is, a significant cost is also incurred when people must go from display to display to find needed information (compared to scanning within a single display). This is particularly true of systems in which menuing and windowing are employed, resulting in the need to page through displays which may be hidden to find needed information. This 'hiding' of information can have significant negative ramifications for SA as it will take longer to find needed information, and important cues may be missed entirely. Therefore, with regard to SA in dynamic systems, there is also a benefit to minimizing the overall number of displays that must be monitored, driving up display density. The solution to this dilemma is to maximize the internal organization provided within a display such that fairly dense information displays remain readable for the user, supporting the need for 'SA at a glance.' Display coherence is maintained through the intelligent grouping of information (see Principles 7 and 24).

Principle 23: Provide consistency and standardization on controls across different displays and systems

A basic human factors principle is to standardize the layout, format, and labeling of controls across systems used by a single operator for similar functions. Consistency across systems aids the operator in forming a reliable mental representation of the system, which allows the operator to predict the meaning and location of information across systems. This predictability reduces the cognitive load on the operator by allowing things learned from one system layout to be applied to other layouts (positive transfer), thereby reducing, among other things, the time an operator must spend searching for frequently used information or functions. If each system were designed independently, then the operator would essentially have to learn the specific terminology and layout for each system and would not only miss the benefit of positive transfer between systems, but negative transfer (e.g., confusing the location of an item on one system with the location of the same item on another system) would result. Designing systems with a standardized approach to all aspects of the system allows the operator to develop one mental model and to build on that mental model to assimilate new or different functionality presented across systems.

Principle 24: Minimize task complexity

Similarly, task complexity (the number of actions needed to perform desired tasks and the complexity of those actions) should be minimized. Requiring the operator to learn and remember a complex series of actions in order to perform a task not only adds to the operator's cognitive load, but also leaves room for error if one or more of the steps is forgotten or performed incorrectly. Reducing the number of steps needed to achieve a particular system state lessens the likelihood that an error will be committed and reduces the complexity of the mental model the operator must develop in order to interact with the system. If 13 steps that must be performed in a fixed sequence can be replaced by a single button press, then the complexity of the system, and the operators need to develop a complex mental model, will be significantly reduced.

CHAPTER NINE

Alarms, Diagnosis, and SA

9.1 AN ALARMING PRACTICE

Alarms or alerts of various types are included in many systems in order to call operators' attention to important information. In aircraft cockpits, Ground Proximity Warning Systems (GPWS) are present to warn pilots when they are too close to the ground or when their vertical profiles show them descending toward the ground too quickly. Nuclear power plant operators are surrounded by alerts that tell them of unwanted states on various parameters for numerous different pieces of equipment. Air traffic control stations are equipped with alerts that flash aircraft symbols when the computer detects they are on a collision course.

Figure 9.1 Alarms are a major part of many systems.

In theory these audio and visual alarms should help to boost operator SA by calling their attention to important information that they may not be aware of due to distraction or lack of vigilance. Certainly that is the intention of the developers of the alerting systems. Reality falls far short of this ideal, however, for several reasons.

First, the tendency to add alarms for different things has grown out of control in many domains. In these cases, there are so many alarms present that the operator cannot adequately process and deal with the specter. In a major incident at a nuclear power plant, for example, between 50 and 300 alarms per minute may be presented to the operators (Stanton, Harrison, Taylor-Burge, & Porter, 2000). Such a large number of alarms can be almost impossible for any person to sort out in order to determine what is happening to create these alarms. Sheridan (1981) reported that during the first few minutes of a serious accident at Three-Mile Island Nuclear Power Plant more than 500 alarm annunciators were activated. Other domains are plagued by alarms as well. A typical medical operating room may have as many as 26 different audio alarms, less than half of which people can reliably identify (Momtahan, Hétu, & Tansley, 1993). When alarms are too numerous, people cannot reliably respond to them as the designers intended.

Second, many alerting systems are plagued by high false alarm rates. In response, operators are often slow to respond or may even ignore those alarms they believe to have low reliability. It is not uncommon to see fire alarms, car alarms, and home security system alarms go completely unheeded. At one workshop for safety professionals, one of the authors witnessed only three people leave a room of approximately 300 people in response to a hotel fire alarm. And that response rate is for people who are very focused on safety! People have simply learned through experience not to believe alarms that often occur but are rarely indicative of a dangerous situation. Breznitz (1983) has called this, quite aptly, the cry wolf syndrome. People simply tune out alarms they don't believe.

Third, in many domains, operators have been found to actively disable alarm systems (Sorkin, 1989). This behavior has been observed in train operations, aviation, and the process control industries. In many cases, this is an all-too-human response to alarms that sound too frequently and have too high a false alarm rate. The alarm becomes a nuisance and a major source of annoyance rather than an aid. The need to respond to the alarm becomes an extra task that distracts the operator from dealing with the underlying problems at hand. In the medical field, for example, doctors and nurses complain that the need to respond to the monitor that is producing the alarm can take their time and attention away from the patient. They end up treating the device rather than the patient.

This sad state of affairs points to a failed philosophy of alarm development and a poor integration of alarms with the mental processing of the operator. With these failings, alarm systems often do not successfully enhance operator SA and in some cases may help contribute to its downfall. Developing a more successful approach to the use of alarms in complex domains requires that we explore a number of factors that contribute to the efficacy of alarm systems in practice.

9.2 PROCESSING ALARMS IN THE CONTEXT OF SA

Alarm system designers often proceed with the assumption that people react immediately to each alarm to correct the underlying condition that caused it. In general, this is often not the case. A study of controlled flight into terrain (CFIT) accidents among commercial airlines over a 17-year period showed that 26% of these cases involved no response, a slow response, or an incorrect response by the pilot to the GPWS alarm (Graeber, 1996). Kragt and Bonten (1983) reported that only 7% of the alarms produced in one plant resulted in action. In studying anesthesiologists' response to alarms in operating rooms, Seagull and Sanderson (2001) noted that only 28% of the alarms resulted in a corrective action being taken. Clearly operators often respond in different ways, and for a variety of reasons.

In actuality, a complex set of mental processes is involved in responding to an alarm (Figure 9.2). The alarm signal (either visual or audio) must be integrated with other information in the environment in order to properly interpret its origin. The meaning and significance of the alarm is interpreted (comprehended as Level 2 SA) based on a number of factors, including the operators' past history with the alarming system, their mental model of what is happening in the underlying system, and their expectancies. Based on their interpretation of the alarm they will decide how to respond to the alarm. Their actions might include correcting the problem, muting the alarm but not changing any underlying system conditions (i.e., waiting), or ignoring it completely in the belief that it is not a valid indicator of a real problem, but an artifact of other situation conditions the alarm system does not take into account. We will discuss the factors that are relevant to this process in more detail.

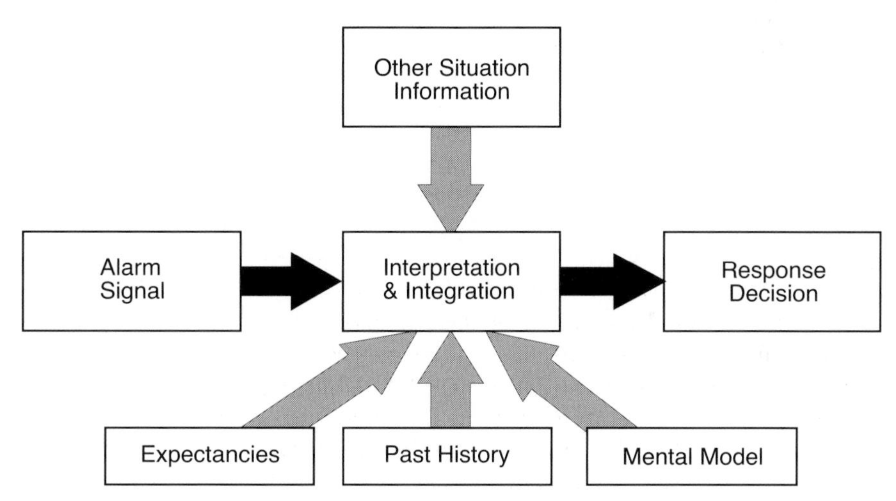

Figure 9.2 Cognitive processing of an alarm signal.

9.2.1 Alarm reliability

The operators' past history with the alarm system in question will have an immense impact on how they interpret the alarm and choose to respond to it. A significant issue affecting their interpretation of the alarm is its reliability. High false alarm rates, leading to the cry wolf syndrome, lead operators to have a low confidence in the alarm. High false alarm rates with the GPWS system, for example, are a key driver of pilots' slow response or nonresponse to alarms with that system (DeCelles, 1991).

False alarms are a considerable problem in many systems. In one study of a typical operating room, on average an alarm sounded every 4.5 minutes, 75% of which were found to be false alarms (Kesting, Miller, & Lockhart, 1988). In another study that reviewed aircraft hazard reports at the U.S. Naval Safety Center, at least 50% of the reports involved false or erratic indications that were at least partially responsible for the mishap (Tyler, Shilling, & Gilson, 1995).

People's confidence in alarms has been shown to be a direct function of their past experience with false alarms for that system and to affect the likelihood they will respond to the alarm (Bliss, 1993; Bliss, Dunn, & Fuller, 1995; Bliss, Gilson, & Deaton, 1995; Lehto, Papastavrou, & Giffen, 1998). Once trust in a system is lost, there can be a considerable lag in regaining that trust (Lee & Moray, 1992). Surprisingly, people's confidence in a particular alarm has also been found to be a function of the number of other concurrent alarms present in the system, even if those parameters are functionally separate (Gilson *et al.*, 2001). Alarms which are closest—spatially, temporally, or functionally—have the most effect. Inactivated alarms nearby also tend to subtract from the perceived confidence that an alarm is valid, even reducing confidence below publicized levels for the device (Gilson *et al.*, 2001). Confidence in the reliability of an alarm appears to be a function of both people's experience with it and the presence of other concurrent alarms, which people apparently take as confirming or opposing evidence of its validity.

A person's reluctance to respond immediately to a system that is known to have many false alarms is actually quite rational. Responding takes time and attention away from other ongoing tasks perceived as important. As they can create a significant decrease in reaction to even valid alarms, however, false alarms seriously degrade the effectiveness of all alarms and should be reduced to the maximum extent possible.

The problem of false alarms is a difficult one for system designers. The degree to which a system produces false alarms must be weighed against the likelihood that it will have missed alarms (situations in which a real problem exists, but for which the system does not produce an alarm). As shown in Figure 9.3, where the designer places the criterion for deciding whether incoming data indicates that a signal (a condition that should produce an alarm) is present produces a trade-off between the probability that a false alarm will be indicated by the system and the probability that a real problem will be missed (*missed alarm*). Under such a scenario, actions to reduce the likelihood of a false alarm (shifting the alarm criterion to the right in Figure 9.3) will also act to increase the probability of a missed alarm, and vice versa.

This is a very difficult decision for designers to make. Often the decision is that, because missed alarms represent a significant hazard, it is best to err on the side of more false alarms. This decision does not recognize the detrimental effect of false alarms on human response to all alarms, however. In new systems being designed to alert air traffic controllers of projected loss of separation between aircraft, for example, missed alerts were designed to be fairly low at 0.2%, compared to false alarm rates of 65% (Cale, Paglione, Ryan, Oaks, & Summerill, 1998). In a different alerting system for the same function, 12% of actual separation losses (aircraft within 5 miles) were missed by the system, while the system produced false alarms 60% of the time for aircraft pairs when they were 25 miles away.

(Paielli, 1998). These are very high false alarm rates that can seriously undermine the effectiveness of the system. High false alarm rates such as these can lead controllers to be slower to respond to, ignore, or turn off the alerting systems. One source shows that reliability needs to be above 95% for systems to be considered useful (Scerbo, 1996), although identifying a precise number for acceptability appears to be elusive.

The only way to resolve this trade-off is to increase the sensitivity of the system—the ability of the alarm to distinguish between true events and nonevents (signal and noise). This in effect is represented as decreasing the amount of overlap between the noise and the signal (Figure 9.4). By creating sensors or algorithms that are more sensitive in their ability to distinguish true events worthy of an alarm from other events in the environment, in theory, both false alarms and missed alarms can be reduced simultaneously.

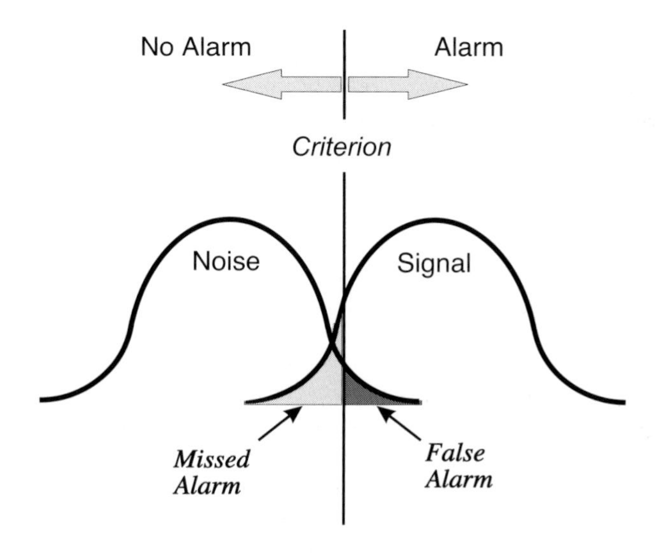

Figure 9.3 Probability of false alarms and missed alarms in system design.

Interestingly, the engineers who create the algorithms behind alarm systems frequently believe their systems do not have false alarms—the systems reliably detect the phenomenon they are designed to. The problem underlying false alarms is often not the functioning of the algorithms themselves, but *the conditions and factors that the alarm systems cannot detect or interpret*. For example, GPWS will often produce alarms at particular airports that involve approaches with flight paths near mountains. The pilot is on the correct path and will clear the terrain with no problem, but the system does not have information about the flight path and thus produces a nuisance alarm (one that is correct to the system, but false to the pilot in terms of warning of an actual hazard). In Cincinnati, Ohio, terrain at a riverbank near the runway rises at a significant rate from the river's water height. This climb produces a GPWS alarm because the system sees rising terrain on the flight path, but has no idea of whether the terrain keeps rising into a mountain or flattens out into a bluff where the runway is located, as is the case. Pilots who fly into this airport become accustomed to frequent false warnings, desensitizing them to it.

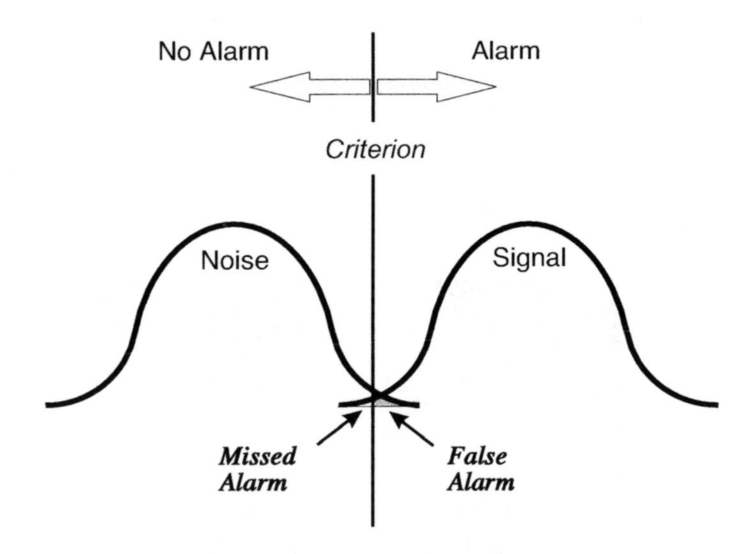

Figure 9.4 Increased system sensitivity leads to both reduced missed alarms and false alarms.

Other systems have similar shortcomings. Anesthesiology alarms, for example, are not able to tell when the physiological parameters they are measuring are changing due to a real patient problem, or due to some act the surgeon has undertaken that poses no real patient risk (such as making an incision which creates a temporary drop in blood pressure), or due to an extraneous artifact (such as patient movement which can disrupt ECG or SpO_2 monitors). Approximately 46% of anesthesiology alarms were found to be ignored because of such artifacts (Seagull & Sanderson, 2001). Air traffic control alerting systems frequently alarm for detected aircraft separation problems, when in fact the controllers know that one of the aircraft will climb, descend, or turn on its route and no problem will occur.

While the alarm systems may reliably perform to detect the signals they receive, these signals only tell part of the story. The fact that such electronic and computerized alarm signals are limited in their awareness of the situation (low system SA) leads to an *experienced false alarm rate* that is often much higher than that expected by its designers. In terms of affect on human responsiveness to alarms, it is this experienced false alarm rate that matters. The development of alarm systems that are capable of taking into account a much larger range of situational factors that reveal whether detected information poses a real hazard or not (i.e., increasing its sensitivity) is much needed to effectively reduce the system false alarm problem.

9.2.2 Confirmation

At least partially because false alarms are such a problem, people do not immediately respond to alarms but often seek confirming evidence to determine whether the alarm is indicative of a real problem. While the designers of GPWS systems created their algorithms under the assumption that human response to the alarms would be immediate, and pilot training stresses just such a response, data show that pilots delay their response to GPWS alarms as much as 73% of the time (DeCelles, 1991). They look for information on their displays or out the window that would indicate why the alarm might be sounding (Figure 9.5). The availability of evidence of nonhazardous factors that might cause the alarm, or the occurrence of the alarm in a situation that frequently causes a false alarm, can easily lead operators to discount that alarm.

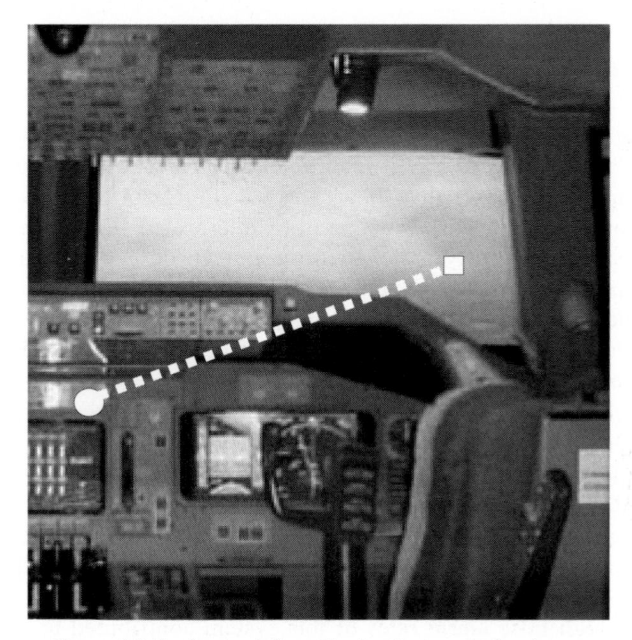

Figure 9.5 Confirming information is often sought for alarms.

In a study of pilot conformance to alerting system commands, Pritchett (1997) found that pilots were more likely to believe visual cues than audio alerts when the two conflicted and that they would be slower to respond to the alert under such conditions. Whereas the philosophy behind alarm systems is that this is the only information that people need to arrive at a decision as to what to do, in reality the alarm signal is only one cue that is processed in addition to all the other information available in the situation. To some degree this is intentional. People are the final decision makers precisely because they can take situational information into account that the alarm systems do not. Otherwise systems could act completely automatically.

There are several downsides, however. First, it takes extra time to look for evidence to confirm or deny whether the alarm is real. In time-critical and high-workload situations, this time is ill afforded. Secondly, people may err in determining whether to heed the alarm. Visual illusions can lead pilots to falsely believe they are not in danger of hitting the ground when in fact they are, for example.

Designers of systems that incorporate alarms must be cognizant of people's need to confirm that alarms represent true hazard situations and must not expect that people will blindly react to the alarms. This will be particularly the case for alarms that have high experienced false alarm rates, and no amount of training or wishful thinking will eliminate this reality. As this confirmation requires time and can be subject to error, the best systems will also seek to minimize experienced false alarm rates as much as possible.

9.2.3 Expectancy

In addition to seeking confirming evidence for an alarm, people's *expectancies* regarding the alarm can affect their interpretation of its cause and significance. In many cases operators report that they expected the alarm to sound due to their actions (or some other event) and therefore interpreted the alarm accordingly. GPWS alarms are frequently anticipated on a steep descent into an airport, for example. Approximately 6% of alarms in operating rooms were found to be anticipated and expected by the anesthesiologist due to the actions of the surgeon (Seagull & Sanderson, 2001). Alarms that sound in expected conditions become a source of nuisance to operators.

When in fact the alarm is due to some other factor than the expected one, these expectations can also lead to a major error in interpretation of the alarm. In one aircraft accident, a DC-9 landed in Houston with its wheels up causing considerable damage to the aircraft. The landing horn alarm was interpreted by the captain as sounding because he put the flaps to 25° before the gear was down rather than as a warning that the landing gear was not down for landing (National Transportation Safety Board, 1997). The National Transportation Safety Board found that the high false alarm rates of this system contributed to this error. The GPWS alerts that sounded in this accident were attributed by the pilot to the high sink rate of the aircraft (which he expected), (Figure 9.6B), rather than the fact that the wheels were not down (Figure 9.6A), as was actually the case. Many similar examples exist, where accidents were not prevented by alerting systems because pilots anticipated and interpreted the alarms according to their expectations rather than the actual conditions.

Fundamentally, such errors in interpretation can only be prevented by increasing the diagnosticity of the alarm information. The cues provided to the operator need to disambiguate each possible causal factor so that incorrectly attributing alarms to one's expectations can be overcome. The Mark II version of the GPWS improves upon the diagnosticity of the factors causing a 'whoop, whoop, pull up' alert shown in Figure 9.6, by adding additional verbiage for 'too low gear,' 'too low flaps,' 'too low terrain,' 'sink rate,' 'don't sink,' and 'minimums.' Some conditions are still difficult to disambiguate, however, such as a sink rate warning, which can be produced by either being too fast or too steep.

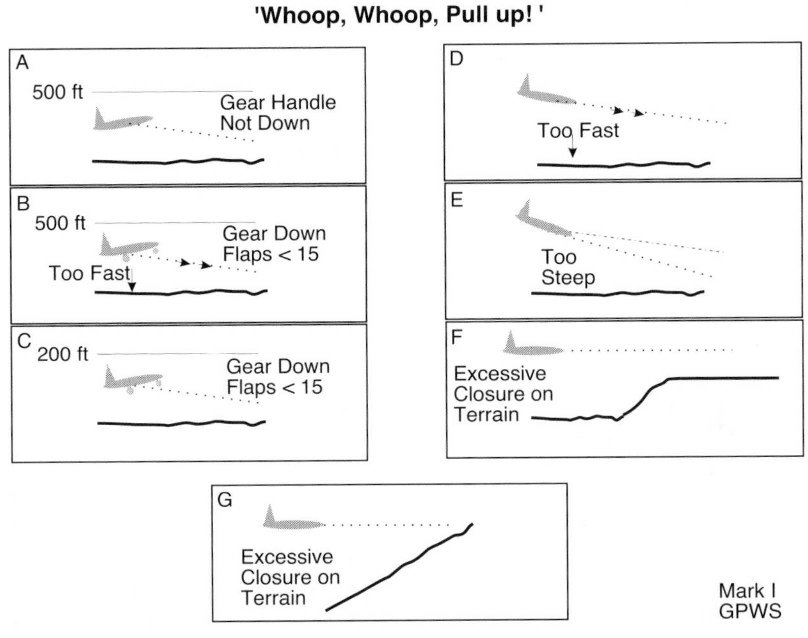

Figure 9.6 Expectations can lead to a mismatch between alerts and belief in the underlying cause of the problem. At least seven different possible conditions can lead to the GPWS alert 'Whoop, whoop, pull up!'

9.2.4 Disruptions and diversions

At the very least, false alarms and those that signal the operator about situations they are already aware of act as a source of distraction and disruption to the individual's ongoing tasks. Depending on the design of the alarm itself, operators may be called upon to acknowledge, clear, or mute the alarm that is flashing on their displays or sounding audibly. This takes time away from the other tasks they are performing, and can contribute to other errors, such as losing their place in performing a procedure or checklist.

Even when no action is required of operators, the alarms can distract them, using part of their limited cognitive resources for processing environmental information. This is particularly so as most alarms are designed to be very attention seeking. Mentally blocking them out, therefore, requires more effort and is harder to do. Banbury and his colleagues, for example, found that the obligatory processing of extraneous auditory information can have a negative effect on attention and monitoring of other tasks, particularly those that require serial ordering (such as determining trend information from displays) (Banbury, Macken, Tremblay, & Jones, 2001). Thus, unneeded alarms can act to degrade ongoing SA processes.

In another investigation of systems that visually cue soldiers as to where to look for targets in a busy visual scene, Yeh and Wickens (2001) found that these cues draw attention away from other targets elsewhere on the display. Systems that visually cue people to look at certain information in order to alert them to potential problems (which may or may not be real problems), can act to effectively divert their attention from other problems they should be attending to. Alerts of traffic conflicts on many air traffic control displays work in this manner, with the system flashing the symbols of suspected aircraft conflicts for the controller's attention. When the high number of missed conflicts and false alarms are considered, however, this diverting of the controller's visual attention may be both unwarranted and inadvisable. They can be distracted from real problems and higher priority tasks they need to attend to. When system reliability is low (i.e., false alarm rates are high), alarm systems that capture the operator's attention can actively contribute to poor SA.

9.2.5 Workload and alarms

So when are alarm systems of use? And how should designers deal with the inevitable trade-offs that must occur between false alarms and missed hazards? Some insight into this problem can be provided by considering the issue of operator workload as it interacts with people's response to alarms (Figure 9.7). Three different conditions need to be considered: underload, normal operations, and overload.

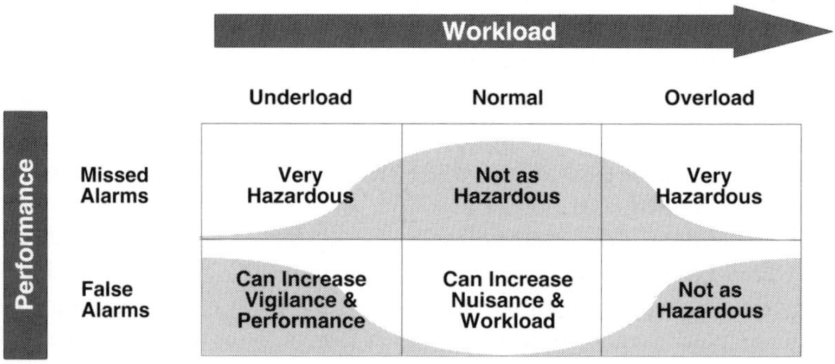

Workload →

Performance	Underload	Normal	Overload
Missed Alarms	Very Hazardous	Not as Hazardous	Very Hazardous
False Alarms	Can Increase Vigilance & Performance	Can Increase Nuisance & Workload	Not as Hazardous

Figure 9.7 Impact of missed alarms and false alarms on performance with different levels of workload.

Underload, or vigilance conditions, can provide a significant challenge for maintaining SA. In situations where people go a long period of time (greater than around 30 minutes) with nothing happening, reaction time and accuracy in detecting significant events can decline significantly (Davies & Parasuraman, 1980; Macworth, 1948, 1970). Typical vigilance situations include radar monitoring and long-haul flight. At very low levels of workload, performance can be very negatively affected by missed alarms as people have a low level of vigilance and an increased likelihood of missing problems that they should respond to. False alarms under these conditions can actually act to increase their vigilance, however, and thus performance (Wilkinson, 1964). Thus in systems in which operator underload and vigilance is a significant problem, we advise shifting the alarm criterion point in favor of more false alarms and fewer missed alarms.

Normal working conditions, with a moderate level of workload, presents a completely different situation. Under moderate levels of workload, people are exerting energy to attend to competing tasks. Under these conditions, false alarms are more likely to become a problem. They become a significant nuisance or annoyance and can act to unacceptably increase workload for the operators who must respond to them. Under these conditions, operators often act to silence alarms or override them, thus losing the potential benefits completely. As operators will be actively involved in performing their tasks, however, missed alarms would not be as much of a problem as in the low workload condition. Operators will be more likely to have detected hazardous conditions on their own (provided they have displays for supporting such diagnostics). False alarms should be severely minimized under normal operations.

Overload conditions can also exist in many situations in which the amount of information, pace of operations, or number of systems to consider far outpaces any human capability. Under conditions of severe overload, people may shift their operational strategy. Instead of trying to perform all operations manually and using the automation as a backup, they may rely on the automation to prioritize their tasks, attending only to the problems alerted by the automation. This strategy allows them to maximize the utilization of their resources. The downside of this strategy is that missed alerts lead to significant performance failures, as people will be less likely to catch these problems on their own with such a strategy. False alarms will not be as great a problem in this case, however, as this strategy insures that many potential problems are attended to.

9.2.6 Alarm formats and compliance

A considerable amount of work has focused on determining the best characteristics of alarms for insuring that they will catch operators' attention and gain an increased probability of compliance with the alarm system. Edworthy (1997) has conducted studies on 'urgency mapping' of audio alarms. The loudness of the alarm is the characteristic that is most likely to lead to a faster response. Other factors that contribute to an alarm's perceived urgency include its frequency, harmonics, speed, repetition, pitch, rhythm, and melodic structure (Edworthy, Loxley, & Dennis, 1991; Hellier, Edworthy, & Dennis, 1993).

Auditory icons, or earcons, in which the sounds presented have inherent meaning associated with the state they are warning about, have been found to lead to better performance than traditional alarm sounds (Belz, Robinson, & Casali, 1999). An example of an auditory icon might be the sound of a tire skidding to alert a driver to an impending collision. In some grocery stores the sound of thunder precedes the activation of automated water sprayers over the fresh vegetables, warning shoppers who might get wet. These alarms have a high information content that supports rapid interpretation of the meaning of the alarm.

In addition to auditory alarms, which are superior for catching the attention of people who may be looking in many different directions in the environment, visual displays of alarms are often provided. Visual displays can provide significant advantages in supporting diagnosis of multiple alarms and support human memory in processing different alarms over time, particularly if they are attending to other information when the alarm is presented.

The best compliance rates have been found for a combination of visual and audio alarm displays (Belz, Robinson, & Casali, 1999; Selcon, Taylor, & McKenna, 1995). The more modalities that are involved in the warning, the greater the compliance. Increasing levels of alarm urgency and clearly publicizing alarm reliability has also been found to increase compliance and reduce the effect of the cry wolf syndrome (Bliss, Dunn, & Fuller, 1995). The proportion of activated alarms (from among multiple possible alarms) has not been found to be as important for alarm detection as the rate at which alarms are presented. Multiple alarms that occur very rapidly in sequence are less likely to be detected and processed properly (Stanton *et al.*, 2000).

9.2.7 Diagnosis of alarms

Correctly diagnosing alarms to determine what is happening with the system is a significant challenge in many cases, even when operators are attending to them and actively attempting to address them. A number of challenges exist for the diagnosis of alarms.

First, operators can fall prey to incorrect expectations, as discussed previously, or to a similar phenomenon called a *representational error*. A representational error occurs when the relevant information in the situation (in this case an alarm) has been correctly perceived, but the significance of that information is misinterpreted (incorrect understanding of the information for Level 2 SA) to fit a different situation representation. Representational errors have been identified in a variety of domains, including the medical field (Klein, 1993), system diagnosis (Carmino, Idee, Larchier Boulanger, & Morlat, 1988), and aviation (Jones & Endsley, 1996).

In each of these cases, operators already possess an internal representation of what is happening in a situation. Subsequent cues are then interpreted to fit that internal model. For example, a nurse may misinterpret symptoms that should indicate that a patient has been misdiagnosed by finding explanations that allow her to fit those symptoms to the initial diagnosis. Carmino *et al.* (1988) found many cases of this misinterpretation of cues in diagnosing alarm conditions in power plant operations. Jones and Endsley (2000b) investigated representational errors in air traffic control operations and found that despite very obvious cues that initial representations were wrong, controllers would find a way to fit new information to that representation, even when fairly far-fetched rationale was required. In their study, nearly two-thirds of all cues did not result in a revision of the representational error.

These representational errors can contribute to the misdiagnosis of alarms by operators. Instead of alerting operators that their internal representation of what is happening is wrong, the alarm may be explained away to fit with that representation. Carmino *et al.* (1988) for example, attributed misinterpretations of the alarms at Three-Mile Island Nuclear Power Plant to a representational error. In numerous accidents in aviation, pilots explained away the GPWS warnings to fit the situation they believe themselves to be in. In such cases, they falsely attributed the warning to their sink rate or another configuration condition, rather than the actual problem to which they should have been alerted. Unfortunately, it can be very difficult to shake someone out of a representational error unless the cues provided are unambiguous as to their meaning. If they can be misinterpreted to fit the representational error, odds are high that the error will persist.

Another major challenge for alarm diagnosis stems from the problem of *multiple faults*. People will often find a simple, single fault explanation for a problem, rather than one involving multiple faults that may actually be causing the alarm (Bereiter & Miller, 1989; Reising, 1993; Reising & Sanderson, 1995; Sanderson, Reising, & Augustiniak, 1995). Moray and Rotenberg (1989) discuss this as a type of cognitive lockup, where people will focus on the problems with a single system and will ignore other subsystems. Successive faults will be detected much later, or not at all (Kerstholt, Passenier, Houttuin, & Schuffel, 1996). Additional system support may be needed to help guide operators to consider multiple faults in their diagnoses.

In some systems, a significant problem exists from the sheer *number of alarms* provided to operators. Although system designers create displays or alerts that may be fine if presented one or two at a time, in situations where hundreds of alarms can be present at once, it is almost impossible for the operator to easily sort out what is happening. A long list of alarms will be shown, many of which are scrolled past the end of the computer screen. Operators must sort through the alarms, prioritize them, determine which are valid indicators of a problem and which are artifacts or minor issues, and attempt to determine the underlying problems that may have caused them. A number of approaches for dealing with these problems are in practice (Figure 9.8), including serial alarms displays, annunciator alarm displays, and mimic alarm displays.

Serial displays, which list multiple alarms, can be useful in retaining information about the order in which alarms occurred (often useful for diagnosis of the underlying problem), but are problematic for operators due to the large number of alarm messages that can build up on the list (Combs & Aghazadeh, 1988). It can also be challenging for operators to find alarms that are related to each other on the list and sort through relationships between them in a complex system (Stanton & Stammers, 1998).

These problems will be exacerbated with *latching alarms*. Latching alarms show a parameter or system to be in an alarm state until the operator takes an active step to clear it (even if the fault activating the alarm no longer exists). Latching alarms are designed to insure that operators are aware of each alarm that occurs. Many alarms, which may only be in an alarm state for a brief period of time associated with some anomaly, can end up on the list. Latching alarms require a significant amount of effort by the operator who must actively try to clear each alarm to be able to tell if the system is still in an alarm state. As such, they are very poor for supporting operator SA.

11:23:12 Lead wire error - main carr
11:17:33 Plt no transmission
10:55:12 Main turbine lock out - clea
10:54:04 Main turbine lock out
10:52:04 Turbine door open
10:51:44 Water heater #5 override
10:49:32 Hi Temp fault clear - Plant
10:49:16 Hi Temp fault - Plant 3
10:47:59 Pressure alert - valve 47
10:47:23 Hi Temp fault clear - Plant
10:47:22 Hi Temp fault - Plant 3
10:45:33 Plt transmission restore
10:44:42 Main carrier fault - dempl
10:44:22 Battery low - backup #3
10:42:39 Heater level low
10:41:47 Carrier lost - ground fault

Serial Alarm Display

Main Turbine #1	South Turbine #2	East Turbine #3	East Turbine #3
Plant Heat Low	Carrier Failure	Trans-mission Error	Main Battery Low
Valve Pressure High	Valve Pressure Low	Heater Override	Backup Battery Low

Annunciator Alarm Display

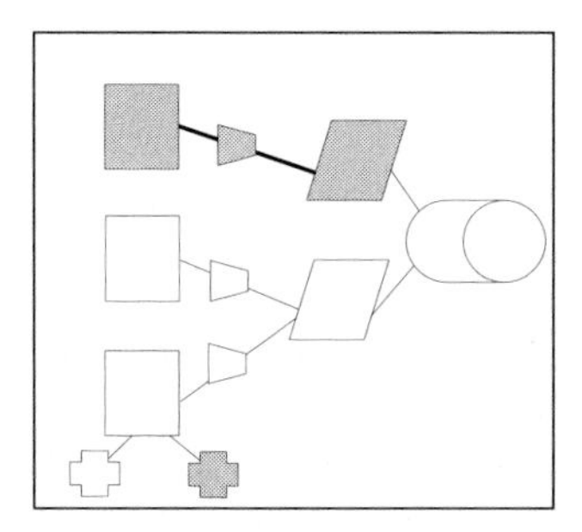

Mimic Alarm Display

Figure 9.8 Approaches for displaying multiple system alarms.

Annunciator-based alarm displays, the middle example in Figure 9.8, show multiple alarms for a system all at once, usually on a panel or display screen. These devices help people assess multiple alarms at the same time. As the same alarm is usually displayed in the same location, they can also support human pattern matching. As a downside, however, they are very poor at helping operators to determine the order of occurrence of multiple alarms, which is often important for diagnosis. When a new audio alarm sounds, it can be hard to tell which is the new system that now has an alarm. If multiple things are happening, integrating the newest alarm into the existing situational model is far more efficient than having to sort through all the alarms to reassess the state of the system. On medical displays for anesthesiologists, searching for the new alarm can take extra time that is needed for attending to the patient. As many non-annunciated alarms may also be present on the display (i.e., non-lit alarms for those items not in an alarm state), people may also have to do more hunting and searching for alarms to find what is happening (Stanton & Stammers, 1998). In complex power plants, for instance, annunciator displays can take up several walls from floor to ceiling.

Mimic displays are also being used in some systems. These displays, as shown in the third example in Figure 9.8, attempt to display alarm signals in a pattern that corresponds to some pictorial representation of the underlying system. For example, it may show a schematic of how each system component relates to the others, with a light or message that can be displayed associated with each component that shows any alarms relevant to that system. Such representations can help operators determine how a particular pattern of alarms might be related to determine underlying fault conditions.

In investigating and comparing the utility of these different approaches to supporting diagnosis of multiple alarms, Stanton and Stammers (1998) found that text-based serial displays were preferable when operators needed to understand temporal order of alarms for diagnosis. Annunciator displays were best for supporting patterns and mimic displays were preferable for those systems where spatial representations are important. In many systems, it may be best to offer multiple approaches for supporting diagnosis of alarms in order to provide the best features of each approach.

9.2.8 Alarm reduction schemes

To help deal with multiple alarms, a number of alarm reduction schemes have been proposed. In some systems, filtering techniques may be used to help reduce the occurrence of spurious alarms—those which can occur due to anomalies in the system that don't really correspond to an actual hazard state. For instance, a medical system may require collecting a certain number of heartbeats before problems in the signals detected set off an alarm. Spikes in the data, such as those due to a loose sensor or noisy data, may be omitted from alarm algorithms. Other schemes involve mode-based conditioning, in which the system does not alarm in certain states when the incidence of false alarms might be high (e.g., when the system is first starting up).

Alarm reduction schemes that reduce spurious and false alarms are likely to help the situation and should be incorporated. Unfortunately they will not significantly reduce the incidence of alarms in many complex systems where alarm overload continues to exist. Alarm diagnosis systems, which lead operators through a sequence of actions to determine the cause of alarms, have been advocated as a method of assisting operators with the problem of too many alarms (Deaton & Glenn, 1999; Noyes & Starr, 2000). Such systems are purported to help operators by sorting information and providing checklists to aid in proper diagnosis of the alarms.

As we'll discuss further in chapter 10, automated systems can also act to undermine diagnosis in some cases. For example, an MD-11 aircraft crashed after overshooting the runway and landed in the water at Subic Bay in the Philippines (Air Transportation Office of the Philippines, 2002). A significant contributor to the accident was the crew's misdiagnosis of alerts related to a faulty airspeed indicator. They followed the checklists the automated system presented to them for diagnosis of the alarms as trained; however, the correct checklist was one of several that was not included in the system's programming. Thus, the system they had been trained to rely on led them to consider problems and factors other than the actual underlying problem. In addition to using up precious time, the crew did not have the support they needed to solve the problem correctly as the diagnostic system led them away from the true problem. The development of 'smart' systems for supporting system diagnosis must be done very carefully to avoid such tragic outcomes.

9.3 PRINCIPLES FOR THE DESIGN OF ALARM SYSTEMS

The use of alarm systems to help support SA is fraught with pitfalls. Operators will view audio or visual alarms through the lens of their expectations and mental model of what is happening in the system, other information available to them on the current state of the system, and past experience with the system's reliability. Designing alarm systems that can deal with these realities and still provide effective support for SA in demanding environments is not easy. Several design principles are offered for designing to best utilize alarms for supporting SA.

Principle 25: Don't make people reliant on alarms—provide projection support

By their very nature, alarms put people in the position of being reactive. When an alarm sounds, they must act to develop an understanding of why it alarmed and what they should do. The alarm itself adds stress to this process. Instead of requiring people to make good decisions under such circumstances, a better strategy is to provide them with the information they need to be proactive. While some systems provide displays that show data on the state variables upon which the alarms are constructed, in many cases this data is not provided in a way that supports SA. Level 3 projection is really required.

Trend information, for example, if displayed directly and graphically can be effective for allowing an individual to recognize impending problems. Rather than being alerted suddenly to a low heart rate, a display showing how the patient's heart rate has changed over time can be invaluable. In addition to allowing medical professionals to anticipate possible problems, this type of display supports their development of an understanding of the problem. Is this a sudden drop, or did it develop more slowly over time? Did the drop co-occur with some other event or change in another physiological parameter? Is it a real drop, or a faulty reading? These types of assessments cannot be met through simple alarms and only current display readings. They can be provided instantaneously through displays that map to Level 3 SA. While alarms can act to back up the process of trend analysis, they should not be the sole means of developing the SA needed for good decision making.

Another way of viewing this principle can be shown by examining the role of alarm systems in the overall SA and decision process. People need to have situation awareness regarding elements 'within a volume of space and time,' per our definition of SA discussed in Chapter 2. This volume of space and time is illustrated for a pilot in Figure 9.9. The volume can vary dynamically based on the pilot's changing goals. That is, pilots have an immediate need for information regarding other aircraft, terrain, or weather that is directly around them. They also, however, have a need to develop SA about 'the big picture.' They need to know what is coming down the line in the intermediate and long-term ranges as well so that they can plan accordingly. At different times their focus may shift between these zones depending on their tasks.

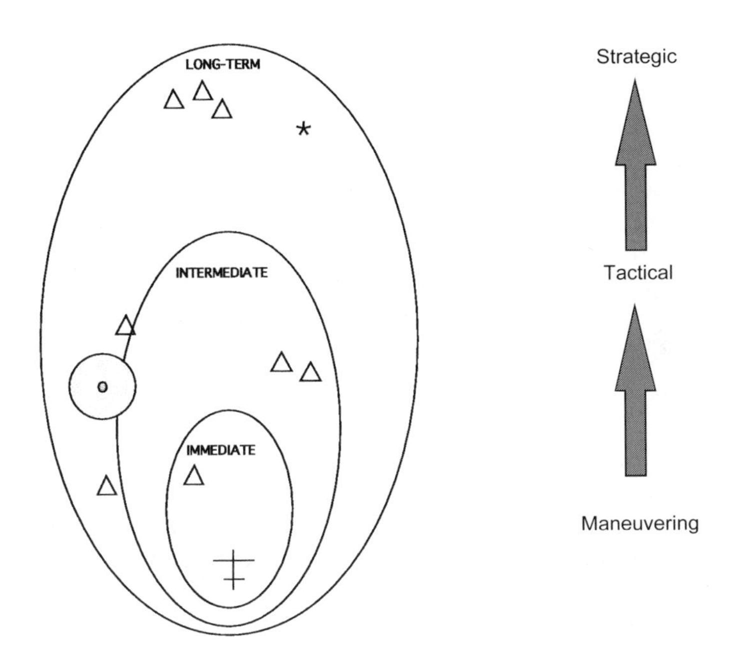

Figure 9.9 SA zones of interest.

Different focuses apply to tasks in these zones of interest. For instance, aircraft maneuvering would fall within the very near-term *immediate zone*. Tactical thinking might involve information within the *intermediate zone* and strategic planning would involve information in the *long-term zone* of interest.

Thinking along these lines, an audio alarm of an impending ground collision would be an information source that would affect the pilot in the near-term maneuvering zone, but would be of little help in the strategic or tactical planning zone. Visual displays supporting pilots' need to understand the relationship between their aircraft and surrounding terrain (including those conditions shown in Figure 9.6), conversely, would be far more useful for long-term planning and tactical assessment than an audio alarm. While audio alarms support the near-term, visual displays that allow projection better support avoiding conditions that might cause an alarm. For making geographical SA assessments, such as planning or re-planning safe routes, frequently pilot focus will be at the intermediate to long-term range zones and could greatly benefit from visual displays of needed information. For near-term maneuvering visual displays would be less helpful than audio alarms if pilots also need to focus on other information (e.g., out-the-window). The use of these two types of information presentations will be quite different and appropriate for different kinds of tasks.

Another way of conceptualizing this issue is shown in Figure 9.10. In most controlled flight into terrain (CFIT) accidents the pilots have made an error of some sort. This may involve violating procedures or airspace, misreading instruments, setting the system incorrectly, or falling prey to spatial disorientation. Projection displays allow pilots to detect that they have made an error and the consequences of that error for producing a potential CFIT accident. They also might be useful for providing the pilots with effective planning tools that may prevent them from making the error in the first place. Audio alarms are most effective further down the line, when it becomes evident that the pilots have not recovered from the error themselves and are close to an accident scenario. These systems are particularly effective when the pilots' attention is directed elsewhere (thus preventing them from noticing or properly comprehending the information on the visual displays) or they are spatially disoriented and unable to understand their situation properly.

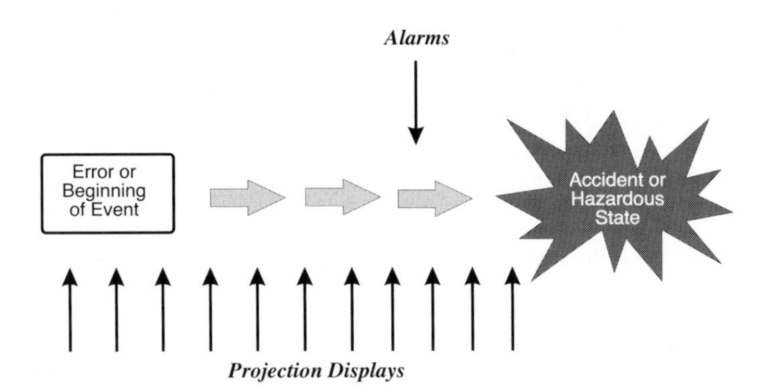

Figure 9.10 Projection displays support strategic and tactical needs while alarms are best for back-up support such as last ditch maneuvering.

An analogous situation exists for other systems. Both alerting systems and projection displays have an important but different role in supporting SA. SA is best supported by providing projection displays to support the strategic needs of the individual, allowing proactive decision making. Alarms can provide a backup to this process if the individuals are unaware that the hazardous condition is building up. However, care must be taken that alarms truly provide new information rather than add tasks to a situation operators are already attempting to deal with.

Principle 26: Support alarm confirmation activities

The idea that people will simply react in a prescribed way to an alarm is seriously flawed. History shows that they will most often seek other information to confirm or deny the validity of that alarm. To best facilitate this process, and the speed and accuracy of responses to actual alarms, information should be displayed that allows them to quickly assess the alarm's validity. This usually requires more than just a display of the value of the underlying system parameter that is causing the alarm. It also frequently requires information on situational factors that can lead the system to alarm, yet not represent a hazardous state.

For example, mechanics who are viewing an alarm on a system in a particular location may want to know of other activities in the area (e.g., routine maintenance or utilities work) that might set off an alarm. A graphical display of the parameter's value over time can help them assess whether the parameter has truly gone out of tolerance, or is simply the result of an anomalous spike in the reading perhaps caused by sensor or transmission error. The SA requirements analysis will reveal what other data operators draw upon to decide on the veracity of an alarm. This information should be used to build confirmation displays that allow a rapid assessment of both the cause of the alarm and of factors that could produce the alarm. Without this data, people will be slower and more error prone in their response to an alarm as they search for this information.

Principle 27: Make alarms unambiguous

Due to the tendency for people to interpret alarms to fit preexisting expectations or mental representations of the situation that are not correct, it is very important that potential causal factors for alarms can be clearly distinguished from the displays provided. Alarms will not be effective if they can be easily misinterpreted. Either through the visual or audio cues provided, different underlying conditions that can lead to the alarm state should be discernable. For example, adding the verbiage 'too low gear' to the 'whoop, whoop, pull up' alarm in Figure 9.6 lessens the ambiguity of the alarm.

Principle 28: Reduce false alarms, reduce false alarms, reduce false alarms

The vast majority of problems with respect to alarms and SA stem from the preponderance of so many systems producing false alarms. Anything that can be done from a system engineering standpoint to reduce the number of experienced false alarms is bound to help. By this, we mean increasing the capability of the sensors and alarm algorithms to detect the difference between actual hazard states that warrant alarms and other situational conditions or anomalies that could produce alarm states. By and large, this is probably the single most important thing that can be done to improve the degree to which system alarms aid SA rather than confuse it.

When false alarms do occur, it is important that operators are able to tell why so that they can calibrate their trust in the system appropriately. It is as important to know when the alarm system is reliable (and why it isn't on occasion) as it is to know how reliable it is in general. Without this information, operators may trust or mistrust the alarms when they should not and will be more hindered by false alarms overall.

One approach to the reduction of false alarms is allowing people to tailor alarm limits, which may be appropriate in some domains. Medical equipment, for instance, often allows the physician to adjust the level at which a particular physiological monitor will alarm—she can set mean heart rate to alarm at 120 for one patient or 150 for another based on her knowledge of physiology (based on age, weight, physical condition and whether the patient smokes) for that patient. This situationally defined tailoring minimizes false alarms. An important thing to consider if such an approach is used is that often more than one person monitors and responds to alarms in many environments. If alarm limits are changed, support should be provided that tells others of such changes so that false expectations are not set up regarding the level at which the system will alarm. In some cases, people may rely on the system to alarm at a certain point to provide them information and they could be caught unaware if their expectations are violated because the alarm limits have been changed.

Principle 29: Set missed alarm and false alarm trade-offs appropriately

Once the ability of the system to distinguish conditions that are truly hazardous from those that are not is maximized, inevitably some difficult decisions will need to be made with regard to the trade-off that exists between a system that doesn't alarm when it should and one that does alarm when it shouldn't. Such decisions are usually made by examining the consequences of the system missing a situation which should be alarmed, without regard to the consequences associated with too many alarms.

The actual consequences need to be weighed with regard to the workload of the individuals involved (Figure 9.7). In vigilance situations, it may be acceptable for more false alarms to occur in order to minimize the possibility of any missed alarms. Even in such cases, it is important that the individual be able to rapidly assess the legitimacy of any alarm that occurs. In moderate workload situations, however, high numbers of false alarms should be avoided. Use great care to create reliable systems where false alarms do not degrade the responsiveness of people to alarms in general, nor where missed hazards lead them to have little trust in the system. Approaches that emphasize providing clear information on developing problems (e.g., projection displays) are more likely to maximize SA than poorly performing alarm systems. In situations where the number or rate of events is so high that people cannot reliably process information themselves, assistance will be needed to alert them to hazards. Effective strategies for performing in this environment lean toward more false alarms and fewer missed hazards. Displays must support the operators' need to make rapid confirmation of the validity and etiology of alarms, however, and operators cannot be held responsible for missed hazards that could occur under these circumstances.

Principle 30: Use multiple modalities to alarm, but insure they are consistent

Compliance with alarms is best when multiple modalities (usually visual and audio cues) are provided together to alert people to the presence of an alarm. People can be slowed down considerably, however, if visual displays and auditory alerts are not consistent, with the visual display usually taking precedence. Visual displays need to do more than simply provide a visual indication of the alarm; they should help people rapidly confirm that the alarm is a valid indication.

In many systems where immediate reaction is needed in stressful situations, the visual indication tells the person what to do about the alarm—for example, a display that reads 'eject' in a fighter cockpit in the case of an engine fire. Command-based displays can be used in environments where the correct action to take is clear and unambiguous. In many domains, however, such as medicine or power plant operations, command-based displays are not advisable as human judgment and decision making are needed to determine the best course of action, often based on information that the system might not monitor.

Principle 31: Minimize alarm disruptions to ongoing activities

It is often difficult to decide what response should be required of people when an alarm sounds. Should they be required to mute it (to stop it from sounding), or to acknowledge it in some way if a visual display is presented (to stop it from blinking, for example)? The idea behind requiring an operator to mute or acknowledge a display is that it ensures they have seen or heard the alarm, and haven't missed it by virtue of being out of the room or busy with another task.

The downside is that the need to respond to the alarm can become a demanding task in and of itself. In a single alarm situation, this may not be such a burden. But in reality, displays end up flashing like Christmas trees and operators can spend a substantial amount of time muting alarms that are audibly annoying. Since oftentimes systems have values that hover near the alarm threshold causing the alarm to sound frequently, the annoyance level and distraction for the operator can be high.

The need to insure that operators are aware of an alarm must be balanced against the disruption that responding to it will have on ongoing tasks (including the act of resolving the situation that may be causing the alarm). Alarms that resound after a period of time, while ensuring that the alarm is not forgotten, end up requiring someone to 'baby-sit' the machine.

Visual displays that do not allow spurious or false alarms to be removed will lose much of their value. When certain subsystems have been disabled, for example, or when a new system is brought online, operators find the entire alarm system to be of little use because the displays become cluttered with alarms that have no value. These circumstances lead people to create workarounds, including disabling alarms, which can lead to negative consequences.

In general, alarms need to inform people of the event in question, without requiring that they be distracted from current tasks unless they choose to be. Prioritization of tasks should, in the end, be up to the person who knows the relative importance of what is happening in the environment. Alarms that sound on a periodic basis (not requiring a person to mute it) may be used as long as visual displays show a consistent alarm state. This prevents the operator from believing between audio alerts that the hazard state has gone away or was due to some spurious condition. Visual displays may blink to show operators a new alarm. Requiring a physical action to acknowledge that alarm (and stop the blinking) can again take time away from other tasks, or can lead to distraction if left unresponded to. Such displays can similarly be set to only blink (or be a different color) for some specified period of time.

The best answer to the question of how invasive alarms should be probably relies on what percentage of the alarms are likely to be false. In a system with a fairly high percentage of false alarms, particularly where people are busy, distracting and invasive alarms are likely to be resented, and thus turned off or disabled. In a system where alarms are rare, particularly false ones, more invasive measures are more acceptable.

Principle 32: Support the assessment and diagnosis of multiple alarms

For highly complex systems where many different parameters on many different components need to be monitored, providing displays that allow operators to reliably oversee the trends and status of all variables may be infeasible. Operators may need to rely on alarm systems to help them ascertain when problems arise. Such is the case in many operations, such as a power plant.

Even in these systems, projection displays can be very useful. Computer support can be provided that shows trends and status for subcomponents on an operator-controlled or as-needed basis. Such displays can be helpful to operators in building their mental model of what is happening with the system and can prevent forcing reactive behavior.

Displays in highly complex systems need to assist operators in discerning (1) which systems are in alarmed and non-alarmed states, (2) which alarms are new, and (3) the temporal order of alarm occurrence. A combination of serial-based and annunciator- or mimic-based displays may be needed to support such assessments. No one format supports all these needs. The format of the displays needs to support people in determining the relationship between multiple alarms (if any) by virtue of the physical and spatial display of the alarms. Subsystems or variables that are related to each other should be shown together, for example. The underlying data that supports the analysis of the alarm should be shown in direct juxtaposition to the alarm displays. Designs are very helpful in supporting the analysis and diagnosis of multiple alarms when the underlying structure of the system is used to arrange the information in a meaningful way.

Principle 33: Support the rapid development of global SA of systems in an alarm state

Finally, it is important that operators are able to ascertain at a glance the status of the entire system to gain global SA. When computerized displays are provided it is critical that alarm messages do not get lost in lists that have scrolled off the screen or become covered with other windows on the display (such as those required to diagnose other alarms, for example). It is all too easy for important information to become lost to the operator in these situations. A central display location that provides the status of all alarms and is never covered up by other information should be provided for maintaining global SA (see Principle 4). This is critical to keep operators from losing track of situations that may be of even higher importance than the task they are working on, or which may have some bearing on it.

Latching displays that show the system in an alarm state until the operator acts to clear it should be avoided for most systems. Not only can they be intrusive for the operator, but they also can act to obscure SA by misleading the operator in assessing which systems are currently in alarm states.

Serial alarm lists can be notoriously inscrutable. They often include arcane abbreviations and messages concerning when a parameter goes in and out of an alarm state. These lists can become very long and require considerable effort to piece together along a timeline. A listing of which systems are currently in an alarm state (and for how long) is much more useful. (A separate log can be provided for the operators' inspection if an exact history is needed.) Methods may be used to insure that this list can be viewed on one screen, without requiring scrolling of information off the screen. An alarm timeline, showing system alarms along a fixed time scale, can also be a very useful way to show the temporal order of alarms. Such displays should also allow operators to superimpose other key events or information that may be useful for analysis of alarms, such as current temperature fluctuations or the starting or ending of some process, for example. Alarm displays must support the operators' need to rapidly understand which systems need their attention and why a problem may exist.

CHAPTER TEN

Automation and Situation Awareness

10.1 AUTOMATION—A HELP OR A HINDRANCE?

The drive to automate has characterized much of the past century. With it has come a mixed bag of both progress and problems. Automation of many different functions formerly performed by people has become common. Elevators carry people without a human operator, cruise control is commonly found in many automobiles, and automated teller machines span the globe in the 21st century. In cases where automation has been developed with a high degree of reliability and fairly simple requirements of the human operator or user of the system, it has been highly successful.

In many systems, however, automation has also been accompanied by much higher levels of complexity, lower reliability levels, and a brittleness (susceptibility to error under varying system conditions) that can lead to tragic consequences. For example, in 1983 a Korean Airlines flight en route from Anchorage, Alaska to Seoul, Korea strayed over Soviet airspace and was shot down. This tragedy was blamed on an error the crew made in programming their path into the navigation computer (Billings, 1997; Parasuraman, Mouloua, & Molloy, 1996). A simple error made large deviations from the intended path possible. In another example, an American Airlines flight crashed in the mountains of Colombia in 1996 killing all aboard due to a mixture of programming error, overreliance on the automation, and poor feedback depriving the pilots of an understanding of what the automation was doing (Endsley & Strauch, 1997). These problems are not limited to aviation, but also can be found in air traffic control, driving, maritime operations, power plants, and oil and gas pipeline systems as they begin to incorporate more and more automation (Parasuraman & Mouloua, 1996). While it is easy to blame the operator for the accident in many of these cases, closer scrutiny reveals that the operators acted in ways that were consistent with the effects of automation on human information processing and decision making.

The goal of this chapter is to examine the complex ways in which different forms of automation interact with human characteristics to produce both benefits and challenges for SA and overall performance. Research evidence shows some significant shortcomings in historical efforts at producing a human/automation symbiosis that can undermine much of the effectiveness of automation initiatives.

Before proceeding with this discussion, it should be pointed out that automation is a broad term applied to a wide class of initiatives, spanning from routing information directly from computer to computer (as is being discussed under the term *datalink* in the aviation system and *digitization* in the military arena), to efforts that provide inputs to people on what they ought to do, to systems that perform certain tasks for the operator. These different types of automation, and numerous factors regarding how they are implemented, can have very different effects on SA and human performance in interacting with the system. Endsley and Kaber (1997; 1999) identify four major aspects of a task that can be automated:

- Monitoring or taking in information,
- Generating options or strategies for achieving goals,
- Selecting or deciding what option to employ, and
- Implementing or carrying out actions.

Each of these roles can be performed by the human, the computer, or by a combination of the two to result in numerous ways of adding automation to a task. Their research shows that the effects of automation on SA, workload, and human performance are quite different depending on these different types or levels of automation (Endsley, 1997; Endsley & Kaber, 1997; 1999; Endsley & Kiris, 1995a; Endsley, Onal, & Kaber, 1997). In this chapter we will attempt to separate the wheat from the chaff; our goal is not to point to automation as either good or bad, but to determine automation types and implementation strategies that are most successful at achieving good human/system performance under a wide variety of conditions.

While some people falsely believe that automation reduces the operator's need for SA, it turns out that SA is not only important for human performance when operating alone, it is just as important when people work to operate systems where they must interact with or oversee automation. In most systems that include even high levels of automation, people still retain important roles as overall system monitors and operators. Although the 'lights out' factory was touted as almost a certainty by the end of the 20th century, today's factories include humans in many roles from system operator to maintainer. In most other domains, the same is true. Human operators maintain important roles in these systems because they are flexible and creative; they can learn in the face of new events in ways that most automation cannot, and can 'plug the holes' in the system's capabilities that its designers did not foresee (Rasmussen, 1980). In this chapter we will focus on the effects on SA of the human operator of a system that incorporates automation of at least some functions.

In order to properly oversee automated systems, to insure that they are operating properly to achieve system goals, operators need to maintain a high level of SA regarding not only the functioning of the automation ('Is it working? Are its outputs meeting operational goals?'), but also the functioning of many of the underlying system parameters that are needed to answer these fundamental questions.

For example, in order for a pilot to know that the Flight Management System (FMS) (automation for navigating the aircraft) is working properly, she must monitor not only the information that the FMS displays provide on its programmed path and current mode of operation, but also information to verify its performance, such as the current speed, altitude, and heading of the aircraft (i.e., raw data). The need for SA does not diminish with the addition of automation to a system, but often increases, as operators must maintain SA over all basic system information as well as information from the new automated system. This job can become much more difficult due to the inherent complexity involved in many automated systems.

There are a number of critical issues with regard to how automation can affect SA that need to be understood in developing sound principles for automation design. These include:

- A tendency for automation of various forms to reduce SA such that human operators have a diminished ability to detect automation failures or problems and to understand the state of the system sufficiently to allow them to take over operations manually when needed. This has been called the *out-of-the-loop syndrome.*
- A problem with operators frequently misunderstanding what the system is doing and why it is doing it, which is necessary for accurate comprehension and projection (Levels 2 and 3 SA). With comprehension and projection compromised, significant errors can occur as operators struggle to insure compliance between what the automation is doing and what they want it to do. This class of problem is typically referred to as a *mode awareness* or *automation understanding problem.*
- A tendency for decision-aiding automation (often called expert systems or decision support systems) to interact with attention and information evaluation processes in such a way as to diminish their intended effectiveness. We'll call this the *decision support dilemma.*

Each of these challenges will be discussed more fully so that we can demonstrate the problems inherent in many typical implementations of automation, and derive principles for implementing automation successfully.

10.2 OUT-OF-THE-LOOP SYNDROME

The out-of-the-loop performance problem is of considerable concern. In most systems, when automation fails or a situation is encountered that the automation is not programmed to handle, human operators need to be able to detect the problem, diagnose what is happening and then take over the task manually. Their ability to do so is often compromised, however. First, people are often slow to detect that there is a problem with an automated system (Ephrath & Young, 1981; Kessel & Wickens, 1982; Wickens & Kessel, 1979; Young, 1969). In addition, once the problem is detected, it can take considerable time to reorient themselves to the situation enough to understand the nature of the problem and what should be done about it. In many cases this loss of SA and the resulting lag in taking over manual performance can be deadly.

For example, in 1987 a Northwest Airlines MD-80 crashed during takeoff from Detroit Airport because the flaps and slats had been incorrectly configured, killing all but one passenger (National Transportation Safety Board, 1988). The investigation showed that a primary factor in the crash was the failure of an automated takeoff configuration warning system. The crew failed to properly configure the aircraft for takeoff because they were distracted by air traffic control communications. The automated system, designed to detect and prevent exactly this type of problem, also failed. The crew was unaware of both the state of the aircraft and the state of the automation.

In another example, in 1989 a US Air B-737 failed to successfully takeoff from New York's LaGuardia Airport, coming to rest in a nearby river with loss of life and destruction of the aircraft (National Transportation Safety Board, 1990). In this accident, the autothrottle was accidentally disarmed without the captain or first officer noticing. The crew was out-of-the-loop, unaware that the automation was not managing the throttles for them. As a result, they had insufficient airspeed to takeoff and ran off the end of the runway into the river.

While most examples of such problems with automation come from aviation (due to both the long history of automation implementation in this industry and from detailed accident investigation and recording procedures required by law), some examples can also be found in other arenas as forms of automation become more prevalent. For example, a recent fratricide incident in the war in Afghanistan was found to be due to the simple automation failure of a global positioning system (GPS) device (Loeb, 2002). The device was being used by a soldier to target opposition forces when its batteries failed. Changing the batteries quickly, the soldier provided the needed coordinates for an airstrike. The soldier did not realize, however, that when the system lost power, it automatically reverted to reporting its own position. He was out-of-the-loop in understanding what the system was telling him. The subsequent bombing killed 3 and injured 20 friendly forces rather than the intended target.

Consider another example from the maritime industry. In 1996, a freighter vessel crashed into a pier in New Orleans, injuring 116 people. It was reported that a lube pump failed, causing the backup pump to come on and a simultaneous reduction in speed. This speed reduction significantly eroded the steering control available at a critical turn on the Mississippi River. As the ship lost engine power at this turn, it crashed into a crowded shopping area on the bank. The automation worked as designed, but it left the boat captain out-of-the-loop and unable to take the control actions needed (Associated Press, 1996).

The out-of-the-loop problem can be fairly prevalent. In one study of automation problems in aircraft systems, 81% of reported problems in cruise were associated with not monitoring the system (Mosier, Skitka, & Korte, 1994). The out-of-the-loop syndrome has been shown to directly result from a loss of SA when people are monitors of automation (Endsley & Kiris, 1995a). This loss of SA occurs through three primary mechanisms (Endsley & Kiris, 1995a):

1. changes in vigilance and complacency associated with monitoring,
2. assumption of a passive role instead of an active role in processing information for controlling the system, and

3. changes in the quality or form of feedback provided to the human operator.

We'll discuss each of these factors in more detail.

10.2.1 Vigilance, complacency, and monitoring

Many people attribute the out-of-the-loop syndrome to complacency or over-reliance of the operator on the automation (Parasuraman, Molloy, Mouloua, & Hilburn, 1996; Parasuraman, Molloy, & Singh, 1993; Parasuraman & Riley, 1997). In general people are not good at passively monitoring information for long periods of time. They tend to become less vigilant and alert to the signals they are monitoring for after periods as short as 30 minutes (Davies & Parasuraman, 1980; Macworth, 1948, 1970). While vigilance problems have historically been associated with simple tasks (such as monitoring a radar for an infrequent signal), Parasuraman (1987) states that 'vigilance effects can be found in complex monitoring and that humans may be poor passive monitors of an automated system, irrespective of the complexity of events being monitored.'

By placing too much trust in the automation, it is believed that they shift attention to other tasks and become inefficient monitors (Parasuraman, Molloy, & Singh, 1993; Parasuraman, Mouloua, & Molloy, 1994; Wiener, 1985). Reliance on automation is related to both operators' subjective confidence in the system's reliability (although imperfectly calibrated with actual system reliability levels) (Wiegmann, Rich, & Zhang, 2001), and to their comparative feelings of self-efficacy or competence to perform tasks themselves (Prinzel & Pope, 2000). Wiegmann, Rich and Zhang (2001) found that people's reliance on automation exceeded their subjective confidence in the system when the automated aid was viewed to be better than their own performance would be without it.

This complacency problem is worse when other tasks are present that demand the operator's limited attention (Parasuraman, Molloy, & Singh, 1993). This has been found to be not a visual problem (inability to look at multiple input sources at the same time), but rather an attention problem. Superimposing the information on the status of the automated task and other concurrent manual tasks does not appear to solve the problem (Duley, Westerman, Molloy, & Parasuraman, 1997; Metzger & Parasuraman, 2001a).

While complacency is often described as a human failure, Moray (2000) makes the convincing argument that people actually employ their attention close to optimally, based on the utility likelihood of different information sources. That is, while people may sometimes miss important cues that the automation has failed, it is not because they are not monitoring various information sources or because they are too trusting. Rather, they miss detecting the problem due to the fact that with an optimal sampling strategy usually reliable sources need not be monitored very often. 'Because an operator's attention is limited, this is an effective coping strategy for dealing with excess demands. The result, however, can be a lack of situation awareness on the state of the automated system and the system parameters it governs' (Endsley, 1996a, p 167).

10.2.2 Active vs. passive processing

A second factor that underlies the out-of-the-loop problem is the inherent difficulty involved for operators in fully understanding what is going on when acting as a passive monitor of automation rather than as someone actively processing information to perform the same task manually. Cowan (1988) and Slamecka and Graf (1978) found evidence that suggests that the very act of becoming passive in the processing of information may be inferior to active processing. Even if monitoring the appropriate information, the operator may not fully process or update that information in working memory. This is analogous to riding in the car with someone on the way to a party at a new place and then finding out at the end of the evening that you must drive home. Your memory trace of how you got there is frequently not the same as if you had actively driven there yourself, illustrating the poor SA rendered by passive processing.

It should also be pointed out that checking the output or performance of a system is often not possible without doing the task oneself. For example, something as simple as checking that a computer is adding correctly is almost impossible to do without performing the calculations oneself. To manually perform every task to check the system is usually not feasible or timely, however. Therefore, efforts at 'checking' an automated system may often be only partial attempts at a manual performance of the same task. Layton, Smith, and McCoy (1994), for example, found that pilots did not fully explore other options or do a complete evaluation of system-recommended solutions for route planning, but rather performed a more cursory evaluation. This makes sense, as it would hardly be worth the trouble and expense of having automated aids if the operator still had to perform everything manually in order to check the system.

Endsley and Kiris (1995a) performed a study involving an automobile navigation task. They found much lower understanding (Level 2 SA) of what had happened in the navigation task when people performed the task with an automated aid, even though their Level 1 SA (ability to report on information in the navigation task) was not affected. Although they were aware of the basic information, they did not have a good understanding of what the data meant in relation to the navigation goals. In this study, complacency and potential differences in information presentation between automated and nonautomated conditions were ruled out as possible contributors to this loss of SA. Lower SA was found to be due to changes in the way information is processed and stored when monitoring the performance of the automation (or any other agent) as compared to doing the task oneself. In other studies, poor SA has been found under passive monitoring conditions with experienced air traffic controllers (Endsley, Mogford, Allendoerfer, Snyder, & Stein, 1997; Endsley & Rodgers, 1998; Metzger & Parasuraman, 2001b). The air traffic controllers had much poorer SA regarding aircraft they were just monitoring as compared to aircraft they were actively controlling.

Thus the out-of-the-loop syndrome may not be just due to simple concepts such as complacency or overreliance on automation, but a fundamental difficulty associated with fully understanding what the system is doing when passively monitoring it. As such, automation may continue to pose a challenge for the SA of human operators, even when they are vigilant.

10.2.3 System feedback quality

Either intentionally or unintentionally, the design of many systems creates significant challenges to SA because it is inadequate for communicating key information to the operator (Norman, 1989). Feedback on system operation is either eliminated or changed in such a way as to make it difficult for a person to effectively monitor what is happening. 'Without appropriate feedback people are indeed out-of-the-loop. They may not know if their requests have been received, if the actions are being performed properly, or if problems are occurring' (Norman, 1989).

Sometimes designers will mistakenly remove critical cues in the implementation of automation without realizing it. For example, development of electronic fly-by-wire flight controls in the F-16 aircraft initially caused problems for pilots in determining airspeed and maintaining proper flight control because important cues vibration normally felt in the flight stick were no longer available through tactile senses. Rather, the necessary information was provided only on a traditional visual display (Kuipers, Kappers, van Holten, van Bergen, & Oosterveld, 1990). It was much more difficult for the pilots to simultaneously determine this key information from visual displays when eyes-out during landing. Without realizing it, designers had inadvertently deprived them of important cues they were using. To correct this problem, artificial stick-shakers are now routinely added to fly-by-wire systems to provide pilots with this needed feedback (Kantowitz & Sorkin, 1983). Other research has also cited the lack of hand movement feedback with automation of physical tasks (Kessel & Wickens, 1982; Young, 1969), and loss of factors such as vibration and smell in the automation of process control operations (Moray, 1986).

In other cases, designers have intentionally removed or hidden information on the status of systems that the automation should be taking care of under the belief that operators no longer need to know this information. Billings (1991) found that in some cases automation has failed to tell pilots when it shuts down engines, directly depriving the pilot of SA needed to avoid accidents. In another incident, an aircraft lost 9500 feet of altitude over the mid-Atlantic when the captain attempted to take control from the autopilot. It was found that the autopilot effectively masked the approaching loss of control conditions, hiding the needed cues from the pilot (National Transportation Safety Board, 1986).

The use of a multitude of different computer windows or screens of information that can be called up through menus or other buttons can also contribute to the problem of poor feedback in many automated systems. If the operator is busily engaged in viewing certain information, key data on another hidden window can be obscured. Unless he knows there is a problem, the operator cannot know that he needs to ask to view this other information. The more complex the situation in which a person works, the easier it is for that person to not notice displayed information and understand its importance.

In addition, there may be a real problem with the salience (i.e., prominence) of key information with many systems. 'The increased display complexity and computerized display format reduces the perceptual salience of information, even if it is available. In a complex environment with many activities going on, it is easy for operators to lose track of such information' (Endsley, 1996a). In many aircraft systems, for example, only one letter or number may be different on a very cluttered display to indicate which mode the system is in. This poor salience can be directly linked to poor awareness of automation modes. Efforts to improve the salience of mode transitions through other modalities (e.g., tactile cues) have been successful in helping to reduce the SA problems associated with poor awareness of mode information on these displays (Sklar & Sarter, 1999).

Of the three factors associated with poor SA leading to the out-of-the-loop syndrome, poor feedback and information presentation are the most easily remedied through the application of good human factors design principles and attention to the importance of information cues related to automation states.

10.3 AUTOMATION AND LEVEL OF UNDERSTANDING

Aside from the out-of-the-loop syndrome, people also can have significant difficulties in understanding what the system is doing, even when it is functioning normally. That is, comprehension and projection (Levels 2 and 3 SA) can be seriously strained, inaccurate, or missing. Wiener and Curry (1980) first noted these problems in studying pilots working with automated systems. 'What is it doing now?', 'I wonder why it is doing that?' and 'Well, I've never seen that before' are widely heard comments. Although understanding seems to improve with experience on an aircraft, problems can continue to exist even after years in working with a system (McClumpha & James, 1994).

A number of factors are behind this poor understanding of many automated systems, including the inherent complexity associated with many automated systems, poor interface design, and inadequate training. As systems become more complex, people find it harder to develop a mental model of how the system works (see Chapter 8). Automated systems tend to incorporate complex logic so that the system can accomplish numerous functions and deal with different situation conditions. Often it acts differently when in different system modes, further complicating the logic structures that a person needs to understand. The more complex the system logic becomes, the harder it can be for a person to attain and maintain Level 2 SA (comprehension) and Level 3 SA (projection). The growing complexity of systems makes it more difficult for a person to develop a complete mental model of a system. Many branches within the system logic, as well as infrequent combinations of situations and events that evoke various rarely seen system states, add to the challenge of fully comprehending the system.

Vakil and Hansman (1998) point to the lack of a consistent global model to drive the design of automated systems in aviation as a major culprit in this state of affairs. As systems evolve over the years, different designers add new functions that may work in ways that are not consistent with other functions or modes, leading to an 'entropic growth of complexity.' Mode proliferation reigns as new modes are added to deal with different requirements of different customers and to provide product differentiation. Complicating this problem, Vakil and Hansman found pilots' training to be simple and rule-based, providing little information on causality or connection to the underlying system structure. With the lack of a clearly presented model of how the system works, pilots tend to adopt their own *ad hoc* models that are often inaccurate or incomplete representations of how the systems work.

Adding to this problem, the state of the automation and its current functioning is often not clearly presented through the system display. For example, when the system changes from one system mode to another, the change is often so subtle that the person using the system may be unaware of the shift, and thus becomes confused when the system functions differently. Without system understanding, the ability of the operator to project what the system will do (and assess the compliance of those actions with current goals) is also lacking, making proactive decision making difficult and error prone.

The display of projected system actions is often insufficient or missing, further limiting system understanding and exacerbating this problem. For instance, while the horizontal flight path is displayed directly in automated flight management systems, information on changes in the vertical flight path and the relationship of that path to terrain and key operational points is missing. When a vertical situation display is provided to pilots, a dramatic improvement in avoidance of mode errors is observed (Vakil, Midkiff, & Hansman, 1996).

While these problems may be specific to poorly evolved automation in the aviation domain, they clearly point to problems to be avoided in the development of automation in any domain.

10.4 DECISION SUPPORT DILEMMA

Some automation has not been focused at doing things for the human operator, but rather at advising the operator on what to do. This broad class of systems has historically been termed *expert systems* or *decision support systems*. The basic underlying premise is that these systems can improve human decision making by compensating for a lack of expertise, or for decision errors that operators might make. The system's advice should be able to boost operator decision making and performance by providing information on the right course of action.

Research on how well these systems perform in improving human decision making using expert systems has frequently not borne out this premise, however. Selcon (1990) performed a series of studies on human performance with expert systems that expressed probabilities associated with the 'correctness' of various options. He found that when those probabilities were close or ambiguous, the time required for the operator to decide what to do significantly increased in comparison to situations in which people made the same decision without system advice. (It is worth noting that it is more likely people would need advice under such circumstances, as opposed to when options are widely separated in terms of their degree of goodness.)

Based on the fact that people often have trouble in dealing with information presented in percentages, Endsley and Kiris (1994a) examined different ways of presenting degrees of confidence in expert system recommendations, including digitally (e.g., 92%), in the form of analog bars, ranking (e.g., 1, 2, 3), and categorically (e.g., *high*, *medium*, and *low*). They found that the expert system advice did not improve performance significantly (even for novices at a task), and it increased decision time for most presentation methods. Only the categorical presentation format resulted in decision times that were slightly faster than when no expert system information was presented.

Kibbe and McDowell (1995) examined the performance of image analysts when presented with the output of an automated target recognition system. They found that while the analysts sought aid more in cases where the images were poor, they did not perform better with that advice, but rather performed slightly more poorly. Overall, the combined human/system performance rarely exceeded the better of the two components when either one or the other was presented with a poor image.

In other research, a problem with decision biasing has been revealed. That is, when the decision support system is wrong, people are found to be much more likely to make an error (follow the expert system), than if they had been presented with no system advice at all (Layton, Smith, & McCoy, 1994; Olson & Sarter, 1999; Sarter & Schroeder, 2001; Smith *et al.*, 1995). As great as a 30% to 40% increase in errors was observed in these studies.

Finally, other research has examined how people respond to automated alerting systems. Pritchett and Hansman (1997) found that when pilots are presented with a system that alerts them to runway incursions by other aircraft, they do not respond immediately. Rather they look to evaluate those alerts, forming an independent assessment of their validity, frequently using different methods than the system's algorithms.

Although drawn from different types of systems and domains, this pattern of results indicates something very important about how people use decision support system recommendations. The premise behind these systems is that overall human/system reliability should be improved by the provision of recommendations on what to do. That is, there should be a synergy induced that leads to more optimal performance than either human or computer could achieve alone (otherwise why have the more poorly performing component involved at all?). In general, this type of improved reliability can only be achieved from two systems that operate in parallel (Figure 10.1a) and only when the best performance of the two is used.

The combined results of these different studies demonstrate a very different pattern of performance, however. They show that people are not operating independently of the decision support system to arrive at their decision, but are highly influenced by the system advice. It becomes one more piece of information that they must take into account along with all the other system information they must normally attend to in order to make a decision. This pattern is much more indicative of two components (the human and the computer) that are operating in series (Figure 10.1b). In this case the overall reliability of the system does not increase, but decreases (Sanders & McCormick, 1992). This shows why fundamentally this model of decision aiding has failed to produce expected advantages in system performance. Decision support systems do not operate on human decision making independently from SA, but rather become part of SA in forming decisions.

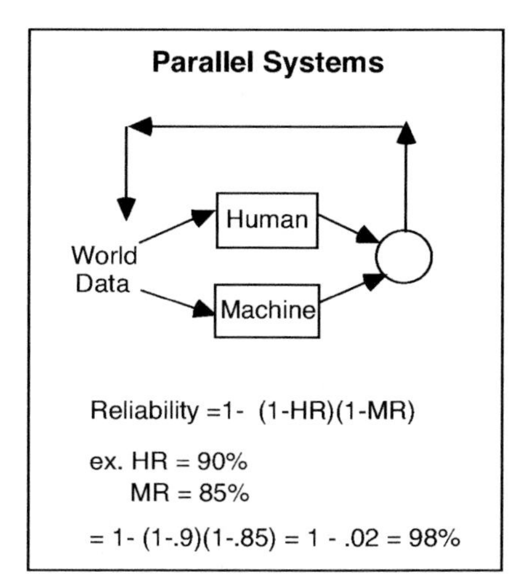

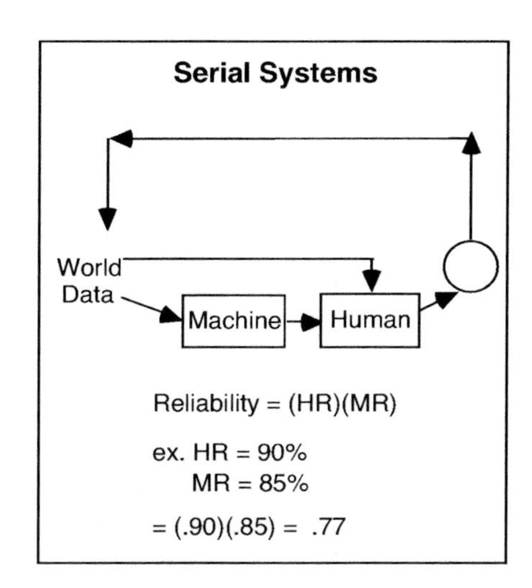

Figure 10.1 Examples of the reliability of a human and machine component (a) when operating in parallel and (b) in series.

In a related approach, rather than telling operators what to do, intelligent target cueing systems have been designed to help direct operator attention to the most important areas of a complex display. This approach seeks to aid SA by directing attention, rather than by aiding decision making *per se*. While these systems have been found to improve detection performance and reduce detection time for designated targets on a display when the system is correct and the target is in the area highlighted, they have also been shown to lead to significant errors when there are additional not cued targets or when the targets are not in the expected area (Wickens, Conejo, & Gempler, 1999; Yeh & Wickens, 2001; Yeh, Wickens, & Seagull, 1999). Furthermore, when the system is incorrect, operators are found to be significantly more likely to select an incorrect but cued target.

These results, showing attention biasing associated with system cueing, are consistent with the decision biasing results found with decision support system approaches. In both cases, while good advice can help performance, poor advice has a large effect on leading the operator astray. Because the output of both system types is most likely processed in a manner consistent with the serial model (Figure 10.1b), this is to be expected. As most systems will not be perfectly reliable, but rather will produce incorrect advice at least a certain percentage of the time, these significant shortcomings in overall performance need to be explicitly considered when implementing such systems.

On the positive side, not all approaches to decision support have encountered these types of problems. For example, Guerlain and her colleagues have explored the use of critiquing systems that provide advice to operators *after* they have made decisions rather than before (Guerlain *et al.*, 1999). These systems typically point out faulty reasoning, inconsistent answers, or violations of constraints. Guerlain *et al.* not only found significantly improved performance associated with use of the system, but also much less susceptibility to biasing in a difficult situation that the system had not been programmed to completely handle; there was a 19% error rate with the system as compared to a 50% error rate without the system. These results indicate that a critiquing approach, by virtue of providing its input after the operator has made an initial decision, is much more in line with a parallel model of interaction (Figure 10.1a), providing better human/automation synergy and leading to improved overall performance of the combined system.

10.5 NEW APPROACHES TO AUTOMATION

New approaches to automation are also being explored with the hope of finding ways of enhancing SA and reducing out-of-the-loop performance problems. These approaches seek to create new ways of combining the human and automation that increase human participation and involvement in system control: either by temporarily transferring control from the human to the automation and back through *adaptive automation*, or by implementing *levels of automation* that increase human involvement and SA.

10.5.1 Adaptive automation

Adaptive automation (AA) recognizes that most automated systems are not under the control of the computer all the time. Rather, at times their human operators need to assume control and operate the system manually. Adaptive automation exploits this opportunity by systematically allocating control of the system to the operator at periodic intervals in hopes of reducing complacency and improving human monitoring performance (Rouse, 1977; 1988).

In many cases the decision of when to pass control from automation to the human has been based on a consistent time interval. It can also be based on a number of other factors (Scerbo, 1996):

- the occurrence of critical events,
- detection of human performance below a certain criterion level,
- use of psychophysiological monitoring to detect losses of arousal or other cues of poor performance (e.g., loss of consciousness), or
- the use of models of human performance to predict the best times to intervene.

Parasuraman (1993) found that human monitoring of automation failures improved when operators were subjected to periods of manual control through adaptive automation that gave them manual control every 10 minutes. In another study, Parasurman, Mouloua, and Malloy (1996) found improved performance with both a model-based AA system and with a performance-based AA system.

Adaptive automation appears to aid human/system performance mostly by reducing workload. Hilburn *et al.* (1997) found reduced workload when AA was used in an air traffic control task (as opposed to full automation or fully manual operation). Using a psychophysiological allocation approach, Kaber and Riley (1999) also found improvements in workload when operators were assigned to intervals of AA, although not when the choice of when to take control was left up to the operator. Kaber (1996) found that the proportion of the time that operators worked under automated conditions through AA (e.g., longer cycle times) contributes most to the reduction of workload, as compared to levels of automation which contributed most to changes in SA.

The development of good methods for implementing AA is ongoing, including establishing methods for activating AA and determining optimal allocation strategies (e.g., duration and frequency of manual versus automated control). If AA is to be successful, it is also important that the interface be enhanced to avoid potential SA challenges associated with AA. Since AA essentially involves the implementation of modes (either manual or automated), assistance is needed to insure that it is very salient to operators which mode the system is in, and when upcoming periods of manual operation are forecasted so that they are not caught unprepared to assume control. These periods of manual control must be seen as relevant work to the operator and not as system unreliability or busywork. Most AA research to date has been conducted under laboratory conditions. The long-term effects of AA in actual operations still need to be determined.

10.5.2 Levels of automation

An alternate approach seeks to find levels of automation that keep the operator involved in system operations on an ongoing basis, rather than subjecting them to periods of passive control. Different levels of automation have been described, relevant to different kinds of systems (Billings, 1991; Endsley & Kaber, 1997; Endsley & Kaber, 1999; Endsley & Kiris, 1995a; Parasuraman, Sheridan, & Wickens, 2000; Sheridan & Verplanck, 1978).

Endsley and Kiris (1995a) showed that SA was improved in system operations that involve intermediate levels of control, and this improvement in SA corresponded to reductions in the out-of-the-loop problem. People were faster to respond to system failures when operating under intermediate levels of control than when operating under full automation.

Endsley and Kaber (1997; 1999) further explored this approach by examining a more detailed level of automation taxonomy that systematically allocates each of four task aspects (monitoring of information, generation of options, selection of actions, or implementation of tasks) to either the human, the computer, or a combination of the two in various ways. They found that when automation aids in or takes over task implementation, overall performance improves. When it is used to generate options, however, performance declines (in agreement with previous discussions on decision support systems). It does not seem to matter whether selection of actions is up to the human or computer, however.

Operators are the most out-of-the-loop (take the longest to recover from automation failure) with levels of automation that employ advanced queuing techniques—those that allow the operator or system to sequence a number of tasks for future implementation. It is worth noting that aircraft flight navigation systems (which have had so many problems) do exactly this; pilots program in a series of waypoints that define the aircraft's route, effectively queuing up future navigation actions. Kaber (1996) found that the level of automation of a task is most predictive of SA (as compared to AA).

10.6 PRINCIPLES FOR DESIGNING AUTOMATED SYSTEMS

Principle 34: Automate only if necessary

As automation can lead to such significant difficulties in lack of understanding, system complexity, decision biasing, and out-of-the-loop performance problems, it should be avoided except in those situations where its assistance is really needed. While this might seem obvious, the rush to automate has characterized a technology-centered design approach for much of the past century. In many cases, system improvements that are generated from implementing automation actually come from merely improving the information and material process flows and reducing unneeded steps, independent of the actual automation itself.

Much of the drive for automation stems from calls for reductions of data overload and high operator workload (Endsley, 1995d). In reality, however, a large portion of these problems stem from disjointed technology-centered systems (see Chapter 1). As a key tenet of SA-Oriented Design, one should *first* seek to improve the user interface to support human situation awareness and decision making. As people are the most creative and flexible problem solvers around (if provided with good SA), this approach creates the most robust human/machine system. Only after the system has been optimized for SA should automation be considered as an aid for specific tasks for which the human operator needs assistance. In addition, it should be recognized that certain circumstances may require automated aiding—hazardous conditions such as mine removal or handling of radioactive materials, for example. Many systems could benefit from a more careful selection of functions to automate, however.

Principle 35: Use automation for assistance in carrying out routine actions rather than higher level cognitive tasks

Computers are typically better than people at carrying out routine, repetitive tasks. Automation of cognitive portions of a task have proven to be the most problematic (see Endsley & Kaber, 1999). Mathematical calculations, psychomotor control (e.g., steering on a fixed course), and moving data from one system to another are examples of repetitive tasks that are probably the most amenable to computer assistance, providing a reduction in errors and workload.

Principle 36: Provide SA support rather than decisions

Due to the problems associated with systems that provide advice or decisions to operators, this form of decision aiding should be avoided. The tendency for decision biasing and slowing of decision making leads to significant reductions in the overall effectiveness of this approach. Much greater improvements and more robust decision making can be found instead with systems that enhance SA.

For example, a typical route guidance system found in a more advanced automobile today will automatically create a map of the best route to an entered destination from the car's current position. It will then direct the driver along that route with visual or aural displays that tell her when to turn. These systems can be quite helpful in unfamiliar cities or when navigating to a new locale. They automate the task of finding the desired destination on the map and determining how to best get there, and act as a memory aid by telling the driver where to turn and how far each segment of the trip is.

They also have some significant shortcomings, however. They essentially create a very narrow picture of the situation and produce very shallow SA. This is not only because the visual displays are typically limited in size, creating a 'soda straw view of the world' and making it difficult to get a big picture of where the suggested route is relative to the larger world. The SA is shallow because it does not support the more robust range of decisions the driver may need to make. She just sees the 'one best way' provided by the system.

For example, one of the authors lives almost equally distant between two major highways, each of which will lead to the city's airport. The two routes are roughly the same distance (less than 1 mile difference over a 30 mile distance). The computer, however, will only plot out the route that is the shortest. On any given day, however, the best route may not be the one recommended by the system, depending on time of day (traffic is very heavy on the route through downtown), or accidents that create traffic jams on one route or the other. These are things the driver would know (by looking at a watch or hearing traffic reports on the radio), but the system would not. It is almost impossible for the driver to add this knowledge to the system's knowledge, however. One has to either accept the system's route or go it on your own.

Route guidance systems are really best in new towns where drivers would have to be reliant on it. In this case, drivers would never know that there is another route that is almost as short, but which would not be so traffic prone. What they really want to know is 'how long is this route compared to others (1% shorter or 50% shorter)?' 'What major factors differentiate the routes that may impact on selection?' (e.g., which route has a toll bridge on it that is known to back up, which goes through a major metropolitan area that is best avoided at 5 p.m. and which goes around, which goes by the stadium where a big football game is likely to create traffic?). This type of knowledge creates robust decision making. If a traffic accident occurs or road construction is encountered (which the system is not likely to know about), 'what is the best side road to use to go around the problem?'. These questions cannot be answered with a system that simply uses a preset optimization algorithm to determine a 'best route.' With a system that provides the SA required for robust decision making, the driver would see this type of comparison data for alternate routings, allowing human knowledge to be more optimally combined with what the system can do. This allows the driver to select situationally appropriate choices and rapidly adapt to changes in the situation.

In one example of such a system, Endsley and Selcon (1997) replaced a route planning system in a cockpit (that had suffered from many of the problems discussed regarding decision support systems) with one that provides pilots with the relevant knowledge they need to select a best course themselves (in this case information on their visibility to threats). This approach resulted in higher levels of SA and better performance in finding good routes through the hazardous areas (see Figure 6.3).

Route guidance systems are but one example of systems that try to provide 'best' solutions to operators. Many other systems can similarly benefit from system designs that provide SA rather than decisions.

Principle 37: Keep the operator in control and in the loop

Avoiding the problems associated with people's reduced ability to detect and understand automation problems is important in SA-Oriented Design. Catastrophic results can follow even simple automation failures when the operator is out-of-the-loop. In an interesting experiment Pope *et al.* (1994) showed a loss of mental engagement (measured through brain activity) to occur as a result of passive monitoring of automation. Endsley and Kiris (1995a) also showed that the act of passive rather than active decision making was behind out-of-the-loop performance decrements in their study.

A key approach to minimizing the out-of-the-loop effect is to increase involvement and control. Intermediate and lower levels of automation are clearly better than higher levels of automation at keeping SA high (Endsley & Kaber, 1999; Endsley & Kiris, 1995a), as long as workload is not a problem. Ensuring that the operator maintains control over the automation (rather than automation that overrides the human), and devising strategies that incorporate the human decision maker as an active ongoing participant with high levels of task involvement are likely to greatly reduce or minimize this problem.

Principle 38: Avoid the proliferation of automation modes

This is an extension of the more general principle of minimizing system modes and logic branches to reduce complexity (see Chapter 8). The use of automation modes increases system complexity, and thus the ability of operators to develop a good mental model of how the system works in all of its possible modes. The training problem is significantly increased, compounded by the lack of opportunity to experience many modes on a frequent basis, making forgetting a problem with rarely seen modes or situations. More automation modes also make it harder to keep up with which mode the automation is in at the present time in order to develop correct expectations of what it will do next. As designers often want to add modes in order to make the automation seemingly more robust (e.g., it can handle more variations in the situation), bucking this trend will be difficult, but persistence in this quest will pay off greatly in system usability and error reduction. A flexible tool that allows user customization of system functioning to fit the current circumstances is probably better than one that simply proliferates modes.

Principle 39: Make modes and system states salient

A key shortcoming of many automated systems is that they simply don't make the current mode salient. One letter on a busy display may be the only cue a busy pilot has that the system will behave differently than expected because it is in a different mode than he thought. Which mode the universal remote control for an entertainment center is in is usually not displayed at all, thus the control inputs made may affect very different systems than expected. In general, many designs do a poor job of making the current mode salient, leading to errors. Mode status is a key piece of information that can affect how other information is interpreted and what expectations for system behavior the operator generates. The current mode should be made salient to the operator whenever automated systems have modes. In addition, the current state of the system should be salient so that any violations of people's expectations will be readily apparent, allowing them to catch any misunderstandings of mode or mode behavior they may have.

Principle 40: Enforce automation consistency

Consistency in the terminology, information placement, and functionality of the system between modes should be enforced. The 'entropic growth of complexity' noted by Vakil and Hansman (1998) has come about due to the lack of maintenance of a consistent model of how the automation is to work. In some systems, different terms will be used for essentially the same function in different modes ('is it start, begin, or run?'), increasing confusion. Controls for the same basic tasks or functions may be distributed in very different ways on the screen, seemingly arbitrarily. Most damaging, the system may work very differently in different modes, driving up complexity and likely errors. The use of a consistent set of principles for system design that is followed across modes and displays can greatly minimize the errors that can come from automation use. The documentation of those principles for use in later system redesigns or revisions can help insure that this consistency lasts longer than individual designers who may come and go (Vakil & Hansman, 1998).

Principle 41: Avoid advanced queuing of tasks

One aspect of automation was found to be most closely related to out-of-the-loop failures: advanced queuing of tasks. Automated systems that allow the operator to set up in advance a number of different tasks for the automation to perform are most likely to leave that operator slow to realize there is a problem that needs intervention. Interestingly, current flight management systems in cockpits and batch processing systems in manufacturing work under just such an approach. While this might seem to be efficient, when the consequences of failures are high this automation approach should probably be avoided. Instead, approaches that maintain operator involvement in the decisions for each task execution should be considered.

For example, in an aircraft flight management system the pilot typically programs in a route consisting of a number of waypoints in three-dimensional space that the aircraft will follow from origin to destination. The computer will follow this route with no additional input from the pilot. If a waypoint isn't captured for one reason or another, however, the pilot can be out-of-the-loop and slow to respond. Altitude busts (exceeding one's assigned altitude) are a common result of this problem and can lead to significant consequences such as loss of separation between aircraft. As an alternate approach that does not rely on advanced queuing, the pilots could be kept in the loop by the necessity to accept or designate each waypoint along the way. This keeps them engaged during generally low workload periods and reduces out-of-the-loop problems.

Principle 42: Avoid the use of information cueing

Information cueing helps direct operator attention to those areas or pieces of information the system thinks are most important. It may highlight gauges on the display or areas of a cluttered visual scene. As this approach is also prone to create significant problems with attention biasing, however, the hazards are significant (Yeh & Wickens, 2001; Yeh, Wickens, & Seagull, 1999). When the system is wrong or incomplete, people are more likely to miss the information they should be attending to. Unless there is a very low probability of error, this type of cueing should be avoided in favor of approaches that allow people to use their own senses more effectively. For instance, instead of a system that highlights which area of a cluttered display to look at, one that allows people to systematically declutter unwanted information or improve picture clarity to better see what they are looking for would be preferable.

Principle 43: Use methods of decision support that create human/system symbiosis

The success of critiquing systems for improving decision quality and reducing decision-biasing problems (Guerlain *et al.*, 1999) should be further explored. In addition, other ways of combining people and computers to create more effective synergy between the two have been insufficiently studied (in favor of traditional decision support systems that supply the human with advice). Alternate approaches for decision support include:

- Supporting 'what-if' analysis, encouraging people to consider multiple possibilities and performing contingency planning that can help people formulate Level 3 SA,
- Systems that help people consider alternate interpretations of data, helping to avoid representational errors, and
- Systems that directly support SA through calculations of Level 2 SA requirements and Level 3 SA projections.

Principle 44: Provide automation transparency

Finally, given that automation will probably be employed in many systems, providing transparency of its actions (current and future) will greatly reduce the problems leading to automation-induced accidents. For example, the crash of an Airbus 320 at Strasburg in 1992 has been attributed to the possibility that the pilots had a mode awareness error. They believed they were in a 3.3 degree descent, but instead were in a 3300 feet per minute descent, a difference that would not be obvious from the mode display panel which would show 33 in either case (Johnson & Pritchett, 1995). Only the mode indication would be different – VS vs FPA in one area of the display. The difference in its effect on aircraft behavior would be very significant however!

Johnson and Pritchett (1995) found that 10 out of 12 pilots they studied failed to detect this mode error in time to prevent crashing the aircraft in a simulated recreation of this accident. Although many detected a faster than expected rate of descent, they were unable to sort out why and take corrective action soon enough. In a related study, (Vakil, Midkiff, & Hansman, 1996) found a significant reduction in pilot susceptibility to this problem when a vertical situation display (which depicts the projected vertical profile of the flight path) was presented to pilots. This display provides pilots with visibility into the expected behavior of the automation that is needed for understanding automation and interacting with it successfully.

In a different approach that also conforms to this principle, Miller, Pelican, and Goldman (1999) created an interface that allows pilots to see what goals as well as tasks the system is working on. By being able to directly inspect these goals, the pilot can easily correct them (e.g., 'no, we are not trying to land now'). While this may seem like an obvious thing to convey, in most automated systems the operator is left guessing as to why the system is behaving the way it is. Approaches such as these that make system behavior clear to the user will greatly reduce many current problems with automation.

Automation, for better or for worse, is likely to form an integral part of many systems being developed now and in the future. The challenge is in formulating effective automation implementation strategies that provide effective human/machine collaboration, and that keep the operator aware of what the system is doing.

CHAPTER ELEVEN

Designing to Support SA for Multiple and Distributed Operators

11.1 TEAM OPERATIONS

In many systems, people work not just as individuals, but as members of a team. Pilots of commercial airliners, for example, typically operate as a team of two or three pilots serving as captain, first officer, and in some cases, a second officer. Air traffic controllers often work in teams of two or three to manage all the aircraft in a sector. In addition, they are part of a larger team of controllers who pass control of aircraft from one sector to another as it moves across large geographical areas. Aircraft maintenance is also performed by teams of mechanics who work together to perform all of the tasks required to service and repair the aircraft. Not only do each of these groups of individuals comprise a team, but together they form a team of teams (Figure 11.1).

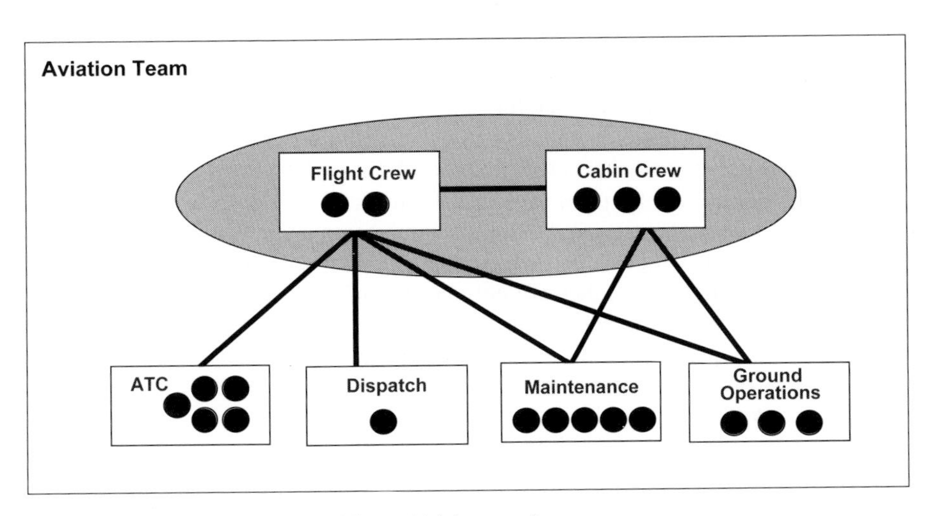

Figure 11.1 A team of teams.

In this aviation example, and in many other systems, it is necessary to design to support the situation awareness of not just individual team members, but also to support the situation awareness of the team as a whole. Understanding how to best accomplish this requires that we understand a few key factors underlying the nature of teams, and also how SA is affected by the presence of team operations.

By examining different factors that teams use to build SA among themselves, we will identify design principles for fostering higher levels of SA within teams.

To begin to understand what is needed for SA within teams, it is first necessary to clearly define what constitutes a team. A team is not just any group of individuals; rather teams have a few defining characteristics. Salas, Dickinson, Converse, and Tannenbaum (1992) define a team as '*a distinguishable set of two or more people who interact dynamically, interdependently, and adaptively toward a common and valued goal/objective/mission, who have each been assigned specific roles or functions to perform, and who have a limited life span of membership.*' This definition brings out several important features of teams.

First, teams have a *common goal*. In the case of the pilots in a cockpit, their common goal is the safe and efficient operation of the aircraft in getting from one point to another. In the case of the broader team of pilots, air traffic controllers, flight attendants, and mechanics, they all share the common goal of safe aircraft operations. As a counter example, the customers in a store each have the individual goal of purchasing items they want, but they really do not share any common goals—they do not form a team. Collocation of individuals is not sufficient (or even necessary) for forming a team.

Second, teams have *specific roles* defined. An aircraft crew has specific duties and goals assigned to the captain, first officer, and flight attendants (although multiple people may be assigned to the same role). In a restaurant, the wait staff, cooks, cleaning crew, and manager each have very clearly defined roles that add up to fulfill the overall team goal of providing food and beverage service to customers for a profit. These roles define the goals for each team member; this is very important for SA, because goals determine the SA requirements for each team member. Referring back to Chapter 5, the goals of the individual drive the determination of SA requirements. By examining the differentiated roles within a team, we can specify just what is required for SA for each member of that team.

Third, the definition of a team says that the roles of the different team members are *interdependent*. The salespeople for the same company may all have a common goal—to sell more widgets for that company—but they do not really meet the definition of a team because they are not interdependent. Each salesperson operates independently to sell more widgets in his or her territory. In a team, the success of each person is dependent on the success of other team members. The success of a pilot in safely operating the aircraft is dependent on the performance of the copilot, air traffic controller, and mechanics, as well as others.

This interdependence between the roles and goals of different team members is key in defining what constitutes team SA. As shown in Figure 11.2, the interdependence between the goals of each team member also implies a certain degree of overlap between the SA requirements of each team member. Within each team, the load is shared in performing a task. In some cases, each individual within a team may perform very similar functions. For example, one person may assist another in performing essentially the same task. In other cases, operations will be divided up so that each team member has very different responsibilities. For example, while the pilots and flight attendants all comprise the crew of an aircraft, their duties are quite different.

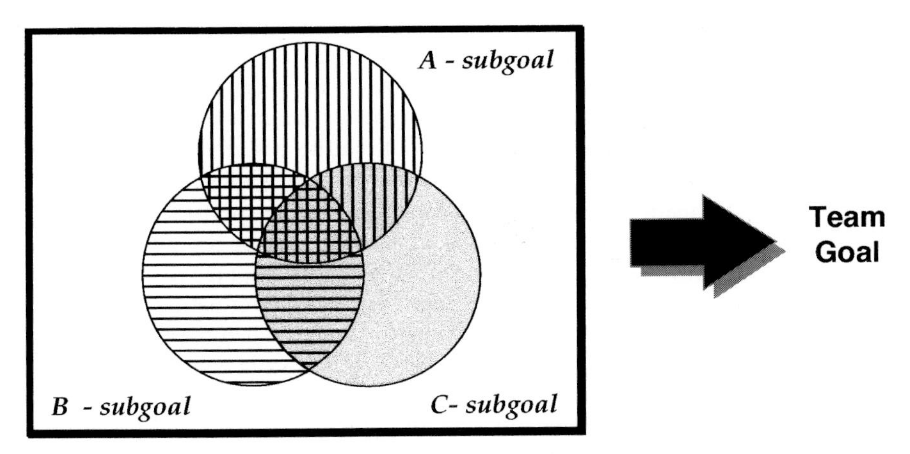

Figure 11.2 Team SA can be determined by examining the goals and SA requirements of all team members (from Endsley & Jones, 1997; 2001). Reprinted with permission from *New trends in cooperative activities: Understanding system dynamics in complex environments*, 2001. Copyright 2001 by the Human Factors and Ergonomics Society. All rights reserved.

If there is a lot of overlap between the goals of two team members (for instance, the captain and first officer in a cockpit), then there will be considerable overlap in the SA requirements they have in common. If there is less overlap in their goals (for instance, the pilot and the flight attendant or the air traffic controller), then there will be a more limited set of SA requirements that they have in common.

It is worth noting that this is true for teams that operate with a fair degree of autonomy, and for those that operate within suprastructures that define clear chains of authority between teams. For instance, military organizations can consist of teams at different levels of the hierarchy (e.g., squads, platoons, companies, battalions, brigades). Yet for each one of these teams, it is possible to define clear goals and SA requirements associated with their goals. SA is needed within teams that are hierarchical as well as those which are more democratically organized.

11.2 SA IN TEAMS

Due to their interdependent nature, in order for teams to be successful they need a high level of *team SA*. The term team SA can lead to certain misconceptions, however. SA, in that it is a cognitive construct, is not something that a team can possess as a whole. There is no 'suprabrain' of a team that can have awareness of anything. Only the individuals within the team can have awareness. Team SA, therefore, must be embodied by the SA of the individual team members.

Team SA is defined as '*the degree to which every team member possesses the SA required for his or her responsibilities*' (Endsley, 1989; 1995c). The success or failure of a team depends on the success or failure of each of its team members. If any one of the team members has poor SA, it can lead to a critical error in performance that can undermine the success of the entire team. For example, on a submarine all crew members must work together for the success and safety of the entire team. If the sonar operator misses a critical cue, the goal of the entire team can be jeopardized when the sub runs aground. If the weapons officer does not understand the status of the torpedoes, other types of failures are possible. In such a team, even a failure of SA by the cook can lead to a serious problem for the entire team if the cook is not aware of a growing fire in the galley.

High levels of team SA are needed for overall team performance. By this definition, each team member needs to have a high level of SA on those factors that are relevant for his or her job. It is not sufficient that someone on the team is aware of certain information if the one who needs that information is not aware. For example, in 1989 a major aircraft accident occurred when the pilots of an Air Ontario flight attempted to take off from an airport in Canada with snow and ice on the wings (Moshansky, 1992). As wing icing is extremely detrimental to the lift capacity of the aircraft, this led directly to a loss of control and crash of the aircraft. In subsequent analysis, it was determined that both the flight attendant and a passenger on board the aircraft were aware of the wing icing, but did not pass that information on to the pilot, whom they assumed was aware of the situation.

In a similar situation, a Boeing 737-400 aircraft crashed in 1989 in Kegworth, UK when the pilots turned off the wrong engine after one of their engines developed a fire (United Kingdom Air Accidents Investigation Branch, 1990). The flight attendants and passengers were aware of which engine was on fire, but did not pass that information forward to the pilots. These examples illustrate that for teams to have high levels of SA, each team member must have the SA he or she needs. It is not sufficient that someone on the team has that information, if the one who needs it does not.

Our goal as designers of systems that include team operations is to ensure that the SA of the overall team is high. To accomplish this, we must ensure that the SA of each individual team member is high. As a significant component of team SA, the degree of *shared SA* between team members must also be high.

11.3 WHAT IS SHARED SA?

In many systems, designers are seeking to support shared or collaborative SA. That is, they want the understanding of the situation by various team members to be the same. In some circles, this is also called developing common ground (Clark & Schaefer, 1989). It is important to point out, however, that we rarely want the SA of two team members to be entirely the same. For instance, should the air traffic controller really have the same picture of the situation as the pilot? Not only would this lead to significant overload, but the comprehension and projection requirements for different team members are often not the same at all.

Shared SA is really dependent not on a complete sharing of awareness between team members, but only on a shared understanding of that subset of information that is necessary for each of their goals. As shown in Figure 11.3, depending on the degree of overlap between the goals of any two team members, only a subset of their SA really needs to be shared. Only this subset is relevant to both team members. Shared SA is defined as '*the degree to which team members have the same SA on shared SA requirements*' (Endsley & Jones, 1997; 2001).

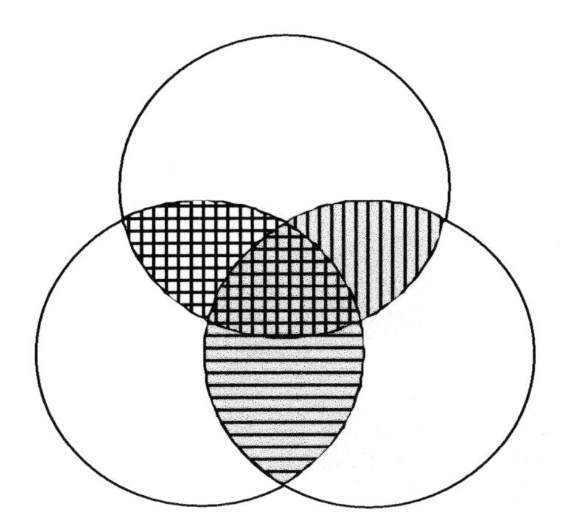

Figure 11.3 The need for shared SA is a function of the overlap in individual goals (from Endsley & Jones, 1997; 2001). Reprinted with permission from *New trends in cooperative activities: Understanding system dynamics in complex environments*, 2001.

The methodologies and background provided in this book provide a useful mechanism for understanding and designing to support shared SA in team operations. By going to the SA requirements analysis developed for each team member (Chapter 5), it is possible to identify exactly which elements are in this overlap (at all three levels of SA) for any pairing of team members. This is the subset of information whose status needs to be shared through training and system design efforts.

There are a number of different possible states of shared SA within a team (Figure 11.4). Two team members can have the same picture of what is happening on a given SA requirement and both be correct. This is clearly the most desired state. For example, both pilots believe the aircraft is at a particular altitude and are correct in that perception.

Or, two team members can have a different picture of what is happening. One may be correct and the other incorrect, or they may both be incorrect in different ways. For example, one pilot may misread a properly functioning altimeter, or may have become distracted and neglected to check the altimeter. Good communications processes and display technologies that support shared SA can be effective at reducing the likelihood of this occurring. Differences in perceptions of information and perspectives on the meaning of that information need to be revealed so that the team can resolve discrepancies.

Alternatively, two team members may have shared SA, but both be incorrect; for example, if two pilots believe they are at a given altitude when the aircraft is at another altitude. This might occur if the pitot tubes are blocked, for instance, causing the altimeters to display incorrect information to the two pilots. Although shared SA is high, team SA would be low in this case. This is a very dangerous state, as it will be less likely that the team will discover that their picture is false as they communicate with each other than in cases where at least one team has the correct SA. In one study, 60% of aircraft incidents were found to occur when both pilots experienced a loss of SA (Jentsch, Barnett, & Bowers, 1997). Very often the team will remain locked into a false picture of the situation until either an accident or incident occurs or some external factor occurs to jar them out of it.

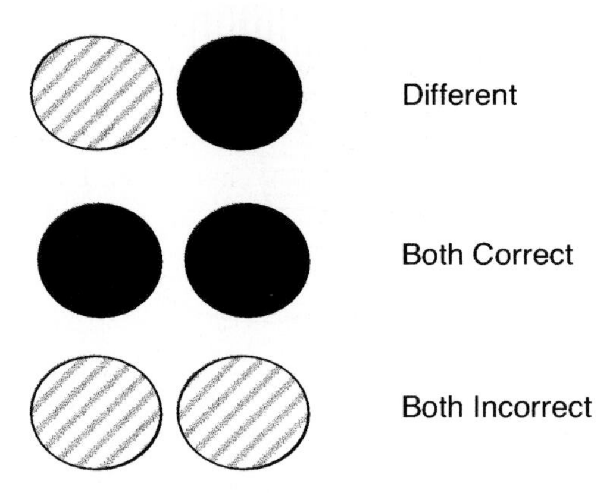

Figure 11.4 Possible states of shared SA (from Endsley & Jones, 1997, 2001). Reprinted with permission from *New trends in cooperative activities: Understanding system dynamics in complex environments*, 2001. Copyright 2001 by the Human Factors and Ergonomics Society. All rights reserved.

Accurate, shared SA is critical for effective team operations. Without it, team coordination can be poor and individual SA needed for good decision making lacking. For example, a significant incident occurred aboard the Russian space station Mir in which the space station lost all power while in orbit, significantly risking the lives of its crew (Burrough, 1998). In this incident, the cosmonauts on board accidentally disconnected the wrong cable during routine maintenance, interrupting power to the central computer and setting the station into a drift. While the crew quickly realized what was happening, they needed the support of ground controllers to help decipher the computer's messages and correct the situation. The ground controller on duty, however, did not understand the situation the cosmonauts were in as voice transmissions between the two were garbled. He continued to treat the situation as routine and let the station drift as he waited for communications on the next orbital pass. This lack of shared SA led to a significant loss of time. No energy saving procedures were put into place and the Mir's batteries became drained so that power to the station was lost.

While one part of the team understood what was happening, the other part of the team did not. Due to the interdependencies involved, this loss of shared SA was critical and life threatening. Effective actions by individuals in these types of team operations often depend on a shared understanding of the situation being developed. It should also be noted that because the individuals on the team are interdependent, the SA of each of the team members often influences the others. The success of a variety of processes and mechanisms that affect this transference of SA between team members therefore becomes of significant concern to trainers and designers of systems to support these teams. It is the goal of the present chapter to discuss the various ways in which this shared SA can be developed and supported through system design.

11.4 CRITICAL FACTORS AFFECTING SA IN TEAMS

Developing methods of creating high levels of shared SA in team operations has become a new design goal. In order to do this we will first discuss the factors that contribute to the development of shared SA in teams, using the model shown in Figure 11.5 (Endsley & Jones, 1997, 2001). This model describes four major factors that are important for the development of good team SA and shared SA in teams.

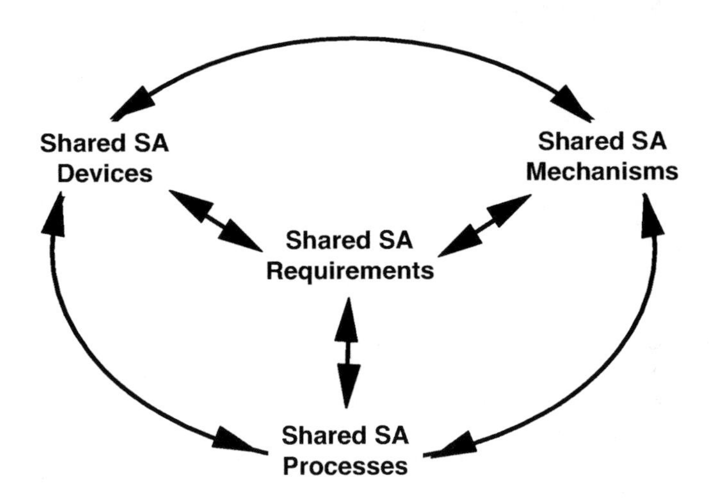

Figure 11.5 Model of team SA.

11.4.1 Shared SA requirements

The elements (at all three levels of SA) that need to be shared between any two team members can be determined from the SA requirements analyses conducted for those team members. The overlap in those requirements specifies the information on which a common understanding is needed. This may include basic information about the system being operated or the environment that affects multiple team members. In addition, how people use and transform that data to form comprehension and projection may also need to be shared. For example,

Table 11.1 shows the aspects of the situation that need to be shared between pilots and air traffic controllers on holding patterns (Endsley, Hansman, & Farley, 1998).

Note that while many aspects are in common (e.g., the aircraft's altitude and current clearance), there are also differences. The air traffic controller (ATC) must keep track of relative information for multiple aircraft approaching an airport in the sector, while the pilot is only concerned about his or her own aircraft. There is also information that only one of the two parties needs to know. Only the controller needs to be aware of the separation between aircraft (horizontal and vertical) and the difference between aircraft arrival rates and the arrival limits for the airport, or the altitudes available for assigning an aircraft to a holding pattern. Conversely, only the pilot needs to be aware of wind direction and speed and their effects on the desired aircraft track. Also, only the pilot knows whether the aircraft has sufficient fuel for continuing in a holding pattern. Only if this fuel sufficiency leads to an emergency state does ATC care to know this information.

Table 11.1 Shared SA requirements between pilots and air traffic controllers: holding.

ATC Only	Both ATC and Aircraft	Aircraft Only
Holding clearance of other aircraft	Holding clearance of given aircraft	
Current track of other aircraft	Current track of given aircraft	
Conformance of other aircraft with ATC clearance	Conformance of given aircraft with ATC clearance	
Relative timing of other aircraft along route	Relative timing of aircraft along route	
Altitudes available	Given aircraft altitude	
		Wind direction and magnitude
Projected clearance timing for all aircraft	Expect further clearance time for given aircraft	
Emergencies affecting safety of flight of all aircraft	Emergencies affecting safety of flight of given aircraft	Impact of hold on safety of flight
Impact of traffic on safety of flight of all aircraft	Impact of traffic on safety of flight of given aircraft	Fuel sufficiency
Aircraft prioritization		Impact of weather on safety of flight
Deviation between aircraft arrival rates and arrival limits		
Relative projected aircraft routes		
Vertical distance between aircraft		
Horizontal distance between aircraft		

While some of the elements that need to be shared qualify as Level 1 SA (e.g., current aircraft altitude), others involve Level 2 or 3 SA. Deviations between current tracks and the assigned holding pattern, for instance, and impact of weather on safety of flight involve comprehension and projection on the part of the controller or pilot. Oftentimes these assessments may be very different by the two individuals (as each has access to different underlying information) and must be actively shared to avoid disparities in SA. A significant task within teams involves the passage of SA (including data, comprehensions, and projections) between team members. Significant team failures occur when needed SA is not transferred successfully.

In addition to the aspects of the situation that are in common to team members, shared SA also often encompasses a consideration of the status of other team members' tasks and their impact on one's own tasks and goals. In many teams, the SA requirements for a particular team member include a consideration of what other team members are doing (e.g., 'What task is he on? Is what she is doing going to affect me? Is what I'm doing going to affect someone else? Has he finished task X yet?'). Each individual has SA requirements that have to do with the system he is operating, and also with the other team members he is interdependent with.

For example, the pilot needs to know when to expect a further clearance from the controller so that she can determine how long she will be holding and whether she has sufficient fuel for that period or needs to seek clearance to an alternate airport. If the controller is not going to meet that expected clearance time, then the pilot must assess what impact that will have. Conversely, if the pilot cannot achieve compliance with a new clearance in a timely manner, that can impact the decisions of the air traffic controller who must make other adjustments to insure aircraft separation.

Shared SA in teams involves knowledge of the status of other team members' tasks *to the degree that they impact on one's own tasks and goals*. Similarly, in well functioning teams, team members often project how their own task status and actions will impact other team members. Aircraft mechanics, for example, need to know what other mechanics are doing on the same aircraft so that they do not take an action (e.g., opening a valve) that might jeopardize the safety of a fellow mechanic working elsewhere on the aircraft (Endsley & Robertson, 1996b).

In teams it is often necessary to project what impact potential changes to a common plan are likely to have on other team members. In military units, when one company is off schedule or makes changes to the common plan, they need to be able to project what impact that will have on other units in order to make appropriate decisions as to the best actions to take. Oftentimes, there is not sufficient time nor the technological capability to verbally communicate; in these cases the projections must be made alone, which has the potential for significantly affecting the functioning of the entire team.

Finally, in teams operating at the highest levels of team SA, team members also project what fellow team members will do. A captain and first officer with a high level of shared SA will be able to project when the other will take certain actions during the approach, for example. Flight attendants also have a set of projections about what the pilots will do that allows them to plan their actions. In many time-critical operations, effective team functioning depends on this type of projection. Xiao, Mackenzie, and Patey (1998), for example, found that operations in a medical trauma unit broke down in cases where team members were not able to anticipate what help would be needed by others.

Shared SA in teams, therefore, involves accurate and timely sharing of system and environmental information that affects both team members. It involves active sharing of the comprehensions and projections of team members (which can be different as each often has information the other does not) on those higher level requirements they share. It also involves understanding the status of other team members (e.g., where they are on tasks, how well they are carrying out their tasks, and whether they are overloaded and need help), and how their progress on those tasks affects one's own goals and tasks. In closely coupled teams, active projection of the actions of other team members may be needed, as well as the ability to assess the impact of one's own actions on others in the team. These shared SA requirements are summarized in Table 11.2. The exact information that needs to be shared is determined through the cognitive task analyses of each team position, and a comparison of the areas of overlap between each team pairing.

A number of problems exist for developing a good common understanding of shared SA requirements in teams. For example, in many teams, team members may not be aware of which information needs to be shared with other team members (Endsley & Robertson, 1996b). In other cases, shared information does not get passed between team members either because they assume the other team member already knows, or because they do not realize that it needs to be passed on. This can often be a problem for Levels 2 and 3 SA, when team members can falsely assume that others will arrive at the same assessments that they do based on the same Level 1 data (Endsley & Robertson, 1996b). There are a number of factors that affect the degree to which teams are successful in achieving shared SA on shared SA requirements: shared SA devices, shared SA mechanisms, and shared SA processes.

Table 11.2 Shared SA requirements in teams (Endsley & Jones, 1997, 2001). Reprinted with permission from *New trends in cooperative activities: Understanding system dynamics in complex environments*, 2001. Copyright 2001 by the Human Factors and Ergonomics Society. All rights reserved.

- Data
  - System
  - Environment
  - Other team members
- Comprehension
  - Status relevant to own goals/ requirements
  - Status relevant to other's goals/requirements
  - Impact of own actions/changes on others
  - Impact of other's actions on self and overall goal
- Projection
  - Actions of team members

11.4.2 Shared SA devices

While many people assume developing shared SA in teams requires verbal communications, in reality there are a number of devices that are available for developing shared SA (Table 11.3). First, teams can communicate verbally and also nonverbally, such as through facial expressions and gestures. People can get a great deal of information about the emotional state of another team member (e.g., overloaded, concerned, asleep), that is important for assessing how well a team member can be expected to perform. Xiao, Mackenzie, and Patey (1998) found that nonverbal communication was very important in medical teams, as did Segal (1994) in studying pilot teams.

Shared displays can also play a critical role in creating shared SA between team members. Different team members may be able to directly view the information that needs to be shared on their displays. For example, datalink initiatives in commercial aviation would pass clearances and other information electronically between air traffic controllers and pilots for visual display to each, off-loading busy radio communication channels. Military units are similarly obtaining displays that will show soldiers key information about the battlefield, communicated across electronic networks. Display of shared information can occur through visual displays (via computers, system instrumentation or paper), auditory displays (e.g., alarms or voice synthesis), or through other senses (e.g., tactile devices).

Table 11.3 Team SA Devices (Endsley & Jones, 1997; 2001) Reprinted with permission from *New trends in cooperative activities: Understanding system dynamics in complex environments*, 2001.

- Communications
  - Verbal
  - Nonverbal
- Shared Displays
  - Visual
  - Audio
  - Other
- Shared Environment

In addition to these devices, many teams also gain shared SA through a shared environment. That is, because they are in the same environment, they do not need to pass information to each other through these other methods. For instance, the driver of a car does not need to tell her passenger that the traffic is heavy. The passenger can obtain this information directly. In a distributed team, however, this shared environment is not present. The driver must tell the person on a cell phone that she cannot talk because she must attend to traffic. In an aircraft, the two pilots build shared SA (at least partially) because they both perceive cues such as the vibration of the aircraft, vestibular perceptions of pitch and roll changes, the sound of the engines, and the feeling of temperature changes. This provides them with a great deal of SA without requiring extra communications or displays.

In different systems and at different times, shared SA may be conveyed through different combinations of these SA devices. In many domains, from medicine to military systems to aviation, engineers are working to develop networks that will transmit and display a great deal of information that can only be communicated verbally today.

Developing effective ways to communicate and display this information remains a central challenge. Bolstad and Endsley (1999b; 2000), for example, provided shared displays to augment verbal communications between team members. They found that certain types of shared displays (those that were tailored to meet the explicit SA requirements of the other team member) could significantly enhance team performance, particularly under high workload conditions. Other types of shared displays (those that repeated all the information of the other team member) had no effect or depressed performance.

It should also be noted that team members may freely trade-off between different information sources. For example, teams have been found to reduce verbal communications when shared displays are available (Bolstad & Endsley, 2000; Endsley, Sollenberger, & Stein, 1999). When a shared environment or verbal communications are not available, a corresponding increase in the need for other information may occur to make up for this deficiency. For example, people communicating across the internet often add *emoticons*—symbols such as smiley faces—to convey emotional content that is present in normal speech and not in e-mail or electronic chat.

In that many new efforts are directed at increasing the capability for electronic and verbal communications between team members and at creating shared displays (often called a *common operating picture* or *visualizations*), it is very important that the effect of these changes on the level and type of information communicated be carefully assessed. For example, it has been found that air traffic controllers gain a lot of information about the pilots they are communicating with, such as experience level, familiarity with an airport, and familiarity with the language, based on the sound of their voices (Midkiff & Hansman, 1992). This information is very useful to controllers, allowing them to rapidly adjust the speed of their communications, their expectations regarding the speed of compliance of the pilot with commands, and the amount of separation they provide with other aircraft in the area (Endsley & Rodgers, 1994). As the types of devices used to provide SA change, great care must be taken to insure that all the information required be provided through some means and that these changes do not accidentally decrease SA in some unanticipated way.

11.4.3 Shared SA mechanisms

Interestingly, teams do not rely solely on these devices for developing shared SA. Some teams also have available internal mechanisms that can greatly aid the development of shared SA. In particular, the presence of shared mental models is believed to greatly enhance the ability of teams to develop the same understanding and projections based on lower level data, without requiring extra communications (Figure 11.6).

Without shared mental models, team members are likely to process information differently, arriving at a different interpretation of what is happening. Considerable communication will be needed to arrive at a common understanding of information and to achieve accurate expectations of what other team members are doing and what they will do in the future.

Shared mental models, however, are believed to exist in some team operations (Orasanu, 1990; Salas, Prince, Baker, & Shrestha, 1995; Stout, Cannon-Bowers, & Salas, 1996). Mental models and schema form the mechanism upon which comprehension and projections are formulated in complex systems. To the degree that two team members have mental models which are similar, the likelihood that they will arrive at the same understanding of the data they process is significantly increased, and the workload associated with doing so significantly decreased. Mosier and Chidester (1991) found that better performing aircrews communicated less often than poorer performing ones, most likely because the use of these shared mental models allowed them to be far more effective on the basis of very limited communications. Bolstad and Endsley (1999b) tested the hypothesis that shared mental models are important by performing a study in which they provided half their teams with shared mental models (through information and training), and half without. They found that the teams with shared mental models performed significantly better on a team task than those without.

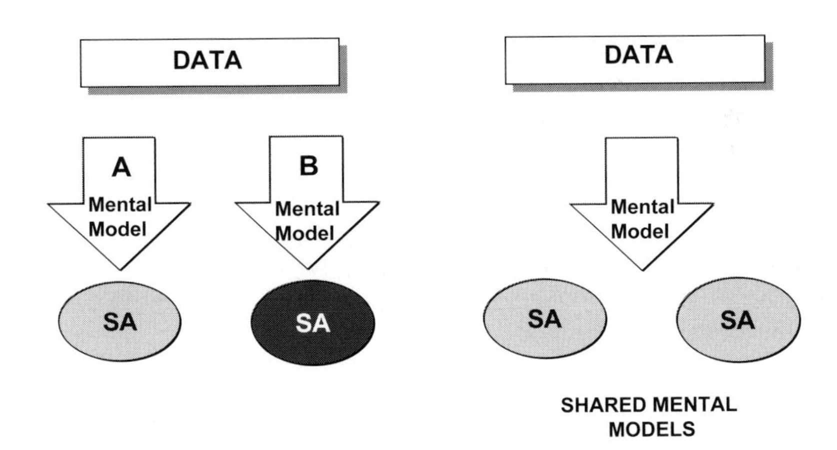

Figure 11.6 Shared mental models provide effective mechanisms for building shared SA (from Endsley & Jones, 1997; 2001). Reprinted with permission from *New trends in cooperative activities: Understanding system dynamics in complex environments*, 2001. Copyright 2001 by the Human Factors and Ergonomics Society. All rights reserved.

Shared mental models can be created through a number of ways. First, they can be developed through common training experiences, such as cross training on each others jobs or joint training exercises that incorporate the entire team. Aircrew, for example, receive standardized training that builds a strong mental model of what other crew members are supposed to do in different situations. Air traffic controllers are encouraged to ride in the jump seat of cockpits to develop a better mental model of what goes on in the cockpit in order to help them communicate with pilots more effectively. Pilots are also encouraged to tour air traffic control facilities to develop reciprocal models.

Second, common experiences are very helpful for building shared mental models. Many teams that have worked together in the past are able to develop a sufficient understanding of each other as individuals, and of the roles and functions of other team members, which helps them communicate effectively, anticipate each others' actions, and perform well as a team. Teams that have not worked together may be at a disadvantage. Some 44% of accidents in commercial aviation have been found to occur on the first leg of a trip with crews that are newly paired (i.e., where the captain and first officer have not flown together before), and 73% occur on the first day of the pairing (National Transportation Safety Board, 1994).

Third, teams can use direct communications to help build a shared mental model in advance of working together. Jones (1997), for example, found that pilots rely on a few key pieces of information about other pilots to form a mental model of them, such as hours of experience in crew position and aircraft type, recent flight experience, and recent experience with particular airports. Information such as this can be provided to team members through databases or verbal communications to help build shared mental models when team members are new to one another. This is a common challenge in many team situations.

Shared mental models are believed to provide several key advantages (Orasanu & Salas, 1993): (1) they can provide important interaction patterns and standardized speech and communications allowing crew members to interact in predictable ways, (2) create an understanding of who has what information, and (3) provide a context for ensuring group resources are directed at common goals. An open question exists, however, as to the degree to which mental models *should be* shared (Duffy, 1993). That is, do mental models need to be identical to be effective (if this is even possible)? Perhaps only certain aspects of the team members' mental models need to be correlated to produce the desired results, just as only some aspects of their situation awareness need to be shared for effective team operations. As mental models are notoriously difficult to measure, these questions remain largely unanswered.

11.4.4 Shared SA processes

Finally, teams are very reliant on good processes for developing shared SA. Orasanu and Salas (1993) reviewed many studies on effective and ineffective teams, which provides a few generalizations on the types of team processes that seem to be successful for building shared SA in teams. Effective teams tend to engage in contingency planning—a process associated with good individual SA—that may also help teams by building good projections of what to expect from other team members. Effective teams also have leaders who create an environment that fosters sharing of information, and who explicitly communicate goals, plans, strategies, and intentions, consider more options, provide more explanations and give more warnings or predictions. These factors are all critical for insuring that the SA of team members are built on the same foundation and expectations.

Effective teams also work to develop a shared understanding of problems before looking for solutions. Citera *et al.* (1995) found this to be particularly important with teams composed of people with very diverse backgrounds, such as is found on design teams. Klein, Zsambok, and Thordsen (1993) also found that effective teams actively encourage the expression of different opinions and engage in a process of convergence to form a common assessment, keep track of team progress and manage time, and work to build a shared understanding of roles and functions of team members, and team goals and plans.

Taylor, Endsley, and Henderson (1996) studied the SA processes of teams in responding to a task in which they were required to adapt to changes in mission goals and plan expectations. They found that effective teams engage in a process of actively checking against each other at each step, coordinating to get information from each other, prioritizing as a group to deal with possible contingencies, and show a group norm of questioning each other (in a positive way) to insure their picture of the situation is accurate. These teams are able to develop the best picture of the situation and establish effective plans in advance of possible problems, even when they do not know the problems actually exist. These processes serve them well when unexpected events occur.

Conversely, Orasanu and Salas (1993) found that teams who perform poorly often accept an incorrect situation model. Group processes that tend to encourage this poor outcome include a reluctance to question a group consensus or powerful group leader. Similarly, there is often a reluctance by individuals to offer novel information that is in disagreement with that held by the group, often in an effort to maintain group cohesion. There often exists the presence of false assumptions that others share one's own opinions, that one knows the goals of others, or that someone is the only one with a different opinion, with pressure to conform acting to repress the willingness of people to check those assumptions. Relevant information offered by team members of lower status is also often rejected by those of higher status. All of these factors directly act to reduce the likelihood of relevant information being communicated between team members in order to create an accurate shared picture of the situation.

Duffy (1993) found three major types of errors in teams: (1) misinterpretation of information on the basis of supposedly shared mental models, (2) acceptance of poor situation models due to social norms, and (3) the presence of poor organizational processes that negatively impact information acquisition. In Taylor, Endsley, and Henderson's study (1996), poorly performing teams exhibited a group norm of not volunteering pertinent information. They failed to prioritize their tasks as a group, with different members pursuing different goals in an uncoordinated fashion. They easily lost track of the main goal of the group. The poorly performing teams relied on their expectations without checking them. This left them unprepared to deal with expectations that proved false. They also tended toward an 'SA blackhole' phenomenon in which one team member with an incorrect picture of the situation would lead others into his or her false picture.

There has been extensive research on why some teams perform better than others that is beyond the scope of this book. Suffice it to say that the processes and norms that teams use can have a significant effect on the quality of shared SA and team SA, even when they have high quality SA devices and mechanisms available to them. Training programs such as Crew Resource Management (CRM) programs developed by Helmreich and his colleagues (Helmreich, Foushee, Benson, & Russini, 1986) have been developed and widely implemented to improve these processes in aviation. CRM has been expanded to similar programs to improve team processes in other areas, such as aircraft maintenance (Taylor, Robertson, Peck, & Stelly, 1993) and medical operating rooms (Bringelson & Pettitt, 1995).

While CRM programs have focused on improving team processes and group norms in general, more recent programs have built on this work to help teams develop better SA. Endsley and Robertson (2000) developed a training program for aircraft mechanics that focuses on developing shared mental models within and across teams, passing key information associated with the higher levels of SA, providing feedback to promote the development of accurate mental models, and mechanisms for dealing with SA challenges. In a different program, Robinson (2000) developed a training program for pilots that teaches the levels of situation awareness and a means of detecting and correcting problems in advance. Both of these programs seek to augment traditional CRM programs that teach teams to communicate more effectively, with information on *what* to communicate to develop good team SA.

11.4.5 Interrelationship between factors

Each of these factors—devices, mechanisms, and processes—act to help build shared SA. Teams can rely on any or a combination of these factors to develop SA, sometimes trading off dependence on one factor for another. For example, Bolstad and Endsley (1999b) found that teams could perform well when provided with either shared displays or shared mental models, but did very poorly when provided with neither. They found that the provision of shared displays was actually very effective at helping teams build up shared mental models, providing good performance even after the shared displays were taken away. Others have found that shared mental models may be used in lieu of explicit verbal communications (Mosier & Chidester, 1991). Overall, as we discuss ways to design systems to enhance the SA of teams, remember that these trade-offs can exist. Care needs to be taken to insure that shifts in the devices, mechanisms, or processes teams use to build SA are detected when evaluating the impact of new design concepts.

11.5 SA IN DISTRIBUTED TEAMS

While we often think of teams as working closely together in the same room, in many arenas teams that work together are increasingly becoming distributed. Cellular telephone networks and the Internet have allowed businesses to distribute their workforce across large geographical areas. Telecommuting has become a common form of business operation, providing flexibility for workers and benefits for companies, but creating new issues in supporting the cooperative work of distributed teams. In many cases, people are distributed not only geographically, but also temporally, with workers in one country taking over software development or customer service functions when workers in other countries are asleep, thus providing around-the-clock operations.

Military operations have always involved operations by multiple units that are geographically distributed. As communications were often very limited, coordination of these operations was tightly controlled by a centralized plan that specified in advance what each unit would do and the areas they could operate in so that activities could be coordinated and unit actions deconflicted. As a downside, however, planning for such operations was typically lengthy and the ability for forces to make effective wholesale changes due to unforeseen enemy actions very limited. Today, military operations are increasingly involving the use of electronic information networks to support real-time planning and coordination, greatly shortening this cycle, and allowing even greater dispersion within teams that traditionally have stayed within eyesight of each other.

Operations in space pose even greater challenges in distributed operations. Communication lags between earth and space stations in low-earth orbit are between 2 and 8 seconds long. A 7 to 20-minute lag will be present for communications each way with astronauts on Mars missions. Yet those working in space, on Mars, and at ground stations on Earth form a team with close coordination and cooperation often being necessary.

Each of these examples show teams who must work in a distributed setting. They may be distributed spatially—in different rooms, different buildings, or different cities. Or they may be distributed temporally, working different shifts, in different time zones or at the same time, or subject to considerable communications delays, as is the case with operations in space. In other situations, while teams may be fairly closely located, they can be blocked from each others' view by the presence of an obstacle. Mechanics working on the same aircraft are a distributed team in the sense that they cannot see each other while working in different areas of the same aircraft.

If the distributed individuals are performing fairly independent tasks, this is not a problem. However, in today's systems we often seek to provide the same levels of team integration for distributed teams as we do for teams that are collocated. In all of these cases, the need for shared SA in these teams is just as great as when they were not distributed. Their SA requirements are still a function of the degree of overlap in their goals (Figure 11.7).

SA for distributed teams is defined as '*SA in teams in which members are separated by distance, time and/or obstacles*' (Endsley & Jones, 2001). The SA requirements are the same, yet the devices available to support the transmission of this information can be quite restricted. In general they will not have available the advantages of a common environment or nonverbal communications. This has traditionally placed the burden of information transmission on verbal communications channels, which have often become overloaded, or on electronic communications to boost the sharing of pertinent information through shared displays. Increasingly, electronic collaboration tools such as electronic chat, file sharing, e-mail, and even Web-based video conferencing are used to fill some of this void. While each of these approaches has certain advantages in supporting team operations, they also have limitations in terms of the types of team processes and information flows they can support in distributed operations (Bolstad & Endsley, 2002). The information that is lost by not sharing a common environment and more restricted nonverbal channels must be compensated for by a greater load on the remaining available SA devices.

As a compounding factor, many distributed teams may have problems with developing common mental models. In that their opportunities for interaction are often more limited, they may have insufficient opportunity in many cases to develop the shared mental models with other team members that will allow them to function effectively in the face of lower information flow. While this may indicate the need for more attention to cross training and team building exercises for distributed teams, it also indicates that display designers may need to pay more attention to supporting comprehension and projection of information on shared displays. Without good shared mental models, it is likely that these aspects of shared SA are likely to become misaligned. More direct display of this information can help to overcome this problem. Designers need to study both the formal and informal processes used for information flow in teams and insure that new methods of information transfer are made available to compensate for those that are lost to distributed teams.

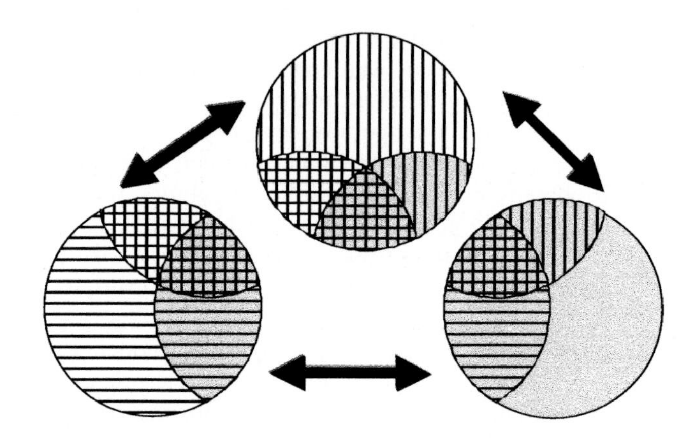

Figure 11.7 SA in distributed teams (from Endsley and Jones, 2001). Reprinted with permission from *New trends in cooperative activities: Understanding system dynamics in complex environments*, 2001.

11.6 SA BREAKDOWNS IN TEAMS

An examination of the ways in which SA can break down in teams can be illustrative for demonstrating the way in which the factors in this model of SA bear on the process, and the features that need to be supported to create high levels of shared SA through system design efforts. The following is a case study, showing the traps for team SA that played a role in the crash of an American Airlines Boeing-757 in Cali, Colombia in December of 1995, that killed all aboard (Aeronautica Civil of the Republic of Colombia, 1996).

The flight to Cali from Miami was initially uneventful until entering Colombia. The first officer was flying the aircraft while the captain was communicating with air traffic control and managing the flight management system (FMS). Upon entering Cali airspace, which was not equipped with air traffic control radar, the following radio transmissions occurred.

ATC: '…is cleared to Cali VOR, descend and maintain 15,000 feet... report, uh, Tulua VOR.'

Cockpit: 'Okay, understood. Cleared direct to Cali VOR. Report Tulua and altitude one, five... is that all correct sir?'

ATC: 'Affirmative.'

In this set of transmissions, the different mental models possessed by the air crew and the air traffic controller led to different interpretations of the clearance. The air traffic controller believed that the aircraft was proceeding on a published route through a navigational point at Tulua towards the airport at Cali. The cockpit crew believed they were on a direct route from their present position to Cali, merely reporting the Tulua fix as they went by it. The word 'direct' to a flight crew means that they can enter a command into their FMS that will navigate the aircraft directly to that point, rather than along a set route.

Although the controller did not initially say 'direct' in his first transmission, he did say 'affirmative' to their readback of the clearance. The air traffic controller, while trained in English as needed for standard cockpit communications, was not fully conversant in English, as it was a second language. It is possible that he did not have the same understanding of the crew's use of the word 'direct' in this context. Indeed, if one is familiar with the area, as the air traffic controller was, then going directly to Cali would make no sense as Tulua is the beginning of the airport approach that the aircraft must be on. The flight crew, however, did not have this mental model of the airport. They were accustomed to getting direct routings in en route airspace. Cleared to Cali VOR sounded to them like a direct routing.

This simple set of transmissions and misunderstandings set the stage for a series of problems that the crew then encountered. When they entered 'direct to Cali' into their FMS, the Tulua fix was dropped from their displays. The air traffic controller then cleared the crew for a different runway than that which they had initially prepared for. This runway used the Tulua VOR as the initial approach fix for the Rozo 1 arrival. The crew could not find Tulua on their FMS, however. Because of their position and altitude, the crew had little time available to descend as needed and perform the steps necessary to prepare to execute the new approach. The crew then elected to begin the approach at the only point on the approach they could find which was Rozo.

Cockpit: 'Can AA 965 go direct to Rozo and then do the Rozo arrival?'

ATC: 'Affirmative. Take Rozo 1 and runway 19, the wind is calm.'

The Cockpit crew then entered 'direct R' into the FMS (sending them off course due to a programming error). In retrospect this seemed like a very poor decision in that Rozo was less than three miles from the runway. To the crew, however, who had little time to gain a sufficient understanding of the new approach and runway, it seemed the best solution. The shared understanding between the aircraft crew and ATC was very low at this point. They continued to talk past each other. ATC still believed the aircraft was on the Rozo 1 arrival, which begins at Tulua. The pilots still believed they were on a direct clearance to the point designated as Rozo.

It is worth noting that in this situation the controller had no shared displays (e.g., radar) showing the aircraft's location or progress toward the airport (radar coverage being poor in many mountainous areas). He was forced to rely strictly on verbal communications to form a picture of what the aircraft was doing. The pilots had very little information about the area they were flying into, terrain information not being incorporated into the FMS displays, and different waypoints not always being shown on the display. Reliant on the navigation display provided by the FMS, paper maps are rarely used en route. Furthermore, the pilots' paper maps used very different nomenclature than the FMS, making integration of the two information sources difficult.

Although there were many causal factors associated with this accident (problems with the automation interface chief among them), the lack of shared SA between the controller and pilots in this accident displays a lost opportunity for accident prevention. Had the controller understood what the pilots were doing he might have been able to intervene or call their attention to their predicament as they crossed out of published airways into areas that had mountainous terrain. As it was, the controller continued to wait for the aircraft to report passing Tulua. When it did not, he asked where they were.

ATC: 'American 965 distance now?'
Cockpit: 'Distance from Cali is 38.'

The difference in the controller's model of what the aircraft was doing led him to miss this clue that there was a problem. Although Tulua was 40 miles from Cali, ATC interpreted the report of 38 miles away as 'they forgot to report passing Tulua along their way.' In reality they never passed Tulua, but were on a completely different path. A minute and a half later, the cockpit reported:

'We're 37 DME at 10 000 feet.'

Again, this was another missed cue. If on direct course to the airport, far more than 1 mile would be covered in one and half minutes. This could only be accomplished if the aircraft was off-course and traversing a path that was orthogonal to the one they were supposed to be on. Shortly afterward, the aircraft crashed into a mountain top as they continued their descent after bringing the aircraft back toward the direct path they thought they were supposed to be on.

This accident illustrates how simple, yet subtle the issues contributing to failures in team SA can be. Simple misinterpretations of the same word can lead to very different pictures of the situation, particularly on multinational teams, or those from very different backgrounds (e.g., mixed teams from marketing and engineering). Different mental models can lead to completely different comprehension and projection of information that is communicated.

In this example, there was little support present for creating good team SA. Very few shared devices were present, loading the verbal communications channel, something that is not optimal in situations where not all participants are fully comfortable in the language spoken. Furthermore, many teams such as this one are formed *ad hoc*; there is no opportunity for cross training or process development.

Often, different teams are not aware of what information needs to be passed to other team members. One team member does not know how much (or how little) the other team member already knows. The controller had little or no knowledge of what the cockpit displays showed to the pilots, or their lack of knowledge about the terrain surrounding the airport.

In these teams, often higher level comprehension or projection is not directly communicated, but only low level data. For example, although the controller was worried that the pilots had not reported passing Tulua yet, he never conveyed this higher level assessment of the situation. Instead, he merely asked for a position report which they supplied, never understanding there was a potential problem. As low level data can lead to such different interpretations by different team members, it is often not sufficient for leading teams to discover these serious disparities in situation understanding.

This example is not out of the ordinary. These same problems leading to poor team SA have been found in other domains as well (e.g., Endsley and Robertson, 2000). In that training opportunities for developing team processes and shared mental models in distributed and *ad hoc* teams are often limited, a higher onus is placed on designing systems that will overcome the problems illustrated here to better support team SA.

11.7 DESIGN PRINCIPLES FOR SUPPORTING TEAM OPERATIONS

Principle 45: Build a common picture to support team operations

Considerable lip service is being provided on the need for a *common operating picture* (COP) to support team operations. How to go about building this common picture, however, is not intuitively obvious. To a certain degree the problem is one of improving the bandwidth and accessibility of electronic networks, particularly for distributed teams. It is very difficult to accomplish the real-time sharing of needed information without such systems. These technological breakthroughs are not sufficient for creating shared situation awareness, however. This also requires that the information shared across these networks be carefully specified to include that information which is truly needed by each team member, and that information be presented in a manner that is compatible with each team member's needs. This is sometimes referred to as a *common relevant operating picture* (CROP).

In most contexts, the body of available data will need to be processed and interpreted slightly differently by different individuals, each of whom has varied and dynamically changing but interrelated information needs, and properly understood by each within the context of a joint mission (for example the pilot, copilot, and air traffic control). Creating information from data is complicated by the fact that, like beauty, what is truly 'information' is largely in the eyes of the beholder. To support the information needs of all the parties on the team and to insure that they are all properly coordinated and 'reading from the same page' is the critical task facing us.

A few key tenets are relevant to the drive to develop the common operating picture. First, one size does not fit all. Although it is important that a shared understanding be developed of that information that different team members share, the ways in which that information is displayed to and used by different team members must differ, based on the unique goals and higher level SA requirements of each team member. This analysis and subsequent system design must be based on an understanding of the unique SA requirements of each team member.

Second, care must be taken to integrate the information sources from which different team members draw their information, particularly when different sources supply similar pieces of information. For example, in aviation, air traffic controllers typically view a picture of where various aircraft in their sector are, based on a network of ground-based radar systems. In the 1980s, new technology called the Traffic Collision Avoidance System (TCAS) became available for aircraft that presents a picture to pilots of other aircraft near to them, based on sensors on board each aircraft. TCAS is meant to provide pilots with aircraft traffic information and provides a last line of defense against potential midair collisions.

These two information sources were never integrated, however. Air traffic controllers have no way of knowing which aircraft a particular pilot may or may not be aware of, or what commands the TCAS may give the pilot for last minute maneuvers to avoid other aircraft. In some cases the TCAS and the controller can provide conflicting guidance to the pilots of aircraft. TCAS can tell a pilot to climb or descend without the bigger picture of other aircraft in the area that will become in conflict in the near future based on these maneuvers causing yet additional maneuvers to be needed.

Controllers, who generally have a bigger picture, may issue very different clearances to separate traffic. A recent midair collision over Germany between a cargo aircraft and a Russian aircraft loaded with passengers was attributed to the fact that the TCAS provided commands to the pilots that conflicted with those provided by ATC (Landler, 2002). One pilot heeded the commands of the controller and the other heeded the command of the TCAS, leading them to maneuver into, instead of away from, each other.

While the exact picture provided to each team member needs to be tailored based on his or her individual needs, some commonality behind the underlying data needs to be provided to insure that such disagreements do not occur.

Principle 46: Avoid display overload in shared displays

While some teams have been forced to try to develop a common picture through voice communications links in the past, the limited bandwidth of these systems has seriously restricted the amount of information that can be passed between team members. Audio overload can be a significant limiter of situation awareness in these systems.

In systems where there is much overlap in SA requirements between team members, visual displays will generally be required to aid in transmitting and visualizing complex and spatial information. In many domains, very large projected displays or data walls are being created to provide a common picture to multiple team members. Simply repeating other team member's displays by these mechanisms does not create shared SA, however. Because other team members' displays are integrated and formatted to meet their needs, the ability to view those displays has been found to overload fellow team members more than help them (Bolstad & Endsley, 1999b). It takes extra time to sort through all the extraneous information to find the parts that are needed by a given team member. Particularly in time-critical tasks, this extra workload and distraction can depress performance rather than aid it.

In teams where the SA requirements of team members are differentiated (i.e., where they have differing roles on the team), shared SA must be created by carefully abstracting out the information that needs to be passed between team members and integrating just that information with the other team members' displays. This approach to creating a common relevant operating picture has been shown to lead to significantly improved team performance without creating workload burdens (Bolstad & Endsley, 2000).

Principle 47: Provide flexibility to support shared SA across functions

The development of a set of shared SA requirements (at all three levels) should direct the selection of information to be shared and the way that it should be presented to support the SA requirements of each team member. Just because the displays will be based on common information does not mean that the information must be displayed in identical ways to every team member. The perspective and information presented to each team member needs to be tailored to the individual's requirements, even though the displays may be created based on a common database.

Several factors need to be considered in this approach. First, in distributed teams, each team member can have different *physical vantage points*. Soldiers on the battlefield, for instance, may look at the same piece of terrain from different locations. Viewing of terrain databases, therefore, may need to be shifted to match the physical vantage point of those using it. Similarly, databases showing mechanical parts can be rotated and shifted to match the physical orientation of the mechanic working on a particular component.

Second, each team member may have different *goal orientations or comparative bases* for the information presented. This is a function of their different SA Level 2 requirements. For example, while many different team members in a military organization may want to know about the surrounding terrain and weather, each has very different higher level assessments to make. The logistics officer wants to know what type of vehicles the terrain can support and how fast the vehicles will be able to travel, or the best area for setting up a supply depot. An operations officer will want to know the effect of the terrain on visibility of his own and enemy troops and on communications capabilities, as well as on rate of movement of troops. The company engineers will want to know the effect of the terrain on potential approach or exit areas they may need to prepare, or on potential barricades or obstacles they may need to create a breech.

While the base information (surrounding terrain) is the same, the relevant comparators that are needed for creating situation understanding are very different. The information needs to be integrated and interpreted differently, depending on the different goals of these different positions. The flexibility required to support comparative shifts needed for different team members requires that information comparators on the display be easily modifiable or tailored to the individual team member. For example, the aspects of the terrain (roads, water features, soil consistency, elevation reliefs or contour lines, etc.), and other situation factors (e.g., troop location, supply vehicles, intended courses of action) being displayed can be controlled by filters that each team member can turn on or off on their own displays as needed for particular assessments. Alternatively, where Level 2 SA information is shown directly on the display, this information can be tailored to the particular position.

Finally, different *semantics* or terminology must often be supported for different team members. Different branches of the military, for example, can have completely different meanings associated with the term 'secure.' To the Navy, securing a building means to turn out the lights and lock the door. To the Army, securing a building means to occupy it and forbid entry to those without a pass. To the Marines, securing a building means to assault it, capture the building, fortify it, and if necessary to call for an air strike. To the Air Force, securing a building may mean merely negotiating a lease. While occasionally terminology differences may be merely humorous, in some situations they can lead to significant miscommunications and misunderstandings that undermine shared SA. Multinational peace-keeping operations and multinational corporations can have particular difficulties in bridging these types of differences in terminology usage. Terminology for a particular team member's displays should be consistent with that used in their particular domain, and bridges made to help understand how others need to be communicated with.

Principle 48: Support transmission of different comprehensions and projections across teams

Although individual team members' displays need to be tailored to their own information needs, they also need to be able to communicate effectively with other team members (each with their own orientation, vantage point and semantics). The solution to this problem lies in creating displays that are flexible. These shared displays should allow physical shifts (viewing the situation from different angles or vantage points), comparative shifts (viewing the information in relation to different goal states and reference information), and with different information filters, allowing different sets of information to be viewed as relevant to different team members or subgoals. Thus while a person might normally want to see the displays tailored to one's own vantage point and goals, the ability to provide a 'quick look' or shift to other team members' vantage points or Level 2 assessments can help to build the shared SA that is needed within the team. In particular, these shifts should help address key issues associated with shared SA including: 'What task is he on?, Is what she is doing going to affect me?, Is what I'm doing going to affect her?, Has he finished task X yet?'. They should also aid in transmitting to team members the ways in which other team members are assessing the situation so that potential disparities can be detected (e.g., two team members planning to use the same resources for different purposes).

This type of flexibility will do much to aid teams in making cross team assessments of information. It will allow them to be effective at achieving a high level of team SA so as to meet their own goals and in obtaining the shared SA across team positions that is necessary to effectively coordinate to meet overall team objectives.

Principle 49: Limit nonstandardization of display coding techniques

Too much individual tailoring of displays can lead to a disruption in team performance, however. If individual team members can modify the color coding or symbology used to represent objects on a display, significant misunderstandings can occur between team members. In many team operations, multiple people within a team will view the same display. Various doctors and nurses in an operating room, for example, commonly share the same displays. In a command and control center there may be more than one person responsible for logistics, with different team members working together at one time, or across several shifts. The same is true for air traffic control and power plant operations.

If one team member significantly changes which information is being displayed or the way in which it is displayed, other team members can misinterpret what they are looking at. Yet, in many systems, programmers who aren't sure how to present information have opted to 'let the user decide,' providing palettes of symbols and colors that can be applied at will by the user. This nonstandardization can be quite dangerous, however, and completely undermines shared SA in these systems. Different team members can easily misinterpret the information they are viewing. In addition, in many systems users communicate in terms of their displays (e.g., 'I've got 2 blue diamonds showing'). If symbology is easily changed by different team members, such communications will be easily fraught with error.

While Principles 47 and 48 promote the tailoring of displays for individual team members' needs, this principle might seem to be at odds. How does one both support tailoring and flexibility and limit nonstandardization? It is possible to meet both goals. The ability to tailor *which* information is shown to different team members cannot be confused with the desire to standardize *how* any given part of it is shown. In these situations take care that the coding used for a piece of information on one team member's display is not used to mean something different on another's. Comparative or vantage point shifts should be clearly indicated as such, so that misunderstandings of what someone is viewing are avoided.

Principle 50: Support transmission of SA within positions by making status of elements and states overt

Finally, in a similar vein, shared SA needs to be supported across team members sharing the same position as well as between positions. In that different people on the team may leave the room periodically, be distracted by other tasks, or work on different shifts, overt means should be used for communicating key information about the status of ongoing tasks and the state of what is happening in the situation.

In many systems this information is not overt, but rather is implicit. For example, who has been contacted about a problem or what steps have been taken are often only in the memory of the person who did them. Errors can occur if this information is not correctly and completely passed to other team members. For example, a nurse who is interrupted when giving medications may not remember to pass on to the nurse on the next shift which patients have not received all their medication. A doctor entering the room may not be aware that another decided to turn off certain monitors that were malfunctioning and be caught by surprise if she expected that an alarm would sound if there were a problem.

In some systems, people have attempted to create markers to help communicate task status information to each other. For example, air traffic controllers have historically used the positioning and tilting of flight strips as well as markings on them to communicate actions taken, plans, and expectations to themselves and others. As many of these systems become increasingly computerized, it becomes incumbent on designers to provide these types of communication aids directly within the system. It should be immediately obvious on a display if certain information has been removed (e.g., sensors turned off or information filtered out), so that a false sense of the picture is not conveyed to others. If something is not being shown that typically is, this needs to be clearly indicated (see principle 9).

The status of ongoing tasks and plans should also be displayed clearly to effect smooth transitions across shifts and team members. For example, in the case study in Chapter 6, a matrix of which organization had been contacted and which permissions had been gained was displayed in a matrix to support communications between team members in the airway facilities maintenance system (see Plate VII). Electronic flight strips that are replacing paper flight strips should allow controllers to manipulate them to show ordering information, expected traffic flows and those aircraft which need attention, just as controllers do with paper flight strips. In some systems aids to allow input of higher level assessments by human operators may be needed (e.g., this patient is experiencing problems and needs to be checked soon), to help foster this communication across shifts and multiple workers.

Supporting the SA required for team operations first requires that we design systems to support the SA of each individual team member. On top of this, supporting SA in teams requires an extra consideration of the design features that are needed to insure that data, comprehension, and projections can be quickly and accurately communicated across team members, as needed for their jobs, without creating overload. Following the design principles described here should avoid many of the pitfalls involved in this process and guide designers towards the additional considerations that are necessary for building high levels of shared SA in teams.

Part Three:
Completing the Design Cycle

CHAPTER TWELVE

Evaluating Design Concepts for SA

Many human factors design guidelines are fairly broad, allowing a wide variety of latitude in their implementation. The SA design principles presented here are probably no exception. In addition, as new technologies develop, solid research on the best way to design their features to enhance SA and human performance will generally lag significantly. These factors lead to a design solution that is often not fully deterministic. Many different design concepts may be considered and designers will need to be able to sort out which are better and which may lead to problems in actual use. Questions may remain in the designer's mind as to the best way to do something. Designers often may be surprised to find that certain design features do not work as well as anticipated. The objective evaluation of system design features, therefore, forms the third major building block of SA-Oriented Design.

Each design should be empirically tested to identify any unforeseen issues that can negatively impact operator SA, and to allow the relative benefits of different design options to be considered. Ideally, this evaluation will occur early in the design process using a prototype of the design, thereby allowing adjustments to be made to the design without unduly affecting cost and production schedules (see Chapter 4). In order to effectively evaluate a design's ability to support SA, appropriate measures must be employed to assess the operator's level of SA when interacting with the system. The ability to measure SA is fundamental to being able to positively affect it through the system design.

Because SA is an internalized mental construct, creating measures to adequately assess and describe it is not an easy task. Metrics of SA generally approach the issue either by inferring SA from other constructs that are easier to assess, or by attempting to obtain a direct assessment of the operator's SA. Four classes of approaches are illustrated in Figure 12.1—process measures, direct measures, behavioral and performance measures. (Detailed information on SA measurement approaches can be found in Endsley and Garland (2000).)

As shown in the figure, the measurement approaches attempt to either directly measure SA, or to infer it from observable processes, behaviors, or performance outcomes. These inferences are not exact, however. In that a number of factors act as moderators between the processes people employ and the level of situation awareness that results (e.g., the strategies they employ and their knowledge and abilities for interpreting information), there is not a complete mapping between behaviors and outcomes. Likewise, many factors can influence the quality of decisions or task execution, independently of SA. Therefore these are only indirect indications of SA and, in fact, may not give a very complete picture of SA.

In general, we believe that direct and objective measurement of SA is the best way to approach the evaluation of a system design (in addition to assessments of

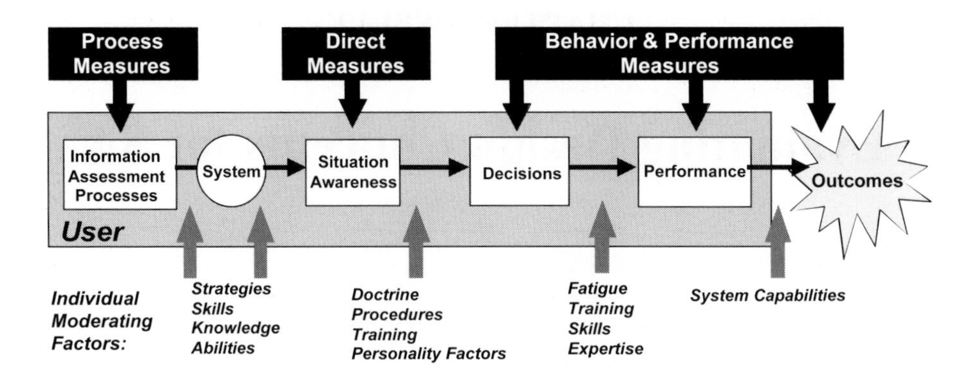

Figure 12.1 Approaches to SA Measurement (adapted from Endsley & Garland, 2000).

workload and human/system performance). We will also review the advantages and disadvantages of different SA measurement approaches that are commonly used for different purposes which you may also want to use at times, depending on your objectives and circumstances. We will then explore a case study that illustrates how these different measurement approaches might be applied and the type of data you could expect to get from them.

12.1 INDIRECT MEASURES OF SITUATION AWARENESS

Measures that infer SA based on an assessment of other constructs are indirect measures. Indirect measures try to infer how much SA a person has by measuring the cognitive processes involved in developing SA or by measuring performance issues related to the operator's interaction with the system. Process measures and behavioral and performance measures are sometimes used to infer SA in this manner.

12.1.1 Process measures

Situation awareness is a state of knowledge that is derived from numerous cognitive processes (see Chapter 2). One approach to assessing SA involves measuring these processes and then inferring from them the person's level of SA. In general, the inability to determine the completeness of the SA inferred of these process measures and the limited degree to which they reflect the entire SA process, constitutes a major limitation of this approach. Advantages, disadvantages, and issues for implementation are presented in Table 12.1. Common examples of process indices include verbal protocols, communication analysis, and psychophysiological metrics.

Verbal protocols. Verbal protocols are running commentaries in which the operators verbally describe their thoughts, strategies, and decisions while interacting with the system. To collect a verbal protocol, the operator is asked to 'think out loud' while performing a task. This information is recorded and later analyzed to identify and categorize information relevant to the individual's SA. The thoroughness of the information collected varies based on the verbal skills of the operator. Some people are much better at voicing their thoughts than others. Although verbal protocols can provide much insight into the way the operator conceptualizes a task and the cognitive processes employed to perform tasks, analyzing verbal protocols can be difficult due to the time intensive and subjective nature of the analysis. As a means of measuring (or inferring) SA, it can only be said that these verbalizations will rarely represent a very complete picture of the entirety of their mental representation of the situation. Verbal protocols typically only provide an incomplete picture of the individual's SA—that part which he or she is actively verbalizing. Information the individual knows about the situation in memory (or has forgotten) is not obvious in the protocols.

Table 12.1 Indirect measures of SA: process indices.

Advantages	Disadvantages	Application Considerations
	Verbal Protocols	
• Provides partial information on data that is used or lacking • Provides information on SA strategies and processes • Provides partial insight into how information is integrated and used • Identifies SA concepts needing more systematic measurement	• Does not provide a complete representation of what is attended/processed • Can slow performance when responding	• Data analysis is problematic • Noisy environments may interfere with data collection • May provide partial assessment of team SA
	Communication Analysis	
• Provides information on: • Unavailable information • Verbal processes • SA strategies • Feedback on actions • Communication types • Operator interactions	• Does not provide a complete representation of what is attended to/processed • Provides partial insight into information integration and use	• Data analysis is problematic • Noisy environments may interfere with data collection • Useful for assessing team SA processes
	Psychophysiological Metrics: Eye Movement	
• Provides indicators of how attention is allocated	• Not conclusive that an object is seen and processed • No information on how information is used or combined	• Equipment is difficult to calibrate in field • Data analysis is problematic • Will not assess team SA

Communication Analysis. Communication analysis focuses on the verbal exchanges between people involved in a task (e.g., between team members or between operators and experiment confederates). Transcripts are made of the entire session, then the transcripts are analyzed and the verbal exchanges are examined and categorized.

This method has been found to be particularly useful in analyzing and identifying team strategies and processes. For example, Orasanu and Fischer (1997) found that pilot teams who performed better increased information gathering earlier in the event, established a range of possible outcomes utilizing worst case rather than best case reasoning, and monitored the evolving information by acquiring information from available sources to determine which possible outcome was becoming most likely.

The applicability of this method for assessing SA as a state of knowledge for design concept evaluation, however, is somewhat less well defined. With this technique, the SA of the individual must be inferred from what the operator does or does not ask for or say. Some people may know a great deal, but be uncommunicative. This limitation can be particularly problematic in non-team situations since an operator does not have to communicate with another team member and may therefore say very little while performing the task. Additionally, although communication analysis can provide information regarding what information the operator does not readily have available, it does not provide insight into what information is being processed or how that information is being integrated and utilized. Like verbal protocols, data analysis of communications can be tedious and time intensive, requiring the categorization and classification of phrases and statements made throughout the experimental session.

Analysis of communications and spontaneous verbalization during a testing scenario may provide insight into the information that is lacking from a display and insight into processes used to gather and interpret data. However, information regarding how well the operator is able to keep up his or her SA (across the spectrum of SA requirements and levels) is not available and must be inferred.

Psychophysiological Metrics. Psychophysiological metrics seek to infer cognitive processes from physical reactions such as eye movements, EEG (electroencephalogram), or ECG (electrocardiogram) data. These types of measures are often used to assess other psychological constructs such as mental workload. Using these measures to assess SA, however, is a fairly new tactic and not much research exists to support their applicability to this construct. Psychophysiological metrics are appealing because they have the potential advantages of being unobtrusive and continuously available, thereby capturing reactions to critical events that are unplanned or uncontrolled.

As an example of their use, Stein (1992) and Smolensky (1993) recorded eye movements of air traffic controllers when processing aircraft on their radar display. They found it took around 5 minutes for eye movements to stabilize after assuming control during a shift, a period during which controllers tend to have higher than normal error rates.

As a downside, psychophysiological metrics can be cumbersome to use because they require specialized equipment and training to administer, as well as considerable expertise and time to analyze the data. At this point, the relationship between SA and most physiological reactions is unknown; more research is needed before any conclusions can be drawn regarding the ability of most of these metrics to reflect SA or changes in SA. (For more information on psychophysiological measures, see Wilson (2000).)

Overall, process indices can be very useful for conducting research on how people develop SA and process information in complex environments. Their ability to infer SA as a state of knowledge for design evaluation is quite limited, however.

12.1.2 Behavioral and performance-based measures

Performance-based measures can be defined as 'any measurement that infers subjects' situation awareness from their observable actions or the effects these actions ultimately have on system performance' (Pritchett & Hansman, 2000). Examples of this class of measures include techniques that assess SA based on the way an operator behaves (behavior measures) and techniques that assess SA based on the operators overall performance (performance outcome measures). An overview of the advantages and disadvantages of this class of measures is provided in Table 12.2.

Behavior Measures. Behavior measures infer the operators' level of SA based on their behavior. For example, if an army commander decides to set up an offensive action at a particular place, the commander's belief that the enemy will be coming down a particular path can be inferred. From this inference, an assessment of the army commander's SA can be made: the commander believes he is aware of the enemy's location and has made a prediction regarding the enemy's future maneuvers. However, the accuracy of this inference is difficult to confirm; a variety of factors other than the state of the commander's SA may be influencing the commander's strategic decisions. Discerning whether people are behaving in a certain manner because of the state of their SA or as the result of lack of skill or poor decision strategies is sometimes difficult. Much SA that does not result in an overt action will also be unobservable and, therefore, unmeasurable.

As an alternative, situations can be created that, given an acceptable level of SA, will cause the operator to react in a predicted manner. Thus, measurements can be taken on expected behaviors that can be used to infer SA. For example, one approach analyzes operators' responses to artificial manipulations of display information (e.g., removing or altering information) in order to assess the operators' SA (Sarter & Woods, 1991). As an example of this approach, Busquets, Parrish, Williams, and Nold (1994) interrupted a flight scenario and placed the pilot's aircraft in a new location. They measured the time required for the pilot to return to his original flight path. Although this method may provide insight into the operators' cognitive processes (i.e., where they are attending or what their expectations might be), the occurrence of an unnatural event can, in and of itself, alter SA and is therefore intrusive. It also only provides a partial picture of their SA (only regarding the manipulated information).

In another approach, unusual (but realistic) events can be incorporated in the simulation and the individual's response or nonresponse to these events analyzed. The events must be established such that a clearly defined and measurable *testable*

testable response is created. That is, the event must be experimentally controlled, unanticipatable through any means other than good SA, and require discernable, identifiable action from the operator (Pritchett & Hansman, 2000). In order for this methodology to be effective, the event must be chosen carefully and have associated with it a highly proceduralized response. For example, if an aircraft descends past its assigned altitude (the event), an air traffic controller would immediately make contact with that aircraft to correct the situation (the testable response).

Table 12.2 Indirect measures of SA: behavioral and performance measures.

Advantages	Disadvantages	Application Considerations
	Behavioral Measures: General	
• Objective and observable measures that are usually non-intrusive	• Assumes what appropriate behavior will be for a given level of SA • Behavioral indices may reflect other processes, such as decision strategy, rather than SA	• Requires operationally realistic scenarios • Behavioral indices must be specific and task relevant • Should be used in conjunction with other measures of SA
	Behavior Measures: Scenario Manipulation	
• Can yield direct observations of processes that may lead to correct assessments or mis-assessments of situations	• Affects attention • Affects SA • Should not be used during concurrent testing of workload or performance • Measures awareness of specific scenario features, not global SA	• Uses realistic scenarios • Manipulations must be mission relevant • Useful for assessing team SA • Need to use with other SA measures
	Performance Outcome Measures	
• Objective and observable measures that are usually nonintrusive	• Global performance measures (e.g., success in meeting a goal, kills in a battle) suffer from problems of diagnosticity and sensitivity • Identifying unambiguous task performance measures may be difficult	• Should be used in conjunction with other measures of SA • Requires operationally realistic scenarios • Performance indices must be specific and task relevant • Subtask performance outcomes should be clearly specified and utilized instead of global performance outcomes

Several examples exist of the use of this approach. Busquets, Parrish, Williams and Nold (1994) measured the time required for pilots to respond to an aircraft that was intruding on their assigned runway. In another study, Jones and Endsley (2000b) measured the time required for air traffic controllers to correct errors that were introduced into an air traffic management scenario. Hahn and Hansman (1992) measured whether pilots detected erroneous flight paths that were uploaded to their flight management system. In all of these cases, the scenarios were designed such that a clear and observable reaction should be elicited from operators if they noticed the key events in the scenario.

By inserting events such as these into a scenario, it is possible to measure the quality of an operator's SA on these specific aspects of the situation and the displays associated with them. The testable response methodology is best suited for examining the sufficiency of operator SA with respect to off-nominal (i.e., nonnormal) or emergency conditions. Situations that involve (1) routine tasks that do not demand a singular distinct response, (2) untrained operators who may not respond predictably, (3) complex tasks and subtle cues that are difficult to artificially manipulate, or (4) too many operators to reliably control throughout the experiment are not good candidates for this technique (Pritchett & Hansman, 2000).

As an advantage, behavioral measures such as these are objective and easy to collect and analyze. As a downside, even with the most careful scenario creation, sometimes people will not respond as anticipated. The researchers are then left to wonder whether their SA was poor, or whether they simply chose to respond (or not respond) differently than anticipated (Farley, Hansman, Amonlirdviman, & Endsley, 2000; Jones & Endsley, 2000b). It should also be noted that great care must be taken to prevent an operator from anticipating these unusual events if they participate in more than one scenario. For example, after one runway incursion when landing, pilots will be primed to respond to any further anomalies of this type (which are extremely rare in real life), thus altering their SA on any future trials.

Performance Outcome Measures. Performance outcome measures infer SA based on how well an operator performs a given task as compared to a predetermined standard, or how good the outcomes of a scenario are (e.g., number of successful missions). While objective and possessing a high degree of face validity, performance metrics suffer from being circular in relation to SA, insensitive to many SA differences, and limited in terms of the ability to collect all the needed measures.

This approach assumes a direct relationship between good SA and good performance. This link, however, is not deterministic, only probabilistic—the odds of better performance are increased with good SA, yet not guaranteed. SA is necessary, but not sufficient for good performance. A person may have good SA, but lack the knowledge, skill and/or training necessary to appropriately solve a problem; consequently performance will likely be low despite a high level of SA. A fighter pilot may be provided with a new display that provides a high level of SA on where enemy aircraft are located. Unless he has good tactics for taking advantage of this SA, however, and the weapons to do so successfully, it is unlikely that performance indicators will show any improvement associated with that SA.

Conversely, sometimes a person with poor SA can still perform well on many performance measures, even if only by chance. For example, the same pilot with generally poor SA may happen upon an enemy aircraft which he is able to shoot down. In other instances, the design provided to an operator may be poor, and consequently the SA afforded by the design poor, but the ingenuity and creativity of the operator still makes it possible to perform well. These confounds cannot be teased out when simply assessing performance.

In a study of SA measurement approaches, Vidulich (2000) found that performance was frequently measured in studies of SA (94%). This approach was sensitive to display manipulations 70% of the time. Due to the circular nature of such measures, however, it is difficult to say if these performance improvements were actually due to SA improvements or other effects (e.g., workload) associated with the designs studied.

In addition, Wickens (2000) notes that one's ability to infer SA from such performance measures depends highly on the scenarios constructed and the type of performance measures. For example, he found that the SA inferred from studies of electronic flight path displays was quite different depending on whether the scenario measured normal performance (e.g., degree of adherence to the path) or nonnormal performance (e.g., ability to find an alternate airport in an emergency). Therefore the range and breadth of scenarios created and aspects of performance evaluated become critical to insuring valid results from performance measurement approaches. As collecting a sufficient number of trials to cover the gamut of possible scenarios is generally infeasible due to the constraints of numbers of operators, time, and economic reasons, it is desirable to collect direct measures of SA that can point to where performance problems are likely to occur in practice, given the right set of circumstances.

12.2 DIRECT MEASURES OF SITUATION AWARENESS

Direct measures of SA attempt to directly assess a person's situation awareness. Both subjective and objective measures have been developed.

12.2.1 Subjective measures

Subjective measures assess SA by asking either the operator or an expert observer to rate the quality of the operator's SA during a specified period. This rating is then used to compare the quality of SA at various points in time or with various systems. Subjective measures may be as simple as asking the operator or observer to rate the operator's SA on a scale of 1 (*low*) to 5 (*high*), or they may be as complex as asking the operator to complete several scales that rate various facets believed to impact SA. Advantages and disadvantages of subjective techniques are presented in Table 12.3.

Subjective metrics have the advantages of being inexpensive, nonintrusive, easy to administer, simple to analyze, and employable in real-world settings without unduly disrupting performance. However, subjective measures have several shortcomings that seriously hamper their ability to provide accurate assessments of SA. First, operators will report what they believe their SA to be, but they may not realize that inaccuracies exist in their SA or that information exists they are unaware of. For example, in comparing several cockpit displays designed to facilitate spatial orientation, Fracker and Vidulich (1991) found that the display that provided the best subjective ratings of SA also resulted in the greatest percentage of inverted recoveries (i.e., the pilots believed they were upright when actually they were upside down). Supporting this finding, Endsley and Selcon

(1997) found a zero correlation between subjective measures of SA and an objective measure of SA. Subjective measures should always be interpreted in conjunction with performance data or direct measures of SA in order to gain a true understanding of self-reported SA versus actual SA.

Table 12.3 Direct measures of SA: subjective measures

Advantages	Disadvantages	Application Considerations
	Self-Ratings	
• Assesses own degree of confidence in SA	• People may not know what information they are unaware of • May be influenced by self assessments of performance	• Data are easily collected in uncontrolled field settings after or during simulations
	Situational Awareness Rating Technique (SART)	
• Measures general constructs • Widely used • Correlates with performance and workload measures	• Workload elements in scale confounded with SA • Not correlated with objective measures • Limited resolution of individual SA elements	• Easily administered in laboratory or field settings • Use with other SA measures • Useful to assess team SA
	SA-Subjective Workload Dominance Technique (SA-SWORD)	
• Good sensitivity and inter-rater reliability • Good face validity • Provides rating of SA for different design features	• May reflect subjective preferences rather than SA • Limited for broader SA construct investigations	• Easily administered in laboratory or field settings • Use with other SA measures • Use to compare design features or concepts
	Observer Ratings (e.g., SARS)	
• Observers may have more complete knowledge of reality than simulation or exercise participants	• Observers have limited knowledge of a person's concept of the situation • Correlated with observer ratings of performance and experience • For teams, multiple observers are needed	• Need observable behaviors to anchor ratings • Multiple raters and opportunities are needed • Scenarios must allow behaviors to be exhibited • Observers should not interfere with task • Confederate observers may be used in the field • Peer ratings may be problematic in the field

Second, operators' ratings of their own SA can be influenced by the outcome of their performance. Operators may rate their SA as good if the simulation scenario has a positive outcome, regardless of whether good SA, luck, or some other factor influenced performance. Similarly, operators may tend to report their SA as poor if the outcome of the trial is less than favorable, regardless of why errors occurred (e.g., poor SA, insufficient skill/knowledge, poor action selection). Endsley and Selcon (1997) found that subjective SA was correlated 0.76 with subjective measures of performance.

Finally, self assessments of SA assume individuals can accurately judge and report feelings, thoughts, and facts from experience. However, errors in judgment and recall as well as limits on working memory can negatively impact the accuracy and sensitivity of subjective metrics (Taylor & Selcon, 1991).

In spite of the limitations, when designed and utilized properly, subjective metrics can provide valuable information regarding SA. Specifically, subjective assessments provide an indication of the operators' confidence level regarding their SA (Endsley & Selcon, 1997), which can directly influence their decisions and action selection (see Chapter 7). Examples of subjective metrics include the Situational Awareness Rating Technique, the Situation Awareness Subjective Workload Dominance Technique, and the Situation Awareness Rating Scale.

Situational Awareness Rating Technique. The Situational Awareness Rating Technique (SART) is one of the best known and most thoroughly tested subjective metrics. It has been found to be sensitive to task difficulty and operator experience (Selcon, Taylor, & Koritsas, 1991), to be reasonably robust (Taylor, 1990; Taylor & Selcon, 1991), and to be sensitive to design variables (Selcon & Taylor, 1990). SART assumes that operators 'use some understanding of situations in making decisions, that this understanding is available to consciousness and that it can readily be made explicit and quantifiable' (Taylor, 1990).

This metric clusters 10 generic constructs, derived from knowledge elicitation techniques with aircrews, into three broad domains: (1) Attentional demand, which includes the constructs of instability of situation, variability of situation, and complexity of situation; (2) Attentional supply, which includes the constructs of arousal, spare mental capacity, concentration, and division of attention; and (3) Understanding, which includes the constructs of information quantity, information quality, and familiarity. (For more information on the development of this technique, see Taylor, 1990.) Although SART was created in the context of the aviation domain, the constructs are domain independent, thereby allowing the scale to be used in other similar domains.

The ratings on the three-dimensional SART scale can be combined to form an overall SA score by using the algorithm SA(calc) = Understanding – (Demand – Supply). However, this formula was developed from considerations of how the dimensions theoretically should interact rather than from statistical or empirical evaluations. Consequently, SA(calc) is somewhat arbitrary and should not be unduly emphasized nor specified as a design requirement; rather, it provides an overall score, which can be useful in comparative system design and evaluation (Selcon, Taylor, & Shadrake, 1992).

One aspect of SART that has caused controversy is its inclusion of constructs generally associated with workload (e.g., supply and demand of attentional resources). From one viewpoint, workload is seen as an integral part of SA and thus should be included (Selcon, Taylor & Koritsas, 1991; Taylor, 1990), while the other viewpoint holds that including elements of workload within the SA scale confounds the measure (i.e., discerning whether decrements are due to workload or SA issues is not possible) (Endsley, 1996b). Because SART considers higher levels of workload to always be detrimental to SA, it may not be applicable to situations in which low workload is a problem for SA (e.g., long-haul flight or vigilance situations). Further, in other than overload situations SA and workload have been found to be independent (Endsley, 1993a). Vidulich (2000) found that SART had the best sensitivity of subjective SA metrics studied. Used in conjunction with objective measures of SA and performance measures, and interpreted with care, SART can be used to complement other data collection efforts.

Situation Awareness-Subjective Workload Dominance Technique. Another scale used to subjectively assess SA was derived from a subjective workload assessment tool called Subjective Workload Dominance (SWORD). SA-SWORD attempts to create a standardized subjective measure of SA (Vidulich & Hughes, 1991). It uses the same data collection and analysis procedures as SWORD (Vidulich, 1989), only the instructions for completing the scale are changed. Operators perform a comparative evaluation of systems, ranking their choice of system based on a 9-point scale. Each level of the scale represents the person's belief in the amount of SA that is provided by each system. Initial evidence suggests that SA-SWORD holds promise to be a sensitive and reliable measure of SA (Vidulich & Hughes, 1991). However, further evaluation of the metric is needed before any conclusions can be drawn regarding its effectiveness for accurately assessing SA.

Situational Awareness Rating Scale. The Situational Awareness Rating Scale (SARS) represents another method for subjectively measuring SA (Bell & Waag, 1995; Waag & Houck, 1994). This metric was developed to assess SA in a tactical air domain. The scale is comprised of eight categories representing 31 behavior elements considered important for mission success. Operators complete the SARS scale by providing a rating on each of 31 elements on a 6-point scale. SARS was used by pilots to rate themselves and others in their units.

The usefulness of this scale for evaluating the SA afforded by a system is limited in several ways: (1) it regards SA more as an innate ability rather than a changeable state of knowledge that can be identified, (2) it evaluates the abilities of the individual rather than the SA supported by a system, (3) it includes decision-making skills, flight skills, performance, and subjective impressions of the person's personality traits as part of the scale (Endsley, 1996b), and (4) it is closely linked to a particular aircraft type and mission and therefore is not easily generalizable to other domains. However, guidelines have been offered as to how to create a similar metric for other domains based on this methodology (Bell & Lyon, 2000).

It should be noted that the subjective ratings of knowledgeable and experienced observers (who theoretically know more about what is actually happening in a scenario, but less about what a particular individual may have in mind) can produce quite different results from self-ratings. For example, the SA ratings of experienced observers were only moderately correlated with air traffic controllers' own ratings of their SA in a study examining display concepts (Endsley, Sollenberger, Nakata, Hough, & Stein, 1999). Bell and Lyons (2000) report fairly low correlations between individual's ratings of themselves via SARS and peers' and supervisors' ratings (<0.50). It is likely these two different approaches to subjective SA measurement are tapping into different things.

12.2.2 Objective measures

Objective measures seek to assess SA by directly comparing an operator's reported SA to reality. This comparison is often made by querying operators about aspects of the environment and then assessing the accuracy of the responses by comparing them with reality. Several challenges exist with respect to collecting objective measures of SA. First, the appropriate timing for the queries must be established. For example, administering the queries at the end of a scenario provides a valid picture of the operators' SA only at the point in time the activity ended. Operators' recall of their knowledge earlier in the scenario will be hampered by reporting problems and influenced by performance outcome (similar to subjective measures). Conversely, administering the queries concurrent with task performance raises other concerns regarding the potentially negative impact of the queries on issues such as performance, workload, attention, and SA.

The second challenge involves determining the appropriate questions to ask in order to obtain a true indication of an participant's SA with respect to the task of interest. For example, asking a pilot how many of the first class passengers are men and how many are women would not provide useful information regarding the pilot's level of SA while operating the aircraft.

The third challenge relates to designing the queries so that they not only address elements essential for good SA, but that they do so in a manner that does not alter the individual's SA. For example, if an air traffic controller were asked 'what is the altitude of the aircraft entering the sector,' the controller's attention would be drawn to that aircraft. If the controller is aware of the aircraft, no harm is done. If not, then the operator will be artificially clued in to the presence of the aircraft, thereby altering the controller's SA.

The fourth challenge concerns the necessity of creating questions that cover the entire range of relevant SA issues. When a limited set of questions is utilized, the participant will be able to identify the elements the experimenter is most interested in and focus on those elements in order to be able to respond to the queries correctly. Consequently, the true nature of SA during task performance will be altered and unavailable for examination.

How well a measurement technique addresses these challenges provides a benchmark for assessing the utility of the tool. General advantages and disadvantages of objective SA measures are provided in Table 12.4. Two examples of objective metrics that contend with many of these issues are the Situation Awareness Global Assessment Technique and Online probes.

Situation Awareness Global Assessment Technique. One widely tested and validated objective metric that addresses these challenges is the Situation Awareness Global Assessment Technique (SAGAT) (Endsley, 1988; 1995a). SAGAT has been shown to be effective across a variety of domains, including aviation (Endsley, 1990a; 1990b; 1995a), air traffic control (Endsley, Sollenberger, Nakata, & Stein, 2000), power plant operations (Hogg, Torralba, & Volden, 1993), teleoperations (Riley & Kaber, 2001; Riley, Kaber, & Draper, in press), driving (Bolstad, 2001), and military operations (Matthews, Pleban, Endsley, & Strater, 2000).

To use SAGAT, operator-in-the-loop simulation exercises are conducted that employ the display or design concepts of interest. At randomly selected intervals, the simulation activity is briefly halted, the displays are blanked, and a battery of queries is administered to the participant. In general, SAGAT questions are administered through a computer, easing administration and scoring of data, although they may also be administered by paper and pencil. Data collected by the simulation computer and by subject matter experts are used to score the participant's responses as correct or incorrect based on what was actually happening in the scenario at that time.

Table 12.4 Direct measures of SA: objective measures.

Advantages	Disadvantages	Application Considerations
	Post-test Questionnaires	
• Less intrusive • Ample time to respond • Does not disrupt mission tempo	• Memories not reliable • Early misperceptions may be forgotten • Captures SA only at the end of an exercise	• Requires detailed analysis of SA requirements • Uses mission-relevant queries • Data collection easily achieved • Administer immediately after exercise completion • Useful to assess team SA
	Situation Awareness Global Assessment Technique (SAGAT)	
• Assesses global SA • Avoids retrospective recall • Minimizes biasing of SA • Good psychometric qualities • Performed in realistic, dynamic environments • Allows assessment of shared SA within teams	• Requires freezes in the scenarios • May negatively impact the pace and flow of real-time scenarios	• Requires detailed analysis of SA requirements • Uses mission-relevant queries • Multiple data collectors for teams • Freezes can last 2-3 minutes • Freezes should not be predictable • Freezes may be problematic in field operations
	Online Queries (e.g., Real-time Probes)	
• Overcomes memory problem associated with post-test queries • Queries can be embedded in task	• May intrude on concurrent task performance • May alter SA by shifting attention • May increase workload • Assesses limited SA requirements	• Requires detailed analysis of SA requirements • Mission-relevant queries can occur as part of normal events • Noise may interfere with data collection • Assessing team SA is limited • Numerous queries must be asked to achieve needed sensitivity

After the participant completes the questions, the simulation scenario is generally resumed from the exact place it was stopped. Although concerns have been raised concerning the intrusiveness of this technique, (Pew, 1995), the preponderance of evidence indicates that these stops do not interfere with operator performance (Endsley, 1995a, 2000a). Participants are generally able to switch from the SAGAT questions back to the scenario fairly easily once they have completed their questions. SAGAT has also been shown to have predictive validity, with pilots with good SA as measured by SAGAT far more likely to have good outcomes than those who did not (Endsley, 1990b). Measurement reliability was demonstrated in a study that found high reliability of SAGAT scores for individuals who participated in two sets of simulation trials (Endsley & Bolstad, 1994).

To be successful it is important that: (1) the timing of the stops not be predictable to the participants (although they may have experience with the types of questions that are asked, and (2) the questions be reflective of a wide range of the SA requirements of the operator.

SAGAT assesses a participant's SA by querying the participant on the pertinent SA requirements relevant for the domain of interest (see Chapter 5 for SA requirement delineation). The queries are developed to cover all levels of SA (i.e., perception, comprehension, and projection issues). Ideally, one question may often cover multiple SA requirements in order to minimize the question set and maximize data collection. Further, the wording of the questions is developed to be consistent with both the terminology of the domain and the manner in which the operator processes the information. Examples of SAGAT questions that have been developed for air traffic controllers are shown in Table 12.5.

After developing a battery of questions covering the entire range of the operators' SA needs, a decision must be made as to how to administer the queries. In general, the questions are administered in a random order at each stop. Administering the entire battery of queries at each stop is often preferable, but not always possible if the question set is very large. Previous research indicates that a stop can last as long as 5 minutes without incurring memory decay on the part of the participants (Endsley, 1995a). Time constraints, task phase, and test-bed limitations can also impact the administration of SAGAT queries.

When time constraints exist (i.e., a shorter stop is desired than can accommodate the range of queries), SAGAT can be customized in two ways to overcome this limitation. The first option is to allow the participant to continue answering questions only until a designated time period (e.g., 2 minutes) has elapsed. Thus, a random subset is administered at each stop. When using this approach, certain questions may exist that are essential to ask each time; in these cases, SAGAT can be customized to ask these questions first, then to proceed to the remaining queries.

A second option for customizing SAGAT when time constraints exist is to only ask a particular subset of the queries at each stop. As attention biasing is a concern, the set of questions should be carefully chosen so that the subset is not too small. When SAGAT queries are appropriately designed and administered, no biasing of operator attention occurs (Endsley, 1995a). In general, it makes sense to limit the queries to subsets that are relevant to the concepts being tested and appropriate for the simulation and scenario.

Several factors are routinely used to limit the question set to those most important for SA: task phase and test bed limitations. In some cases, phase of task may be a driving factor for query administration. Administering the entire battery of SAGAT questions may not be desired if the task has clearly defined phases with distinct information needs. For example, a pilot who is in the landing phase of flight will not be interested in information relevant only to takeoff operations. SAGAT can be customized to present only information relevant to the current phase of the task in such a case.

Table 12.5 SAGAT queries for Air Traffic Control (Endsley & Kiris, 1995b).

1. Enter the location of all aircraft (on the provided sector map)
   aircraft in track control
   other aircraft in sector
   aircraft will be in track control in next 2 minutes
2. Enter aircraft callsign (for aircraft highlighted of those entered in query 1)
3. Enter aircraft altitude (for aircraft highlighted of those entered in query 1)
4. Enter aircraft groundspeed (for aircraft highlighted of those entered in query 1)
5. Enter aircraft heading (for aircraft highlighted of those entered in query 1)
6. Enter aircraft's next sector (for aircraft highlighted of those entered in query 1)
7. Enter aircraft's current direction of change in each column (for aircraft highlighted of those entered in query 1)

Altitude change	Turn
climbing	right turn
descending	left turn
level	straight

8. Enter the aircraft type (for aircraft highlighted of those entered in query 1)
9. Enter aircraft's activity in this sector (for aircraft highlighted of those entered in query 1)
   en route, inbound to airport, outbound from airport
10. Which pairs of aircraft have lost or will lose separation if they stay on their current (assigned) courses?
11. Which aircraft have been issued assignments (clearances) that have not been completed?
12. Did the aircraft receive its assignment correctly?
13. Which aircraft are currently conforming to their assignments?
14. Which aircraft must be handed off to another sector/facility within the next 2 minutes?
15. Enter the aircraft that are experiencing a malfunction or emergency that is affecting operations.
16. Enter the aircraft which are not in communication with you.
17. Enter the aircraft that will violate special airspace separation standards if they stay on their current (assigned) path.
18. Which aircraft will weather currently have an impact on or will be impacted on in the next 5 minutes along their current course?
19. Which aircraft will need a new clearance to achieve landing requirements?
20. Enter all aircraft that will violate minimum altitude requirements in the next two minutes if they stay on their current (assigned) paths?
21. Enter the aircraft that are not conforming to their flight plan.
22. Enter the aircraft and runway for this aircraft.

Second, the limitations of the test bed may necessitate customizing the list of SAGAT queries. For example, in situations where the simulation has no capacity to take weather into account, asking questions about weather would not be a good use of time even if weather information is a relevant SA requirement.

SAGAT is designed to avoid cueing the operator to information they may or may not be aware of. For example, in aviation this is achieved by first presenting the pilot with a map of the area (complete with appropriate boundaries and references), on which the pilot indicates other aircraft she knows about. Future questions are then referenced to the aircraft the participant initially indicates on this map. Thus, she is not specifically asked to answer questions about 'the aircraft behind you' that she may not have been aware of before the question, thus affecting her SA on later trials. Utilizing a computer program facilitates this type of questioning. In situations where a computer program is not available and SAGAT is collected via pencil and paper, the questions should be designed to cover every possible event of interest and the entire range of questions should be presented at each stop with the understanding that not all questions will always be relevant.

Several guidelines have been developed that can help maximize the effectiveness of SAGAT data collection efforts (Endsley, 2000a).

1. The timing of SAGAT stops should be randomly determined. SAGAT stops should not occur only at times of increased activity. This insures a fair and unbiased assessment of SA is obtained for each concept under evaluation.
2. If more than one person is participating in the study (e.g., a team or competing force), the SAGAT stops for all participants should begin and end at the same time. This insures that unfair advantage is not provided to some participants.
3. SAGAT stops should not occur within the first 3–5 minutes of the study to allow participants to develop an understanding of what is going on in the scenario.
4. SAGAT stops should not occur within 1 minute of each other, to insure SA is regained following a stop.
5. Multiple SAGAT stops can be included in a single scenario. Although incorporating too many stops can be counterproductive (and potentially bothersome to the participant), as many as three stops within 15 minutes have been used with no adverse effects (Endsley, 1995a).
6. In order to obtain the needed number of data points for effective analysis, each SAGAT query should be administered 30–60 times (across subjects and trials in a within-subjects experimental design) for each experimental condition (e.g., the design concept being evaluated).
7. Other measures of interest (e.g., other measures of SA, performance measures, or workload measures) can be incorporated into the same trial as SAGAT.

SAGAT is a particularly useful tool in design evaluation because it provides diagnostic information regarding how well the system in question supports the operator's various SA requirements. For example, finding queries that are frequently answered incorrectly can give the designer insight into areas where the operator's SA needs are not being supported. Thus, the designer has specific guidance as to ways to improve the interface design in order to better support the operator's SA.

The main limitation of SAGAT is that it is most amenable to simulation environments and situations where activity can be halted for the queries to be administered. This can be hard to do in some real-world settings or field-testing. Additionally, it requires time and effort to delineate the SA requirements for developing appropriate queries for a new domain, and access to simulations that can be stopped and restarted. For more information on SAGAT, see Endsley (1995a; 2000a).

An example of the use of SAGAT for evaluating the impact of new system concepts can be found in Endsley, Mogford, Allendoerfer, Snyder, and Stein (1997). In this example, a totally new form of distributing roles and responsibilities between pilots and air traffic controllers was examined. Termed *free flight*, this concept was originally developed to incorporate major changes in the operation of the national airspace. It may include aircraft filing direct routes to destinations rather than along predefined fixed airways, and the authority for the pilot to deviate from that route, either with the air traffic controller's permission or perhaps even fully autonomously (RTCA, 1995). As it was felt that such changes could have a marked effect on the ability of the controller to keep up as monitor of aircraft in these conditions, a study was conducted to examine this possibility (Endsley, Mogford, Allendoerfer, Snyder & Stein, 1997; Endsley, Mogford, & Stein, 1997).

Results showed a trend toward poorer controller performance in detecting and intervening in aircraft separation errors with free flight implementations and poorer subjective ratings of performance. Finding statistically significant changes in aircraft separation errors during ATC simulation testing is quite rare, however. Analysis of the SAGAT results provided more diagnostic detail about the problem controllers were having in keeping up with the situation and backed-up the finding of a risk of separation loss associated depending on controllers to detect problems under free flight conditions.

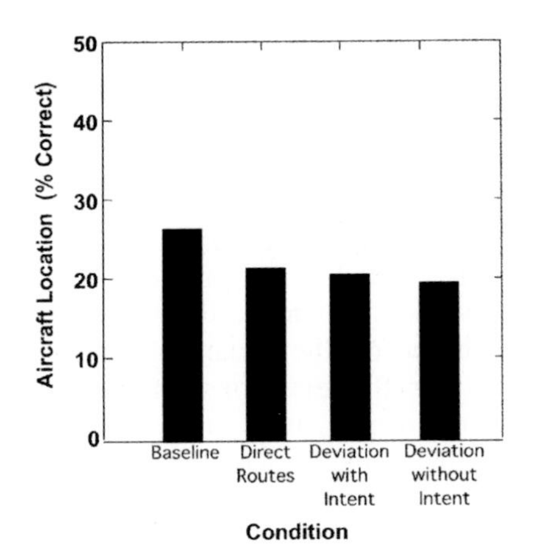

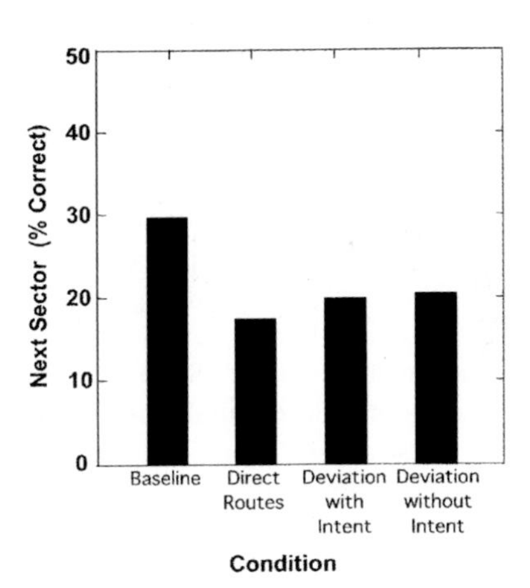

Figure 12.2. Example of SAGAT results (from Endsley, Mogford & Stein, 1997).

As shown in Figure 12.2, controllers were aware of significantly fewer of the aircraft in the simulation under free flight conditions. Attending to fewer aircraft under higher workload has also been found in other studies (Endsley & Rodgers, 1998). In addition to reduced Level 1 SA, controllers also had a significantly reduced understanding (Level 2 SA) of what was happening in the traffic situation, as evidenced by lower SA regarding which aircraft weather would impact on and a reduced awareness of those aircraft that were in a transitional state. They were less aware of which aircraft had not yet completed a clearance, and, for those aircraft, whether the clearance was received correctly and whether they were conforming. Controllers also demonstrated lower Level 3 SA with free flight. Their knowledge of where the aircraft was going to (next sector) was significantly lower under free flight conditions.

These findings were useful in pinpointing whether concerns over this new and very different aviation concept were justified, or whether they merely represented resistance to change. The SAGAT results showed not only that the new concept did indeed induce problems for controller SA that would prevent them from performing effectively as monitors to backup pilots with separation assistance, it also showed in what ways these problems were manifested. This information is very useful diagnostically in that it allows one to determine what sorts of aids might be needed for operators to assist them in overcoming these deficiencies.

For instance, in this example, a display that provides enhanced information on flight paths for aircraft in transitionary states may be recommended as a way of compensating for the lower SA observed (Endsley, Sollenberger, & Stein, 1999). Far from just providing thumbs-up or thumbs-down input on a concept under evaluation, this rich source of data is very useful in developing iterative design modifications and making tradeoff decisions.

Online Probes. Another objective method of assessing SA involves querying participants concurrent with their activity. For example, the *Situation Present Assessment Method (SPAM)* (Durso, *et al.*, 1998) asks questions of operators concurrently with activities and with their displays in full view. Response time to the questions is considered to be the measure of the operators' situation awareness. Theoretically, participants who have good SA will answer more quickly, because they would know where in the environment to look for a particular piece of information and thus could answer the question more quickly. Durso *et al.* found some sensitivity of the metric.

The validity of online probes was investigated with a methodology derived from SAGAT (Endsley, Sollenberger, Nakata, & Stein, 2000; Jones & Endsley, 2002a). This metric utilizes queries called *real-time probes*, which are developed in a similar manner to the SAGAT queries (i.e., based on the situation awareness requirements delineated through a GDTA), but the questions are verbally administered one at a time while people are performing their tasks. Since the queries are administered concurrently with the activity, the displays remain in view of the participant.

Similar to the SPAM metric, reaction time and accuracy to the real-time probes are recorded and used to assess the participant's SA. This metric may provide an alternative to SAGAT in situations where SAGAT cannot be employed (e.g., where the activity cannot be temporarily halted). However, since the queries are presented one at a time, fewer queries can be provided during a trial with real-time probes than with SAGAT, thereby requiring that more data collection points and trials be included in the experimental design.

Initial research on this technique found a weak correlation between the measure and SAGAT, indicating that the technique holds some promise as a way to assess SA. To build up the validity of the measure, adequate sampling of each query or probe is important (Jones & Endsley, 2000a). However, initial evidence also indicates a weak correlation between real-time probes and workload, thereby raising the concern that real-time probes may reflect measures of workload as well as SA. This possible relationship should be taken into account when utilizing real-time probes until such time as further evidence either confirms or does not confirm this relationship. More testing is still needed before the measure is accepted as a valid and reliable metric of SA.

12.3 Measuring team SA

As more emphasis is being placed on understanding team SA (see Chapter 11), metrics are needed to assess this phenomenon. Although ongoing efforts are examining team interactions and team SA, these approaches are not necessarily easily translated to a method for system evaluation. For example, Childs, Ross, and Ross (2000) describe a method to assess team situation awareness by categorizing errors via a breakdown analysis. Prince and Salas (2000) discuss team SA and crew resource management and how those constructs can provide training insights, and Cook, Salas, Cannon-Bowers, and Stout (2000) provide an overview of efforts aimed at measuring and understanding the content of team knowledge.

In order to gain an understanding of the SA afforded to a team by a particular system design, metrics and analysis techniques need to quantify team SA and the shared SA of the members within the team. *Team SA*, the degree to which each team member has the SA required for his or her job, can be determined by measuring SA for every team member during a simulation. *Shared SA*, the degree to which team members have a consistent picture of what is happening on their shared SA requirements, can be determined by directly comparing their perceptions of what is happening on those items (see Figure 12.3).

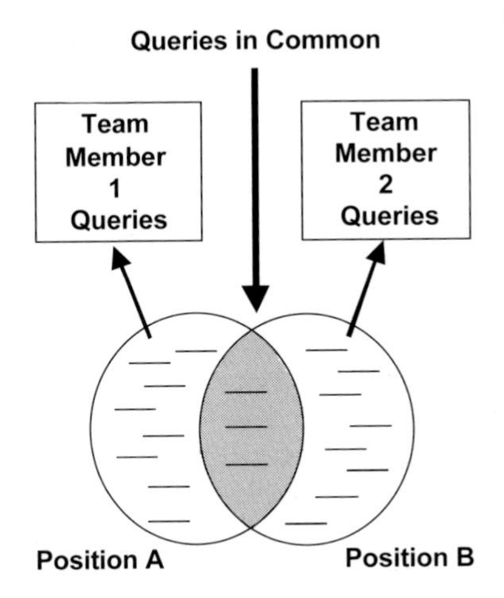

Figure 12.3 Assessment of shared SA.

One approach to meeting the need for measurement of team SA involves utilizing SAGAT and extending the analysis techniques to include multiple team members. Using the same guidelines described in Section 12.2.2, SAGAT can be administered to multiple team members during a test scenario. The team members must complete the test battery by themselves; assisting one another or sharing information is not allowed. The battery of questions provided to each team member is tailored to his or her SA requirements—based on an SA requirements analysis of each position. For example, pilots would get one question set and air traffic controllers would get another. Two team members working side by side in very similar roles (e.g., a radar and a planning controller) would get very similar question sets. Team members whose jobs are very different (e.g., the pilot and the controller) would get very different questions. Team SA can be assessed by evaluating the SA of each team member to determine any weaknesses or SA problems that could undermine team performance.

Shared SA can be examined by comparing the team members' responses to the SAGAT queries *on those questions they have in common* (determined by the degree of overlap in their SA requirements). An initial study examining shared SA for a two-member team identified five categories of responses in analyzing team responses to SAGAT queries (Jones & Endsley, 2002b), shown in Figure 12.4.

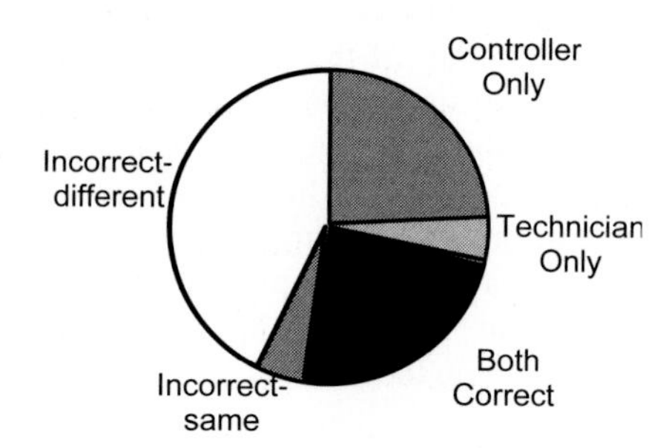

Figure 12.4 Shared SA in an Air Operations Center (in percentage of cases in each category).

Potential categories of shared SA include:

1. Both team members answer correctly. This category represents accurate shared SA—the ideal state.
2. Team member A answers correctly, but team member B answers incorrectly. This is non-shared SA. Unless good team processes are in place to help uncover the discontinuities, this situation could lead to an error.
3. Team member B answers correctly, but team member A answers incorrectly. Unless good team processes are in place to help uncover the discontinuities, this situation could lead to an error.
4. Neither team member answers correctly and their wrong answers are different. In this case team members have different SA and neither is correct. This situation is also undesirable. While good team processes

may help uncover the discrepancy, there is also a possibility that the team will settle on one of the incorrect interpretations.

5. Neither team member answers the query correctly and their wrong answers are essentially the same. This case represents inaccurate but shared SA. This situation is the worst case since the team members are reinforcing each other's inaccurate SA and there is less chance of discovering the error.

Analyzing these categories of shared SA can provide valuable insight into the impact the design has on the ability of the team to develop the shared SA needed to perform effectively and efficiently (see Figure 12.4). It can be used to identify not only what data sharing is not being supported through the system design, but also differences in comprehension and projection that can lead to large disparities in the SA of teams, and thus their performance.

12.4 CASE STUDY

The following example illustrates how some of the measures described in this chapter can be incorporated into a system evaluation and how the results of these measures can provide feedback to the designer regarding the strengths and weaknesses of the design. (This is for illustration only; normally, not all measures would be collected in the same study.) A summary of these measures is shown in Table 12.6.

Table 12.6 Summary of case study metrics.

Metric	When Collected	Collection Method
	Indirect	
Communication Analysis	Continuously	Audio recording for later transcription
Psychophysiological Measures	Continuously	Specialized physiological equipment
Testable Responses	Discrete, preprogrammed events, carefully selected and scripted to ensure limited responses	Participant responses to the subtle cues are recorded by experimenter; system actuation recorded by simulator
Performance Outcome	Dependent on task. Typically some indicators of performance collected during the task and others result from final outcome	Simulation collected data or Subject Matter Expert rated
	Direct	
SART	Post-trial	Pencil and paper or computer
SAGAT	During predetermined temporary halts in activity	Pencil and paper or computer
Real-time Probes	Concurrent with activities	Verbally administered and answers written down by experimenter (Subject Matter Experts should record correct answer)

12.4.1 Test case and testing description

Scenario: A new aircraft navigation display has been designed and a prototype created. The designer scrupulously followed all relevant design guidelines for achieving high levels of SA and performance. Now the designer wants to test the design to ensure that no unforeseen negative interactions or consequences result from the design. Additionally, the designer wants to compare the level of SA supported by the new design with the level of SA afforded by a currently used design. The designer has selected a variety of SA measures to employ during the evaluation process.

Task: Eight pilots are tasked separately with flying from point A to B in a simulator. A confederate posing as the copilot for each of the eight pilots will also participate in the study. During the flight, the pilots will be given a reroute. The course and reroute are designed so that the pilot must interact with the navigation display. Each pilot will participate in two 45-minute scenarios (one with each navigation display).

Training. The pilots are given training on the navigation display prior to the experimental scenario in which they will interact with the system (i.e., they train on the new display, fly the scenario with that display; then they train on the other display and fly the scenario with that display). They will receive training until they reach a specified level of performance with each display.

Prebrief. Prior to beginning the experimental trials, the participants are given instructions on how to complete SAGAT queries, how to answer the real-time probes, and how to complete the SART scale. Additionally, the use of the psycho-physiological equipment is fully explained and the participants are instructed that their overall performance will be analyzed. No directions are given regarding the other measures of SA that will be collected (e.g., communication analysis and testable responses), as these measures are to be 'hidden' aspects of the study.

12.4.2 Implementation and analysis of SA measures

Communication analysis. The entire session is audio recorded and later transcribed for analysis. The analyst must go through the transcript and identify communication exchanges in which information was requested and transmitted. The analyst must then perform an analysis to categorize the information into meaningful categories in order to make qualitative inferences regarding the state of the participants' SA. The patterns of communications will then be compared across the two display types as well as against other measures collected during the trial. Results will tell designers whether pilots communicate more or less with the new navigation display and how these communication patterns change; for example, the frequency of information requests versus the frequency of discussing navigation options.

Psychophysiological metric: Eye tracking data is obtained through the use of disposable electrodes positioned on the top and bottom of the eye. The electrodes were connected to the participant prior to each test scenario. These devices stay in place throughout the scenario and continuously collect and store data for later retrieval and analysis. Although advances in computer technology are making data collection, storage, and retrieval easier, expertise is still needed to interpret the data. The markers of where the pilot is looking is superimposed over a video of the screen. Researchers must create a time log of how long the pilots look at each display or display feature. The data from these measures are then compared to events that took place during the scenario (as noted by the experimenter or captured by the simulation) to look for patterns associated with the eye tracking data and events within the simulation. A comparison is then made between eye gaze measures collected during the two scenarios and other measures collected during the study in order to identify relevant patterns. Results will tell designers how long pilots looked at each display feature and the order in which information was acquired. This data can be linked to specific events of interest to provide information on how these information acquisition times and patterns may be different with the new display as compared to the previous display.

Testable responses. Testable responses must be carefully scripted into the scenario. An example event would be the failure of the autopilot to follow the prescribed course (e.g., it does not turn when it should). If the pilot notices the error, she should take immediate action to correct the course heading (the testable response). Even when tightly scripted, operator actions can preclude an event from occurring. The operators' responses (or nonresponses) to these events and the time at which they make them are recorded either by the experimenter or by the simulator. Results from data collected from the two systems will be analyzed to determine if the results vary with display design. A comparison will be made of the time required for pilots to respond to the autopilot failure with and without the new navigation display. Faster response times can be inferred to correspond to higher SA.

Performance outcome. Measures of performance outcome will vary by task and domain. In this scenario, performance outcome measures include items like 'were all ATC instructions adhered to?', 'At any time was the safety of the flight in jeopardy?', and 'How many errors did the pilots make interacting with the navigation system?' These measures will be collected throughout the study and analyzed at the completion of each trial. Performance measures are important complements to other measures of SA. Statistical analysis will be performed comparing performance outcome between the two navigation system designs. Results will show whether the pilot performs better on these measures with the new display.

Situation Awareness Rating Technique. The Situation Awareness Rating Technique is typically administered at the completion of the scenario and can be administered via pencil and paper or computer program. The participants provide a rating for each of three or ten dimensions (depending on the version) by drawing a line at the appropriate point in the scale (Figure 12.5). The line markings are converted to percentages to obtain the SART scores for each dimension. These can then be combined to form the overall SART SA rating. These measures (i.e., the three dimensions and the overall score) are statistically compared to determine whether pilots feel their SA was better with the new navigation display, providing a subjective assessment.

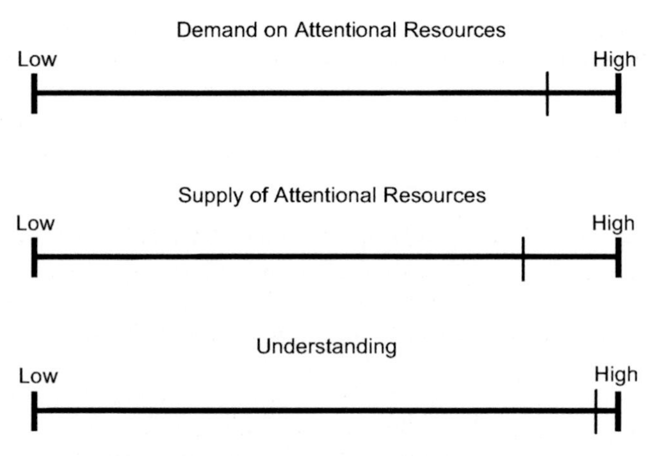

Figure 12.5 3-D SART (adapted from Taylor, 1990).

Situation Awareness Global Assessment Technique. Four SAGAT stops of 2 minutes each will occur in each trial at predetermined times selected at random. The SAGAT queries were developed from a goal-directed task analysis performed for commercial aviation (see Appendix A). A subset of these SAGAT questions are shown in Table 12.7. The first question presented to the participant is, 'What is your current phase of flight?' Questions appropriate for that phase of flight will then be presented in their entirety at each stop via a program on a personal computer. An example of the computer presentation of a SAGAT query is shown in Figure 12.6.

Table 12.7 Example SAGAT queries for commercial aircraft.

Query 0	Phase of Flight
Query 1	Current Fuel vs. Planned Fuel
Query 2	Current Altitude vs. Planned Altitude
Query 3	Current Speed vs. Planned Speed
Query 4	Current Heading vs. Planned Heading
Query 5	Clearance Conformance
Query 6	Traffic Conflict
Query 7	Traffic Location
Query 8	Traffic Conflict Type
Query 9	Special Use Airspace
Query 10	Hazardous Weather

The participants' responses to the queries are scored as correct or incorrect in accordance with operationally relevant tolerance bands as compared to the correct answers collected by the simulator and a subject matter expert. The frequency of correct answers on each question when using the new navigation display is then compared to the frequency of correct answers without it.

This type of analysis provides specific direction to designers as to what aspects of SA are successfully being supported by a particular design, and unforeseen problems that may occur. Results will show if pilots are more aware of whether they are conforming to their assigned altitude, heading, and airspeed with the new navigation display. In addition, it will show whether the new display has affected SA in some other unanticipated way, such as decreasing the degree to which they are aware of their distance from terrain and special use airspace as this information is not represented on the display and they may pay less attention to other information sources that do contain it.

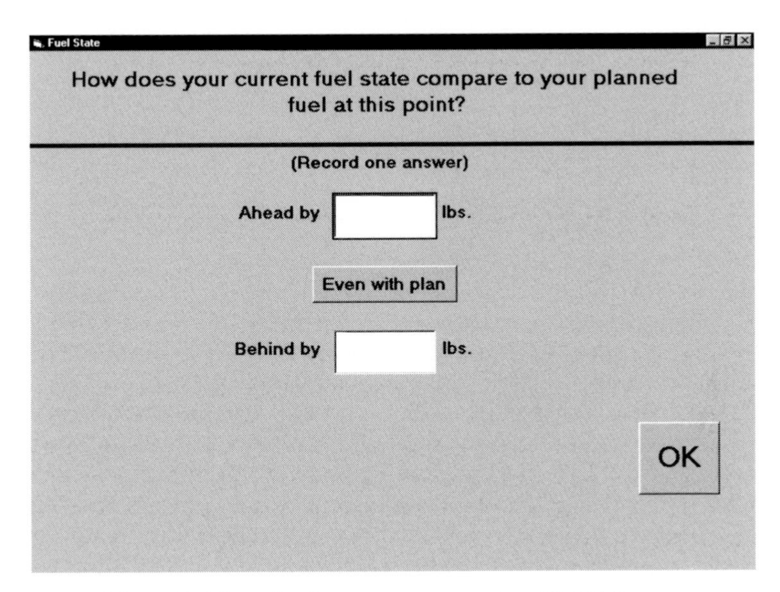

Figure 12.6 Example of a computer presentation of SAGAT query.

12.5 SUMMARY

The choice of metrics for a specific design evaluation depends largely on the goals of the evaluation (e.g., diagnostic evaluation of specific design features versus participants' confidence in their SA when using the system), the characteristics of the medium available for testing (e.g., a simulator or field test), and the expertise of the personnel performing or analyzing the study (e.g., psychophysiological metrics require specific training to analyze).

The application of sound testing metrics and procedures is a bedrock of good design. SA-Oriented Design specifically incorporates the use of objective SA measurement to insure that desired results are obtained in the design process. The results of the evaluation are then used to refine the design concept. Diagnostic measures are important for determining not only whether a given design concept provides better or worse SA than others, but also in what ways it may be better or worse. Many design concepts may help SA in some ways, but hurt SA in other ways. For example, the navigation display discussed here may result in better awareness of horizontal information, but worse awareness of vertical information. Aspects of the design that are found to be problematic can be redesigned. So in this example, a refined navigation display may incorporate a vertical situation display (showing projected altitude changes) as well as a horizontal (top-down) situation display. Information on local terrain may be added to compensate for changes in attention to paper maps with electronic flight displays. The refined design concept would then be retested to evaluate the success of the redesign. In this way, SA-Oriented Design provides specific tailored information to the designer for building system concepts that support high levels of SA.

CHAPTER THIRTEEN

Applying SA-oriented Design to Complex Systems

As a result of technological advances, enormous amounts of data can be made available to operators. Unfortunately, the ability of technology to provide data surpasses the human's ability to effectively process the information. This divergence creates a significant information gap between the data that is available and that which is needed, limited by the human's ability to sort through and make sense of the data, particularly when it changes rapidly in dynamic systems. Simultaneously, technological advances have put at the fingertips of system designers an almost infinite number of possibilities for presenting data. However, taking full advantage of all that is offered by technology is often difficult. The mere existence of so many options can end up making the system designer's job more difficult. *Which among the vast quantities of data is the information that the operator needs? What is the best way to present that information to the operator to maximize performance and minimize frustration? How should the information be organized and what should be emphasized?*

Many of these questions can be addressed by designing systems to maximize the person's ability to perceive needed information, comprehend what that information means, and use it to project the future state of the system – in other words, by designing systems to support the operator in achieving and maintaining a high level of SA. Taking the operator's SA needs into account at the beginning of the design process is key to this approach. It provides the foundation for dealing with system design on the basis of information content, rather than just at a surface feature level (e.g., colors, symbology type, and size of font). Design guidelines for supporting SA based on a model of how people gain, maintain, and in some cases lose SA is tied to this critical foundation, providing the basis for transforming SA requirements into systems that will deliver that SA effectively. Finally, iterative rapid prototyping in association with objective SA measurement plays a key role in identifying unintended interactions within the design, and in insuring that the goals of usability, SA, performance, and operator acceptance are achieved.

The SA-oriented Design process (Figure 13.1) is a simple three-stage progression for addressing SA within the system design process. It supplements other human factors analysis techniques, design guidelines, and measurement approaches to provide a critical linkage to the cognitive foundation of performance in systems operations—the ongoing representation of the current situation that drives decision making and behavior aimed at keeping that environment aligned with operator goals. This approach is aimed not at system design at the component level, so much as system design at an overall level of integration. Its goal is to insure that the dynamic information presentation shown to operators matches with what is required for their SA and with the way the brain processes that information to achieve SA on an ongoing basis in a complex system.

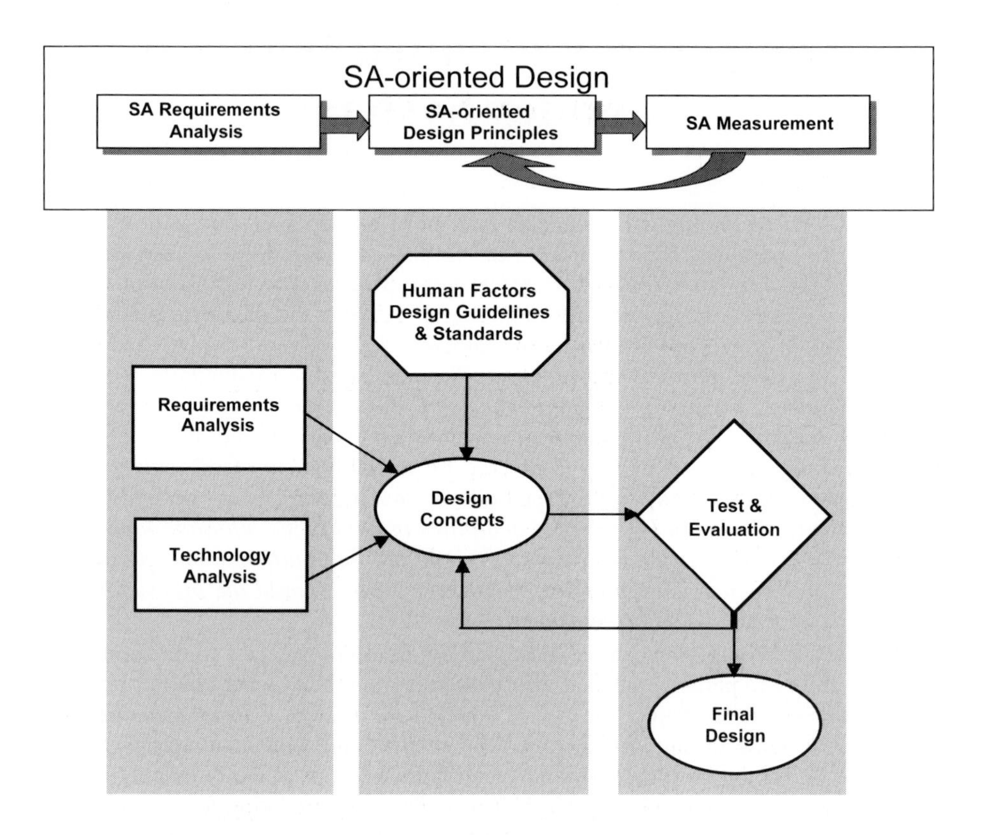

Figure 13.1 SA-oriented Design as a part of the overall system design process.

Fifty design principles are presented in this book to help the system designer create interfaces that effectively support operator SA. These design principles are summarized in Table 13.1. They were developed based on the best of what is known to date on the mechanisms, strengths, and weaknesses of human situation awareness. The basic design principles stem from a theoretical model of SA, and a wide range of research on SA and on the impact of significant system features on SA. The SA-oriented Design principles also cover a wide range of issues that impact on SA in today's complex systems, including certainty or confidence level of information, system complexity, alarms, automation, and multioperator operations. While it is doubtful this list is conclusive by any means (far more research is needed on many topics that impact on SA in complex systems), it does provide a fundamental approach for designing systems with operator SA as an objective. These principles can be applied to a wide range of systems from a variety of domains where achieving and maintaining SA is challenging.

Table 13.1 Summary of SA-oriented Design principles.

No.	*SA Design Principles*	*Page*
	General	
1	Organize information around goals	83
2	Present Level 2 information directly—support comprehension	83
3	Provide assistance for Level 3 SA projections	85
4	Support global SA	86
5	Support trade-offs between goal-driven and data-driven processing	86
6	Make critical cues for schema activation salient	86
7	Take advantage of parallel processing capabilities	87
8	Use information filtering carefully	87
	Certainty Design Principles	
9	Explicitly identify missing information	123
10	Support sensor reliability assessment	125
11	Use data salience in support of certainty	126
12	Represent information timeliness	128
13	Support assessment of confidence in composite data	129
14	Support uncertainty management activities	130
	Complexity Design Principles	
15	Just say no to feature creep—buck the trend	144
16	Manage rampant featurism through prioritization and flexibility	144
17	Insure logical consistency across modes and features	145
18	Minimize logic branches	145
19	Map system functions to the goals and mental models of users	145
20	Provide system transparency and observability	146
21	Group information based on Level 2/3 SA requirements and goals	146
22	Reduce display density, but don't sacrifice coherence	146
23	Provide consistency and standardization on controls across different displays and systems	147
24	Minimize task complexity	147
	Alarm Design Principles	
25	Don't make people rely on alarms—provide projection support	164
26	Support alarm confirmation activities	167
27	Make alarms unambiguous	167
28	Reduce false alarms, reduce false alarms, reduce false alarms	167
29	Set missed alarm and false alarm trade-offs appropriately	168
30	Use multiple modalities to alarm, but insure they are consistent	169
31	Minimize alarm disruptions to ongoing activities	169
32	Support the assessment and diagnosis of multiple alarms	170

No.	SA Design Principles	Page
33	Support the rapid development of global SA of systems in an alarm state	170
	Automation Design Principles	
34	Automate only if necessary	186
35	Use automation for assistance in carrying out routine actions rather than higher level cognitive tasks	186
36	Provide SA support rather than decisions	186
37	Keep the operator in control and in the loop	188
38	Avoid the proliferation of automation modes	188
39	Make modes and system states salient	188
40	Enforce automation consistency	189
41	Avoid advanced queuing of tasks	189
42	Avoid the use of information cueing	190
43	Use methods of decision support that create human/system symbiosis	190
44	Provide automation transparency	190
	Multioperator Design Principles	
45	Build a common picture to support team operations	213
46	Avoid display overload in shared displays	215
47	Provide flexibility to support shared SA across functions	215
48	Support transmission of different comprehensions and projections across teams	216
49	Limit nonstandardization of display coding techniques	217
50	Support transmission of SA within positions by making status of elements and states overt	218

13.1 COMBATING THE ENEMIES OF SITUATION AWARENESS

Effective system designs are those that afford operators a high level of SA. One facet of achieving this ideal is to avoid known pitfalls that can degrade SA. These problems include the SA nemeses discussed in Chapter 3:

- attentional tunneling,
- requisite memory trap,
- workload / anxiety / fatigue / other stressors (WAFOS),
- data overload,
- misplaced salience,
- complexity creep,
- errant mental models, and
- the out-of-the-loop syndrome.

Many of the 50 SA-oriented Design principles are squarely directed at combating these SA demons.

13.1.1 Attentional narrowing

SA is compromised when conditions exist that incline people to stop attending to all the relevant aspects of the environment in favor of focusing on a particular subset of the environment. With attentional narrowing, some aspect of the environment grabs people's attention and draws their focus almost exclusively to that particular information. In some cases, people are aware that their attention is focused on a subset of the information, but they believe that this information is the most important and is thus deserving of such focused attention. Other times, operators simply become engrossed in what is going on with one aspect of the situation and forget to monitor other information. In either case, attentional narrowing negatively impacts SA as the operator's knowledge of other not attended to facets of the situation quickly become outdated and thus inaccurate. They are unaware of new information that should shift their priorities.

A number of SA-oriented Design principles should minimize the likelihood that operators will succumb to attentional narrowing. For example, designing a system to support global SA (Principle 4) will help the operator maintain a high level of understanding of what is happening across the full range of goals and SA requirements, and help lessen the likelihood that one aspect of the environment will become overly dominant. When new information becomes available that should shift operator priorities, it is far less likely to be missed.

Further, making critical cues for schema activation salient (Principle 6) increases the probability that the system design will encourage the operator to notice key triggers for redirecting goal priorities and help prevent the operator from becoming too absorbed in a subset of the information. Taking advantage of parallel processing mechanisms (Principle 7) can also help avoid attentional narrowing, as information from different modalities can more easily be time-shared. Directly supporting the alternating of goal-driven and data-driven processing (Principle 5) helps insure that the process does not get stuck on any one aspect of the situation.

Conversely, design features that restrict the information to which the operator has access can negatively impact SA by obscuring information that should signal to the operator that other information needs to be attended to. Minimizing information filtering helps ensure that the operator is not deprived of relevant information, but rather has easy access to all the information required to prioritize goals and remain proactive (Principle 8).

While alarms have traditionally been relied upon to jolt people out of the poor situation awareness caused by attentional narrowing, some of the limitations of this approach were discussed in Chapter 9. Principles 25 through 33 provide guidelines that consider the effect of alarms on SA in its totality. Unless alarms are used carefully and in accordance with the cognitive processes people employ in processing alarms, they will not be fully successful at combating this SA demon.

13.1.2 Requisite memory trap

Working memory provides the repository where disparate pieces of information in the current situation are brought together and processed into a meaningful picture of what is happening. As such, in many environments it can easily become overloaded, hampering the person's ability to maintain a high level of SA. Thus, designing the system to minimize the load on working memory is very important for supporting SA.

The SA-oriented Design approach overcomes this demon in two ways. First, the SA requirements analysis insures that the information needed is identified early in the design process. Based on this analysis, system designs can be created that directly provide the information needed, rather than relying on human memory to process or maintain the desired information.

Second, the design principles help to accomplish the goal of minimizing memory load through the system design. For example, the demands on working memory are lessened when information is organized around goals (Principle 1) because the operator does not have to remember and mentally manipulate information from various system components or multiple displays to address any given goal or subgoal. Additionally, the integration of Level 1 information requirements into the form required for Level 2 comprehension is generally performed in working memory with many systems today. Presenting Level 2 information directly to support comprehension (Principle 2) dramatically minimizes the mental gymnastics that the operator must perform, and frees working memory capacity for other needs. Similar gains are achieved by supporting Level 3 SA projections in the system design (Principle 3), minimizing working memory demands overall.

Other principles aimed at reducing complexity (Principles 15 through 24) and insuring certain information needs are directly supported, such as designing to make information confidence levels explicit (Principles 9 through 14) will also help insure that working memory loads are kept to a minimum.

13.1.3 Workload, anxiety, fatigue, and other stressors (WAFOS)

The ability of a person to maintain SA can be negatively impacted by both cognitive (e.g., high workload, time pressure) and physical (e.g., heat, noise, cold) stressors. Because stressors often cannot be avoided or eliminated, designing the system to support the operator's decreased mental capacity under stress is imperative.

Stressors can have a variety of effects on the cognitive processes that are required to achieve and maintain SA. For example, when exposed to stressors, people tend to become less efficient at gathering information—they may pay less attention to peripheral information, become more disorganized in scanning, and may be more likely to succumb to attentional tunneling. The design principles that reduce attentional narrowing should be extra effective under conditions of stress. Design features that support efficient information gathering patterns will also help reduce SA decrements associated with these stressors. Grouping information based on Levels 2 and 3 SA requirements and goals so the operator can easily acquire the needed information (Principles 1 and 21), making alarms unambiguous thereby limiting the amount of time a person must spend decoding the alarm (Principle 27), taking advantage of parallel processing by distributing information across modalities instead of overloading a particular modality (Principle 7) and using redundant cueing (Principle 30), and assisting proactive as opposed to reactive behaviors (Principle 25) are all effective strategies for minimizing the negative impact of unavoidable stressors. In that people's ability to execute well-learned behaviors from long-term memory is better under stress than their ability to generate novel solutions, techniques for supporting scheme activation (Principle 6) should also be effective under stress.

13.1.4 Data overload

Data overload is a by-product of advances in technology. The amount of data that can conceivably be presented to an operator can be staggering. The capacity of technology in many systems to present data far surpasses the ability of the human to process it. SA-oriented Design provides a mechanism for the designer to take a close look at the real information needs of the operator and to carefully tailor system designs to those needs creating user-centered systems rather than technology-centered systems.

For example, by organizing information around goals (Principle 1) and Levels 2 and 3 SA requirements (Principle 21), the designer can double check to insure that the information presented is truly information that the operator needs. Furthermore, data overload in many systems exists not as a function of too much data, but as a function of the disorganization of that data or the degree to which its organization fails to map to the organization needed for cognitive processing. These principles, together with the SA requirements analysis, provide the organizing principles needed for directly reducing the data overload induced by technology-centered information presentation.

At times, even reducing data to needed information will not keep the operator from being overwhelmed with available information. In this respect, the data overload problem is similar to the workload problem, and can have negative consequences for SA. Methods for reducing display density and creating coherence (Principle 22) assist the operator by minimizing the need to sort through data to find needed information. Directly presenting higher level SA needs (Principles 2 and 3) can also minimize the amount of lower level data that needs to be presented in some cases.

13.1.5 Misplaced salience

Salience refers to the ability of a particular form of information presentation to grab a person's attention, regardless of the importance of the task or goal in which the person was previously engaged. Salience is largely determined by the physical characteristics of the information, such as blinking lights, buzzers, and alarms, which tend to be highly salient. When these attributes are applied inappropriately or excessively, the operators' attention can be drawn from things they should be attending to, negatively impacting SA.

Salience should be employed in such a way that it does not unnecessarily draw operators' attention away from high priority goals and information. For example, when a recurring false alarm commands operator attention by emitting a loud sound or demanding alarm acknowledgement, other relevant ongoing tasks and information processing will likely be interrupted as the operator addresses the false alarm. Thus not only do false alarms not accomplish their intended goals, they also fail to be benign; they can actively reduce SA.

By restricting the use of perceptually salient features to carefully selected information presentation needs, the disruption to ongoing information acquisition that is important for SA will be minimized (Principle 6). Designing systems to minimize false alarms (Principle 28) also acts to minimize the problems associated with misplaced salience, in that any highly salient alarms that occur will be appropriate in how they direct operator attention. Minimizing the disruptiveness of alarms minimizes the likelihood those salient alarms will interfere with SA (Principle 31).

We also use higher levels of salience to represent higher levels of information accuracy and recency (Principle 11) to support assessments of how much confidence to place in displayed information. Information salience is also used to support the operator in making appropriate shifts between goal-driven and data-driven processing by cuing the operator to attend to certain aspects of the environment (Principle 5), and to support the operator's need for explicit identification of missing information (Principle 9). Thus these design principles work to employ salience in ways that support information processing needs, carefully eliminating its use in ways that are inappropriate or gratuitous.

13.1.6 Complexity creep

Complexity creep emerges as features are routinely or periodically added to a system. These additions may be due either to the need for increased functionality to support a particular task or to the perceived value added by the feature. A superfluous adoption of new features without carefully considering the negative, as well as positive, ramifications, however, can result in a variety of negative consequences for SA. Rampant featurism significantly increases the amount of complexity the user must deal with, taxing working memory, and making it difficult for users to develop appropriate mental models needed for SA.

Careful design can insure that needed system functionality is provided without torpedoing SA. SA-oriented Design ensures that potential features are carefully evaluated for their real benefits in comparison to the costs to the user by virtue of their presence. Each feature is examined and tested through rapid prototyping to ensure that the feature is truly necessary (Principle 15). Existing features are prioritized and organized to minimize the impact of infrequently used features on system complexity (Principle 16). Further, it insures that the logical functioning of a system across modes or features is consistent, minimizing the difficulty of creating a mental model of how the system can be expected to perform in each mode (Principle 17). Similarly, complexity is minimized by reducing the conditional operations associated with a system (Principle 18), making mental models easier to form, and by designing systems that directly minimize task complexity (e.g., how many steps in sequence must be performed to achieve a desired result) (Principle 24).

Automation forms a particular problem, significantly increasing complexity creep in systems it is a part of. Operators who must interact with or oversee automation face a considerable challenge in developing good mental models of automated systems that have any degree of complexity for handling different conditions. By using automation only when absolutely necessary (Principle 34), avoiding the proliferation of automated modes and minimizing the rule sets the operator must remember (Principle 38), enforcing automation consistency (Principle 40), and providing automation transparency (Principle 44), SA-oriented Design seeks to directly limit this avenue for complexity.

13.1.7 Errant mental models

The ability to develop an accurate mental model of how a system functions is essential for SA. Without this framework, correctly assimilating and processing information to form accurate Level 2 and 3 SA is difficult, if not impossible. Poor mental models or mismatches of expectations to system functioning based on the selection of the wrong mental model can be devastating for SA.

Effective system design can go a long way toward assisting the operator in developing these mental models. Mapping system functions to the goals and established mental models of users (Principle 19), providing consistency and standardization across controls on different displays and systems (Principle 23), and enforcing automation consistency so the operator can expect the automation to function similarly in different modes (Principle 40) are all effective strategies for minimizing the problem of errant mental models and the hazards they can create. Further, strategies such as providing system and automation transparency and observability (Principles 20 and 44) and making modes and system states salient (Principle 39) effectively promote the selection of an accurate mental model by allowing the operator to see how the system is performing its task and its current mode. Minimizing the development of inadequate mental models and the incorrect matching of mental models to system functioning will directly help increase SA at the higher levels.

13.1.8 Out-of-the-loop syndrome

The out-of-the-loop syndrome is increasingly problematic for SA in that automation has become a mainstay of many systems. This problem occurs when the operator suffers from a reduced awareness of the state of the automation and the system parameters it is controlling. Numerous factors can contribute to this problem, including vigilance, complacency, monitoring, passive processing, and inadequate system feedback that deprives the operator of needed information.

The careful application of SA-oriented Design principles can combat this significant problem. For example, consideration should be given to the decision as to whether to automate at all (Principle 34). Out-of-the-loop syndrome can be completely avoided if automation is not used.

Recognizing that some automation will continue to be employed in many systems, numerous design principles also seek to use automation in ways that minimize out-of-the-loop syndrome effects. As the human is the most valuable and flexible component of the system and can bring to bear knowledge that is beyond the scope of the automated system, SA-oriented Design seeks to maintain a better synergy between automation and human operators than traditional approaches have provided. Automation should be employed to support the operator, rather than automating to the fullest extent made possible by technology and assigning to the operator the leftover tasks.

Along these lines several principles are directed at how to best design automation: automation should assist operators in carrying out routine actions rather than trying to perform higher level cognitive tasks (Principle 35), support SA rather than provide decisions (Principles 36 and 43), and allow the operator to be in control and in the loop (Principle 37). These approaches have all been found to reduce, if not eliminate, out-of the-loop problems.

Helping the operator to fully understand what the automation is doing, how it is doing it, and what it will do next further reduces the out-of-the-loop performance problem (Principle 44). Automation features that have been found to compound out-of-the-loop problems, such as advanced queuing (Principle 41) and traditional decision support systems (Principle 36), are avoided.

Even with well-designed systems, a person can get out-of-the-loop, so SA-oriented Design recommends systems be designed to bring the operator up to speed quickly in the case of loss of SA. Examples of features that support rapid buildup of SA include supporting global SA (Principle 4), avoiding information filtering schemes (Principle 8), providing enough feedback to support the rapid development of global SA of systems in an alarm state (Principle 33), and supporting the assessment and diagnosis of multiple alarms (Principle 32). Minimizing the out-of-the-loop syndrome is of utmost importance as automation proliferates within systems.

These principles, and others described in this book, provide guidelines for designing complex systems to support the situation assessment processes that operators use and for avoiding known pitfalls in this process.

13.2 SA-ORIENTED DESIGN SYNERGY

Designing to support SA is a holistic approach rather than a set of rules that must be met to create a satisfactory design. Just as no aspect of a system works in isolation, the principles for SA-oriented Design do not work in isolation. Many design principles support SA in more than one way.

For example, avoiding the proliferation of automation modes (Principle 38) not only decreases task complexity and minimizes complexity creep, it also supports selection of the appropriate mental model and minimizes the out-of-the-loop syndrome by assisting the operator in keeping up with the functioning of the system. Appropriate use of salience can be used (1) to support the operator's knowledge of the current system state and mode (Principle 39), thereby reducing out-of-the-loop performance issues as well as encouraging the selection of the appropriate mental model, (2) to assist the operator in shifting between goal-driven and data-driven processing (Principle 5) by drawing the operator's attention to specific information that requires immediate attention, and (3) to support the operator's understanding of the certainty of information (Principle 11) thereby lessening cognitive workload and memory demands.

At times, however, these design principles may also seem to be at odds with each other. In these cases, the designer will have to make a decision as to which design principle should take precedence and then select measures to appropriately evaluate the design (Chapter 12). Understanding how SA-oriented Design principles support SA enables the designer to make design decisions that allow the operator to achieve and maintain a high level of situation awareness, which will, in turn, maximize user-system performance.

13.3 SYSTEM EVALUATION

The application of SA-oriented Design principles provides a solid foundation for system design. However, these principles, like much of design, provide an incomplete recipe. Deciding how best to implement the design principles will at times be nondeterministic. Decisions will have to be made as to how to present the information to the operator. Consequently, rapid prototyping and simulations for experimentally testing the display concepts form an essential backbone for human interface design. Additional benefits are gained from such testing in that unexpected interactions among design features and unintended effects on the operator can be identified and dealt with early in the design cycle.

As a key component of SA-oriented Design, we provide a means of measuring the SA afforded to the operator by the design. Both indirect (process measures and behavioral/performance measures) and direct (subjective and objective metrics) approaches have been described. Often, the resources available for testing a design are somewhat limited and the SA measures must be chosen with financial and time constraints in mind. High returns can be achieved by carefully selecting a subset of available metrics to evaluate the design.

Detailed evaluations of system design concepts should be conducted with representative users and should employ realistic scenarios that represent the challenges that will be placed on operators in real-world system use. In addition to evaluating SA with system concepts when things go well, these scenarios should evaluate that SA when things go wrong; automation failures, missing information, and overload situations should be included in realistic system evaluations to determine where the flaws are in the system design and where operators need additional support to function well.

While performance measures should almost always be included in system evaluation efforts, significant advantages in terms of sensitivity and diagnosticity can be gained by including direct measures of SA as a part of the evaluation process. As it is impossible to test under all possible conditions, the additional benefits of including SA measures are significant for detecting system problems. SAGAT data provide detailed diagnostic information with respect to specific features of the design and can highlight differences in effectiveness of the design to support SA under varying conditions such as differing workload levels or scenario events.

When the SA-oriented Design approach has been utilized from the beginning of the design process, employing SAGAT is an easy proposition because the SA requirements analysis will have been completed prior to the start of the design process, and this same requirements analysis provides the basis for development of SAGAT queries. Thus, SAGAT can provide information to point out areas where unintended interactions influence operator SA (either positively or negatively), where trade-offs in SA are occurring across different conditions, and identify the strengths and weaknesses of different design concepts. Most importantly, this information can be used to modify designs to correct for any deficiencies and achieve design objectives. Iterative design and testing form a core tenet of SA-oriented Design.

13.4 FUTURE DIRECTIONS

SA-oriented Design provides a general process for designing systems to support SA. Based on supporting the key driver of decision making in dynamic systems, this design approach provides one method for overcoming the problems of technology-centered design and creates user-centered systems. The design principles included here form a good start toward the goal of creating systems that support SA. Undoubtedly, they are not complete. As more research is conducted on SA, and the factors that effect it in a wide variety of system domains, we will develop a basis for augmenting these design principles, creating additional levels of detail on system features that will enhance or undermine SA in different operational settings and with new technological advances.

The future lies in conducting the research needed to ascertain just which technologies and interface approaches succeed in providing higher levels of SA under ever increasing challenges. Solid empirical testing and objective measurement of SA is needed to transform this research into usable results for system designers.

The future also lies in the application of this methodology to ongoing system design efforts in a wide variety of system initiatives. Many new technologies are being developed for automobiles, capable of significantly transforming the task of driving. Military operations are changing significantly, as computers and networking technologies change the nature of command and control, and the way in which soldiers, sailors, and pilots conduct operations. They will be faced with never before seen arrays of data and technologies that can transform both war and peace. Air traffic control, power plant operations, space missions—no area is immune from the charge of technology, with all the potential benefits and hazards that come with it. Harnessing these new advances in successful ways depends, more than ever before, on ensuring that the information they provide and the way in which they function meshes with the cognitive capabilities and needs of their human operators. SA-oriented Design provides a blueprint for moving in that direction.

Appendix A: Goal-directed Task Analysis for Commercial Airline Pilots

(from Endsley, Farley, Jones, Midkiff, & Hansman, 1998)

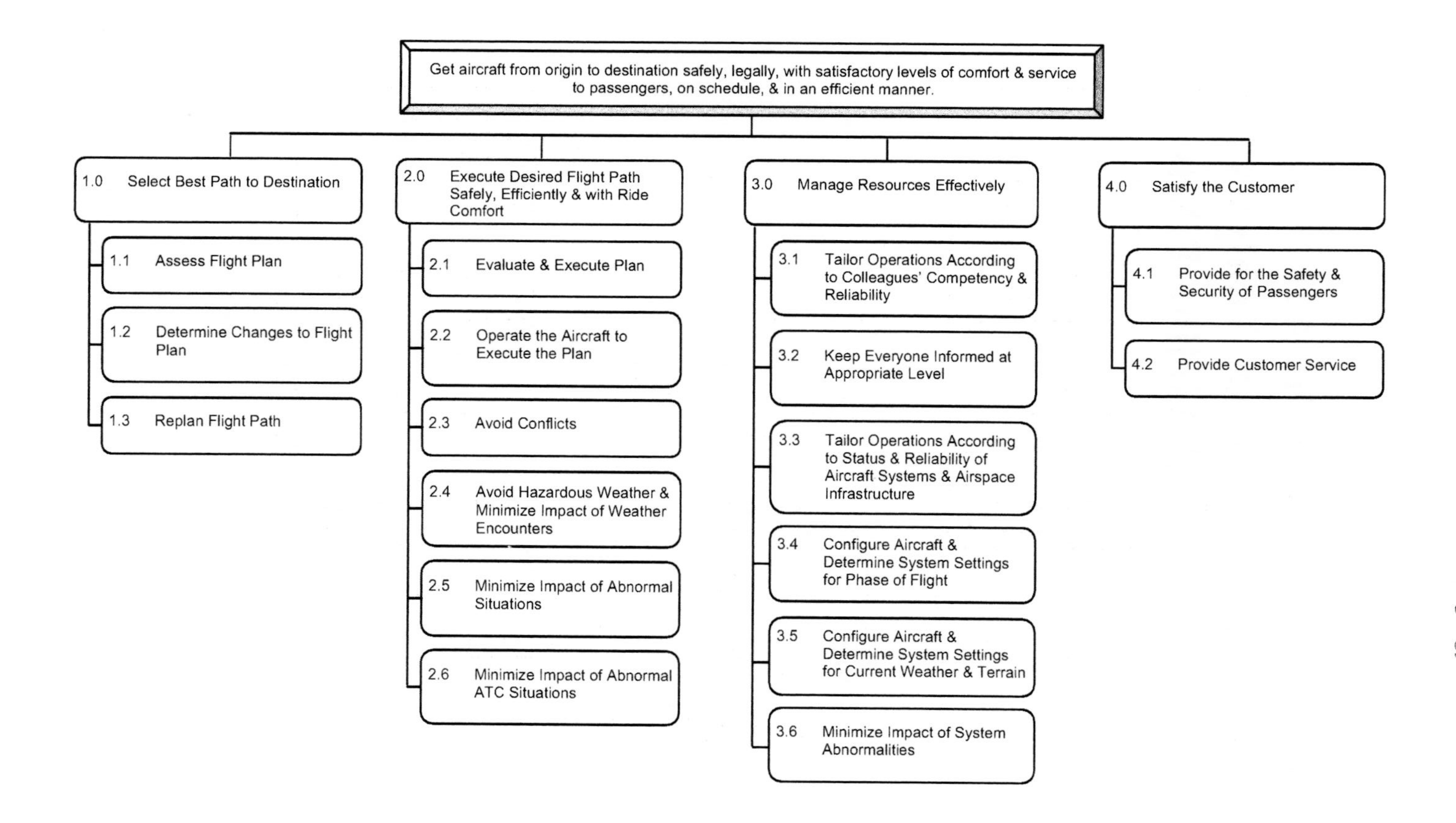
Get aircraft from origin to destination safely, legally, with satisfactory levels of comfort & service to passengers, on schedule, & in an efficient manner.
1.0 Select Best Path to Destination
1.1 Assess Flight Plan
1.2 Determine Changes to Flight Plan
1.3 Replan Flight Path
2.0 Execute Desired Flight Path Safely, Efficiently & with Ride Comfort
2.1 Evaluate & Execute Plan
2.2 Operate the Aircraft to Execute the Plan
2.3 Avoid Conflicts
2.4 Avoid Hazardous Weather & Minimize Impact of Weather Encounters
2.5 Minimize Impact of Abnormal Situations
2.6 Minimize Impact of Abnormal ATC Situations
3.0 Manage Resources Effectively
3.1 Tailor Operations According to Colleagues' Competency & Reliability
3.2 Keep Everyone Informed at Appropriate Level
3.3 Tailor Operations According to Status & Reliability of Aircraft Systems & Airspace Infrastructure
3.4 Configure Aircraft & Determine System Settings for Phase of Flight
3.5 Configure Aircraft & Determine System Settings for Current Weather & Terrain
3.6 Minimize Impact of System Abnormalities
4.0 Satisfy the Customer
4.1 Provide for the Safety & Security of Passengers
4.2 Provide Customer Service

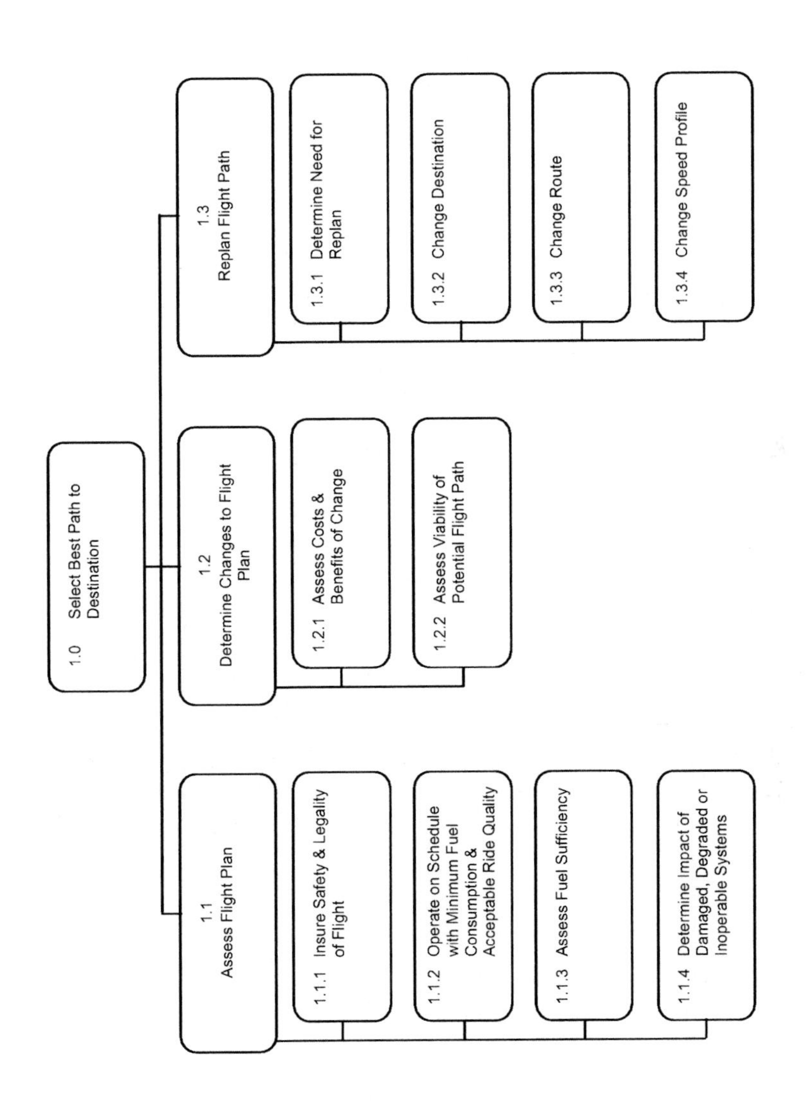
1.0 Select Best Path to Destination
1.1 Assess Flight Plan
1.1.1 Insure Safety & Legality of Flight
1.1.2 Operate on Schedule with Minimum Fuel Consumption & Acceptable Ride Quality
1.1.3 Assess Fuel Sufficiency
1.1.4 Determine Impact of Damaged, Degraded or Inoperable Systems
1.2 Determine Changes to Flight Plan
1.2.1 Assess Costs & Benefits of Change
1.2.2 Assess Viability of Potential Flight Path
1.3 Replan Flight Path
1.3.1 Determine Need for Replan
1.3.2 Change Destination
1.3.3 Change Route
1.3.4 Change Speed Profile

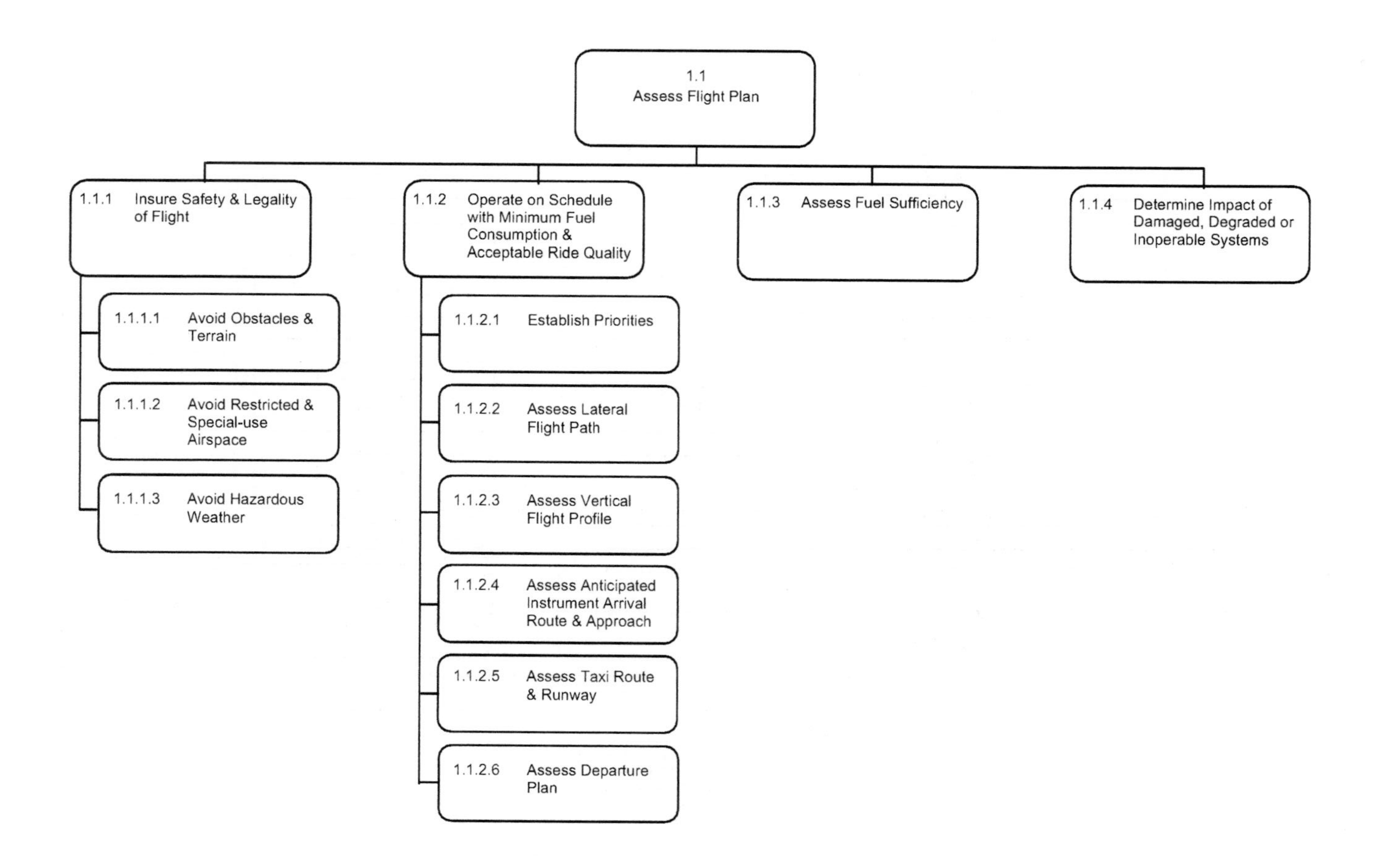
1.1
Assess Flight Plan
1.1.1 Insure Safety & Legality of Flight
1.1.1.1 Avoid Obstacles & Terrain
1.1.1.2 Avoid Restricted & Special-use Airspace
1.1.1.3 Avoid Hazardous Weather
1.1.2 Operate on Schedule with Minimum Fuel Consumption & Acceptable Ride Quality
1.1.2.1 Establish Priorities
1.1.2.2 Assess Lateral Flight Path
1.1.2.3 Assess Vertical Flight Profile
1.1.2.4 Assess Anticipated Instrument Arrival Route & Approach
1.1.2.5 Assess Taxi Route & Runway
1.1.2.6 Assess Departure Plan
1.1.3 Assess Fuel Sufficiency
1.1.4 Determine Impact of Damaged, Degraded or Inoperable Systems

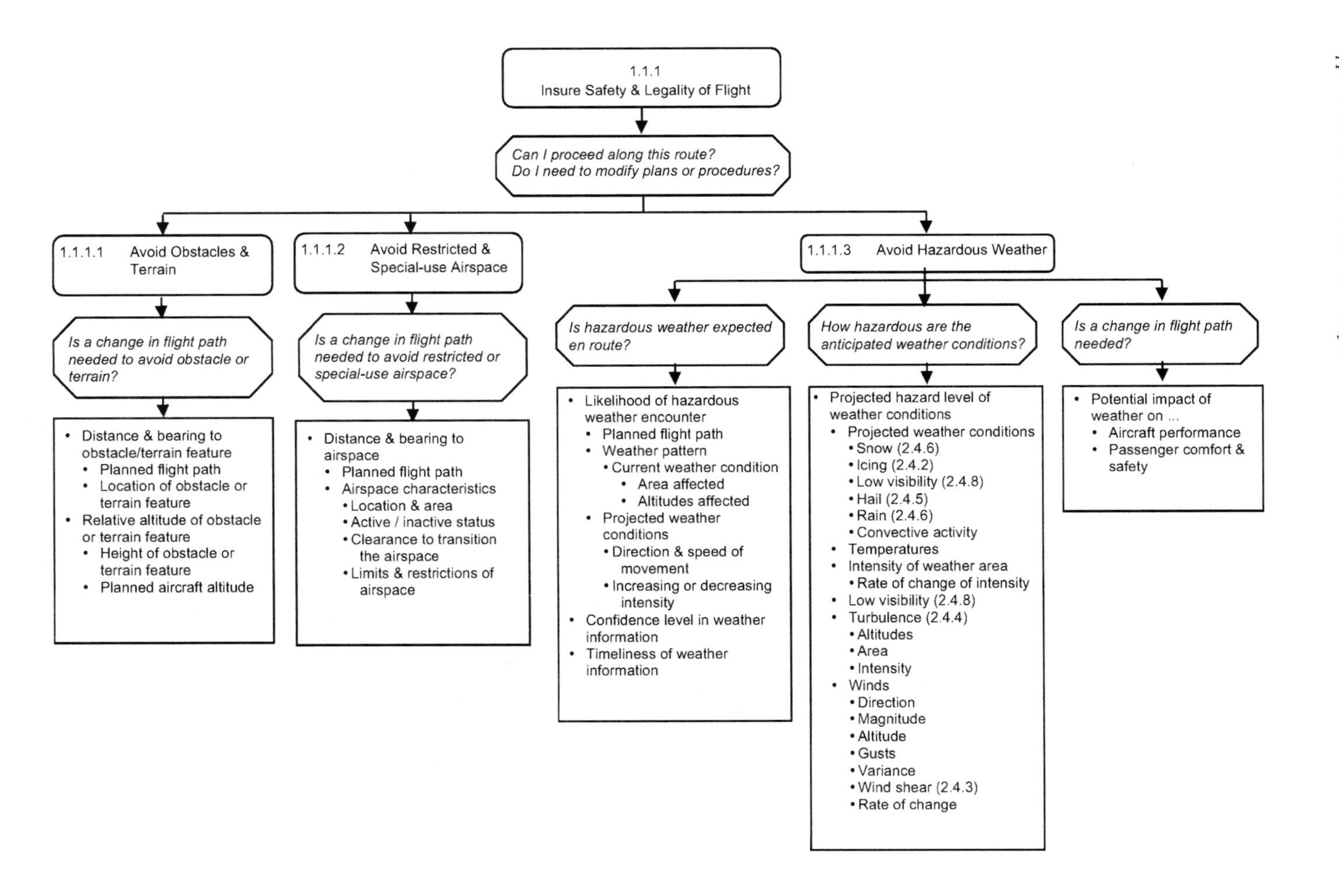

1.1.1
Insure Safety & Legality of Flight
Can I proceed along this route?
Do I need to modify plans or procedures?
1.1.1.1 Avoid Obstacles & Terrain
Is a change in flight path needed to avoid obstacle or terrain?
• Distance & bearing to obstacle/terrain feature
• Planned flight path
• Location of obstacle or terrain feature
• Relative altitude of obstacle or terrain feature
• Height of obstacle or terrain feature
• Planned aircraft altitude
1.1.1.2 Avoid Restricted & Special-use Airspace
Is a change in flight path needed to avoid restricted or special-use airspace?
• Distance & bearing to airspace
• Planned flight path
• Airspace characteristics
• Location & area
• Active / inactive status
• Clearance to transition the airspace
• Limits & restrictions of airspace
1.1.1.3 Avoid Hazardous Weather
Is hazardous weather expected en route?
• Likelihood of hazardous weather encounter
• Planned flight path
• Weather pattern
• Current weather condition
• Area affected
• Altitudes affected
• Projected weather conditions
• Direction & speed of movement
• Increasing or decreasing intensity
• Confidence level in weather information
• Timeliness of weather information
How hazardous are the anticipated weather conditions?
• Projected hazard level of weather conditions
• Projected weather conditions
• Snow (2.4.6)
• Icing (2.4.2)
• Low visibility (2.4.8)
• Hail (2.4.5)
• Rain (2.4.6)
• Convective activity
• Temperatures
• Intensity of weather area
• Rate of change of intensity
• Low visibility (2.4.8)
• Turbulence (2.4.4)
• Altitudes
• Area
• Intensity
• Winds
• Direction
• Magnitude
• Altitude
• Gusts
• Variance
• Wind shear (2.4.3)
• Rate of change
Is a change in flight path needed?
• Potential impact of weather on ...
• Aircraft performance
• Passenger comfort & safety

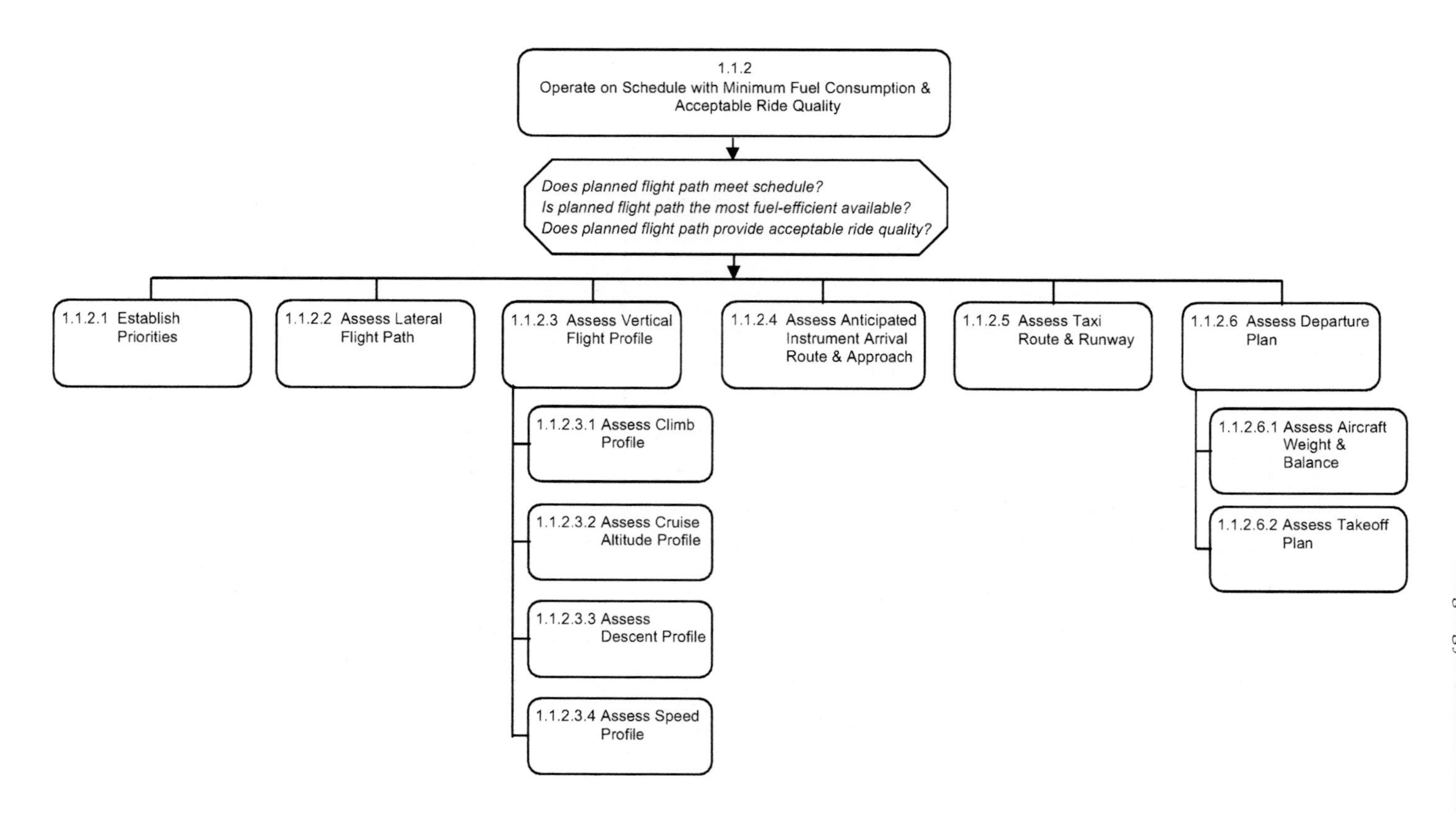
1.1.2
Operate on Schedule with Minimum Fuel Consumption & Acceptable Ride Quality
Does planned flight path meet schedule?
Is planned flight path the most fuel-efficient available?
Does planned flight path provide acceptable ride quality?
1.1.2.1 Establish Priorities
1.1.2.2 Assess Lateral Flight Path
1.1.2.3 Assess Vertical Flight Profile
1.1.2.3.1 Assess Climb Profile
1.1.2.3.2 Assess Cruise Altitude Profile
1.1.2.3.3 Assess Descent Profile
1.1.2.3.4 Assess Speed Profile
1.1.2.4 Assess Anticipated Instrument Arrival Route & Approach
1.1.2.5 Assess Taxi Route & Runway
1.1.2.6 Assess Departure Plan
1.1.2.6.1 Assess Aircraft Weight & Balance
1.1.2.6.2 Assess Takeoff Plan

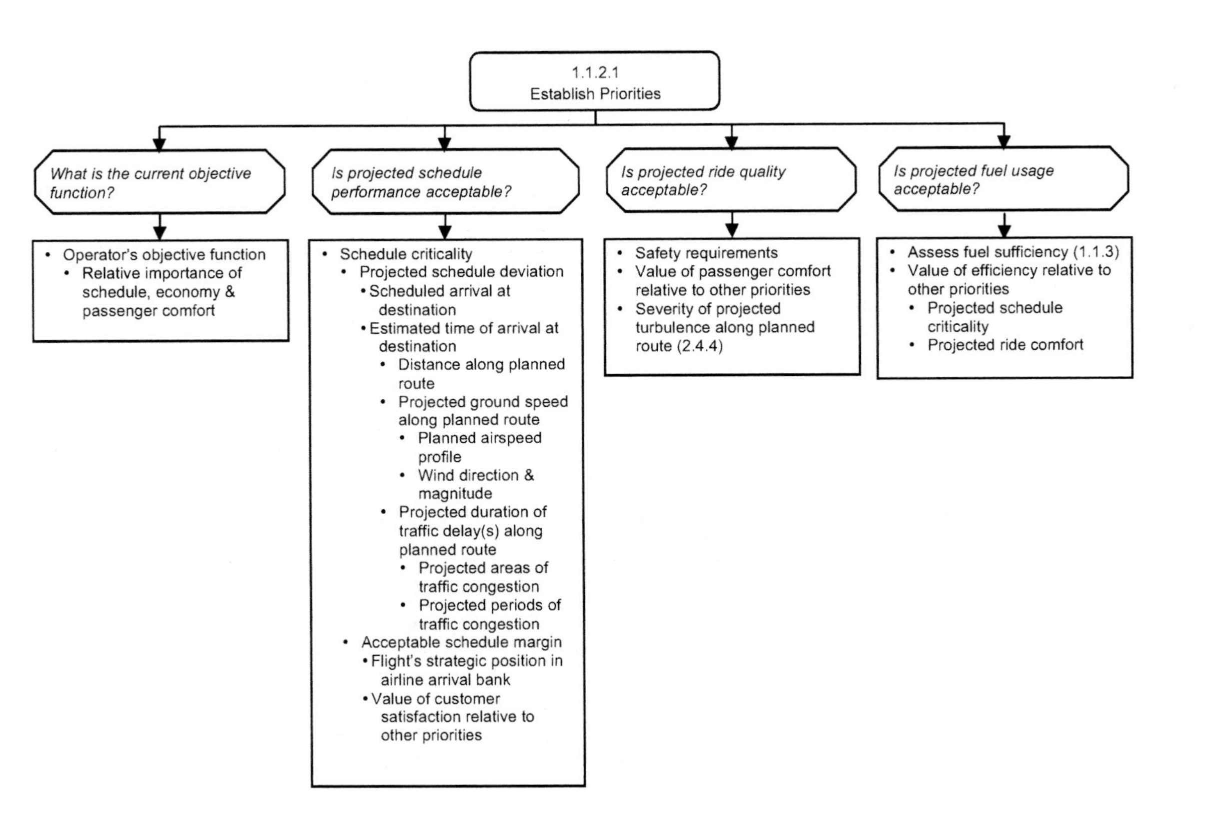
1.1.2.1
Establish Priorities
What is the current objective function?
• Operator's objective function
• Relative importance of schedule, economy & passenger comfort
Is projected schedule performance acceptable?
• Schedule criticality
• Projected schedule deviation
• Scheduled arrival at destination
• Estimated time of arrival at destination
• Distance along planned route
• Projected ground speed along planned route
• Planned airspeed profile
• Wind direction & magnitude
• Projected duration of traffic delay(s) along planned route
• Projected areas of traffic congestion
• Projected periods of traffic congestion
• Acceptable schedule margin
• Flight's strategic position in airline arrival bank
• Value of customer satisfaction relative to other priorities
Is projected ride quality acceptable?
• Safety requirements
• Value of passenger comfort relative to other priorities
• Severity of projected turbulence along planned route (2.4.4)
Is projected fuel usage acceptable?
• Assess fuel sufficiency (1.1.3)
• Value of efficiency relative to other priorities
• Projected schedule criticality
• Projected ride comfort

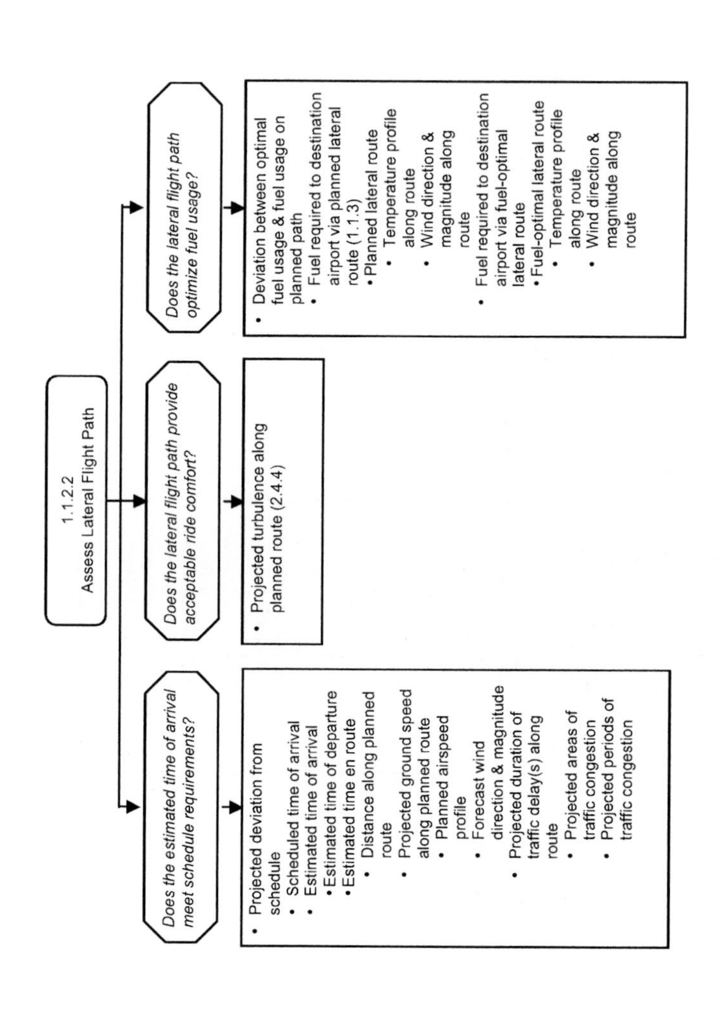
1.1.2.2
Assess Lateral Flight Path
Does the estimated time of arrival meet schedule requirements?
Does the lateral flight path provide acceptable ride comfort?
Does the lateral flight path optimize fuel usage?
• Projected deviation from schedule
• Scheduled time of arrival
• Estimated time of arrival
• Estimated time of departure
• Estimated time en route
• Distance along planned route
• Projected ground speed along planned route
• Planned airspeed profile
• Forecast wind direction & magnitude
• Projected duration of traffic delay(s) along route
• Projected areas of traffic congestion
• Projected periods of traffic congestion
• Projected turbulence along planned route (2.4.4)
• Deviation between optimal fuel usage & fuel usage on planned path
• Fuel required to destination airport via planned lateral route (1.1.3)
• Planned lateral route
• Temperature profile along route
• Wind direction & magnitude along route
• Fuel required to destination airport via fuel-optimal lateral route
• Fuel-optimal lateral route
• Temperature profile along route
• Wind direction & magnitude along route

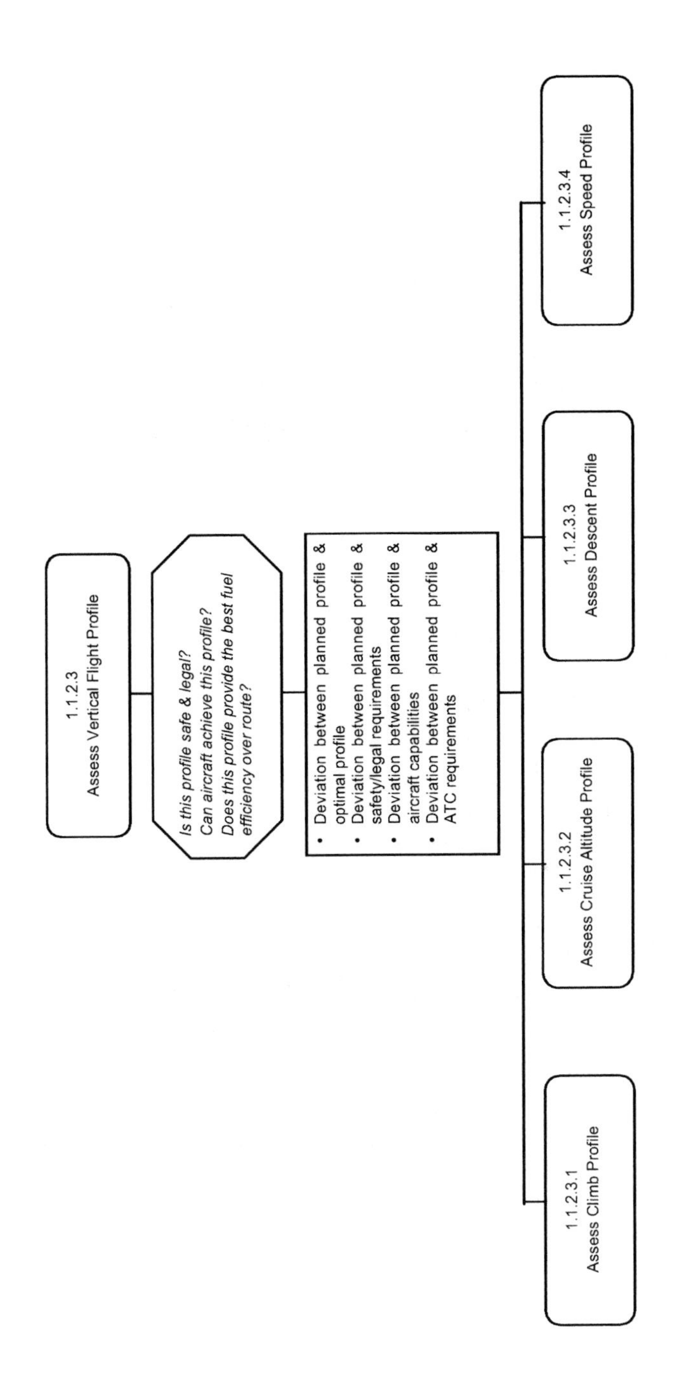
1.1.2.3
Assess Vertical Flight Profile
Is this profile safe & legal?
Can aircraft achieve this profile?
Does this profile provide the best fuel efficiency over route?
• Deviation between planned profile & optimal profile
• Deviation between planned profile & safety/legal requirements
• Deviation between planned profile & aircraft capabilities
• Deviation between planned profile & ATC requirements
1.1.2.3.1
Assess Climb Profile
1.1.2.3.2
Assess Cruise Altitude Profile
1.1.2.3.3
Assess Descent Profile
1.1.2.3.4
Assess Speed Profile

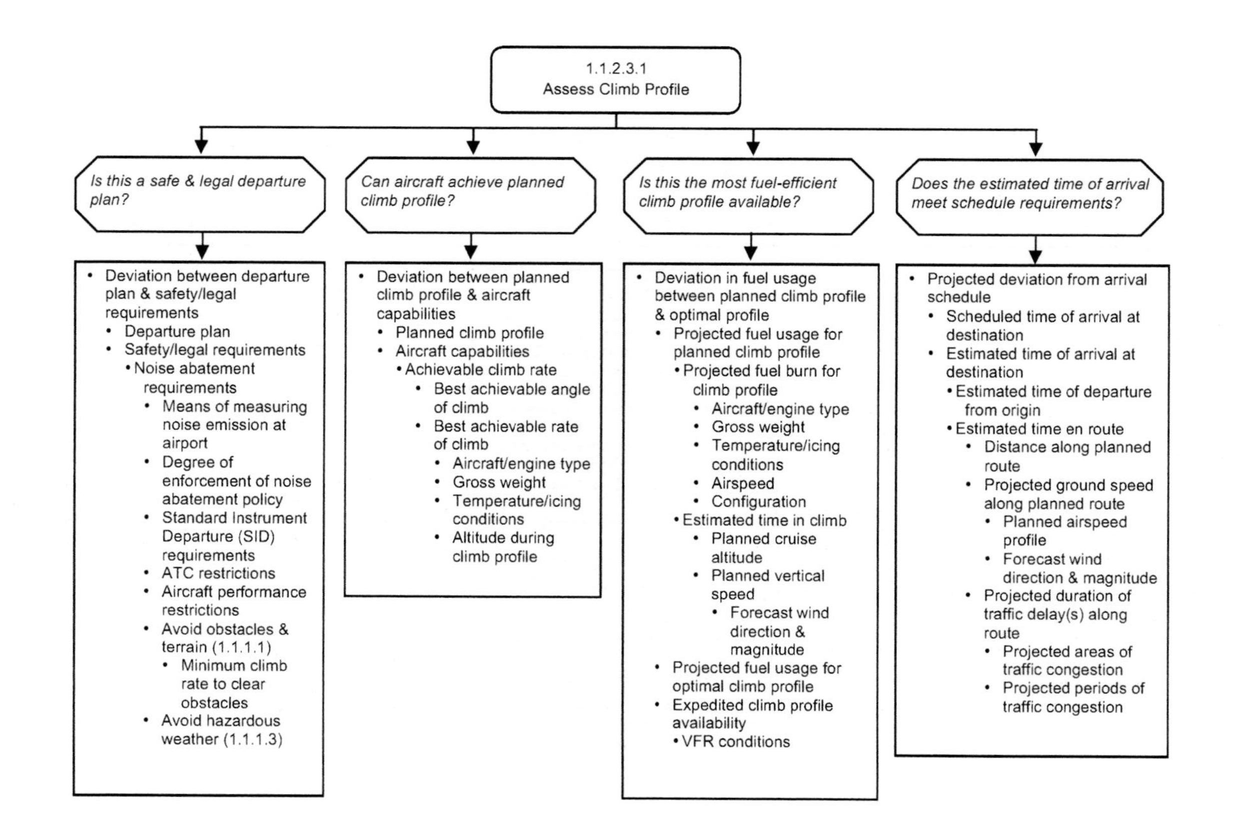
1.1.2.3.1
Assess Climb Profile
Is this a safe & legal departure plan?
• Deviation between departure plan & safety/legal requirements
• Departure plan
• Safety/legal requirements
• Noise abatement requirements
• Means of measuring noise emission at airport
• Degree of enforcement of noise abatement policy
• Standard Instrument Departure (SID) requirements
• ATC restrictions
• Aircraft performance restrictions
• Avoid obstacles & terrain (1.1.1.1)
• Minimum climb rate to clear obstacles
• Avoid hazardous weather (1.1.1.3)
Can aircraft achieve planned climb profile?
• Deviation between planned climb profile & aircraft capabilities
• Planned climb profile
• Aircraft capabilities
• Achievable climb rate
• Best achievable angle of climb
• Best achievable rate of climb
• Aircraft/engine type
• Gross weight
• Temperature/icing conditions
• Altitude during climb profile
Is this the most fuel-efficient climb profile available?
• Deviation in fuel usage between planned climb profile & optimal profile
• Projected fuel usage for planned climb profile
• Projected fuel burn for climb profile
• Aircraft/engine type
• Gross weight
• Temperature/icing conditions
• Airspeed
• Configuration
• Estimated time in climb
• Planned cruise altitude
• Planned vertical speed
• Forecast wind direction & magnitude
• Projected fuel usage for optimal climb profile
• Expedited climb profile availability
• VFR conditions
Does the estimated time of arrival meet schedule requirements?
• Projected deviation from arrival schedule
• Scheduled time of arrival at destination
• Estimated time of arrival at destination
• Estimated time of departure from origin
• Estimated time en route
• Distance along planned route
• Projected ground speed along planned route
• Planned airspeed profile
• Forecast wind direction & magnitude
• Projected duration of traffic delay(s) along route
• Projected areas of traffic congestion
• Projected periods of traffic congestion

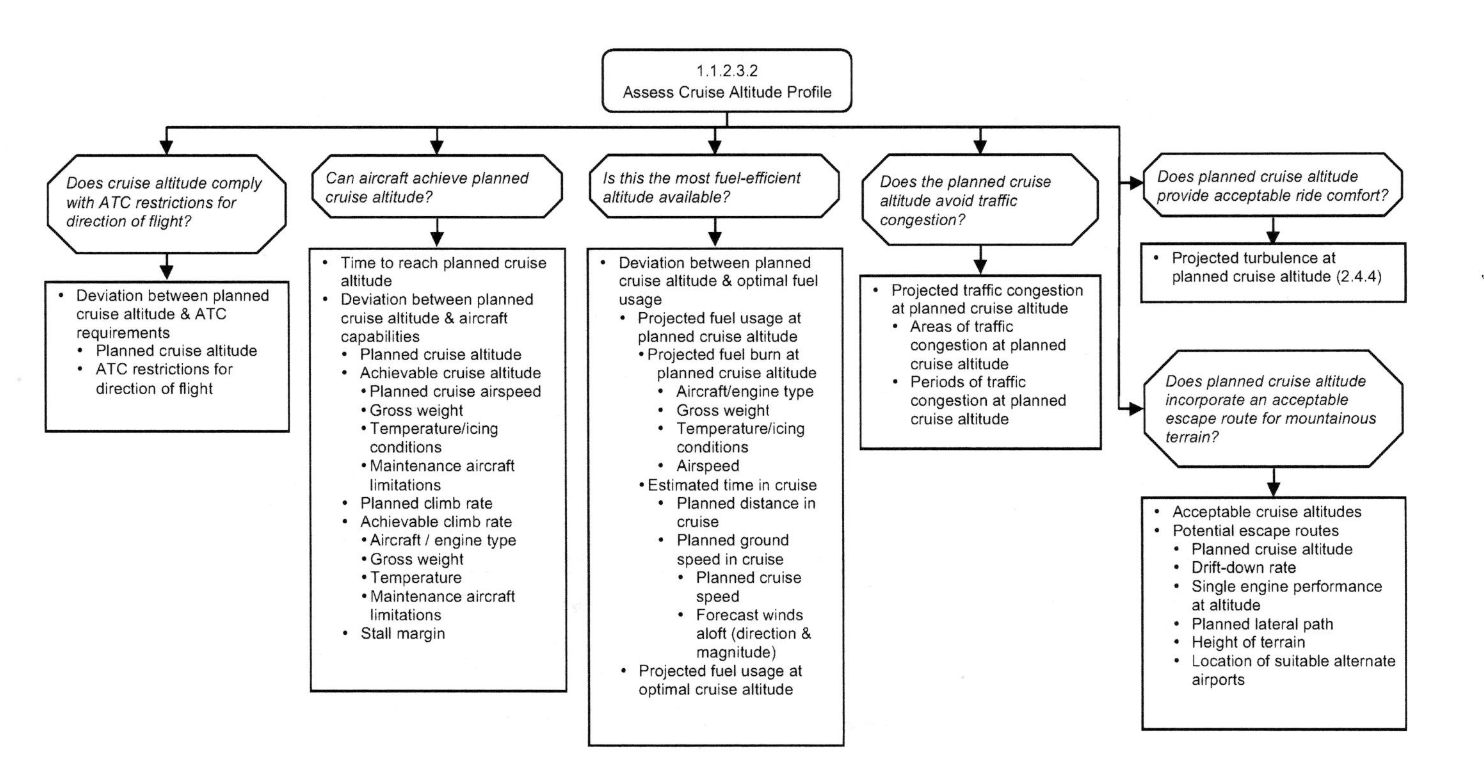
1.1.2.3.2
Assess Cruise Altitude Profile
Does cruise altitude comply with ATC restrictions for direction of flight?
• Deviation between planned cruise altitude & ATC requirements
• Planned cruise altitude
• ATC restrictions for direction of flight
Can aircraft achieve planned cruise altitude?
• Time to reach planned cruise altitude
• Deviation between planned cruise altitude & aircraft capabilities
• Planned cruise altitude
• Achievable cruise altitude
• Planned cruise airspeed
• Gross weight
• Temperature/icing conditions
• Maintenance aircraft limitations
• Planned climb rate
• Achievable climb rate
• Aircraft / engine type
• Gross weight
• Temperature
• Maintenance aircraft limitations
• Stall margin
Is this the most fuel-efficient altitude available?
• Deviation between planned cruise altitude & optimal fuel usage
• Projected fuel usage at planned cruise altitude
• Projected fuel burn at planned cruise altitude
• Aircraft/engine type
• Gross weight
• Temperature/icing conditions
• Airspeed
• Estimated time in cruise
• Planned distance in cruise
• Planned ground speed in cruise
• Planned cruise speed
• Forecast winds aloft (direction & magnitude)
• Projected fuel usage at optimal cruise altitude
Does the planned cruise altitude avoid traffic congestion?
• Projected traffic congestion at planned cruise altitude
• Areas of traffic congestion at planned cruise altitude
• Periods of traffic congestion at planned cruise altitude
Does planned cruise altitude provide acceptable ride comfort?
• Projected turbulence at planned cruise altitude (2.4.4)
Does planned cruise altitude incorporate an acceptable escape route for mountainous terrain?
• Acceptable cruise altitudes
• Potential escape routes
• Planned cruise altitude
• Drift-down rate
• Single engine performance at altitude
• Planned lateral path
• Height of terrain
• Location of suitable alternate airports

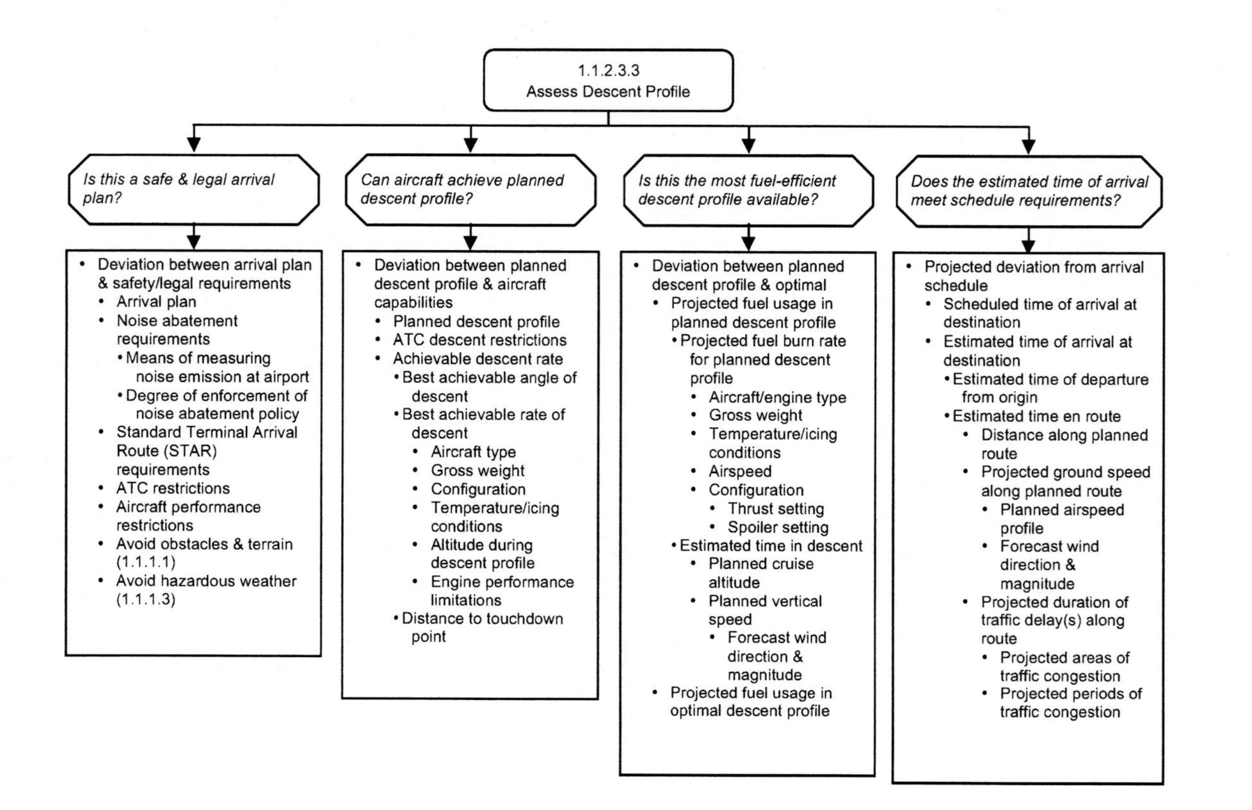
1.1.2.3.3
Assess Descent Profile
Is this a safe & legal arrival plan?
Deviation between arrival plan & safety/legal requirements
Arrival plan
Noise abatement requirements
Means of measuring noise emission at airport
Degree of enforcement of noise abatement policy
Standard Terminal Arrival Route (STAR) requirements
ATC restrictions
Aircraft performance restrictions
Avoid obstacles & terrain (1.1.1.1)
Avoid hazardous weather (1.1.1.3)
Can aircraft achieve planned descent profile?
Deviation between planned descent profile & aircraft capabilities
Planned descent profile
ATC descent restrictions
Achievable descent rate
Best achievable angle of descent
Best achievable rate of descent
Aircraft type
Gross weight
Configuration
Temperature/icing conditions
Altitude during descent profile
Engine performance limitations
Distance to touchdown point
Is this the most fuel-efficient descent profile available?
Deviation between planned descent profile & optimal
Projected fuel usage in planned descent profile
Projected fuel burn rate for planned descent profile
Aircraft/engine type
Gross weight
Temperature/icing conditions
Airspeed
Configuration
Thrust setting
Spoiler setting
Estimated time in descent
Planned cruise altitude
Planned vertical speed
Forecast wind direction & magnitude
Projected fuel usage in optimal descent profile
Does the estimated time of arrival meet schedule requirements?
Projected deviation from arrival schedule
Scheduled time of arrival at destination
Estimated time of arrival at destination
Estimated time of departure from origin
Estimated time en route
Distance along planned route
Projected ground speed along planned route
Planned airspeed profile
Forecast wind direction & magnitude
Projected duration of traffic delay(s) along route
Projected areas of traffic congestion
Projected periods of traffic congestion

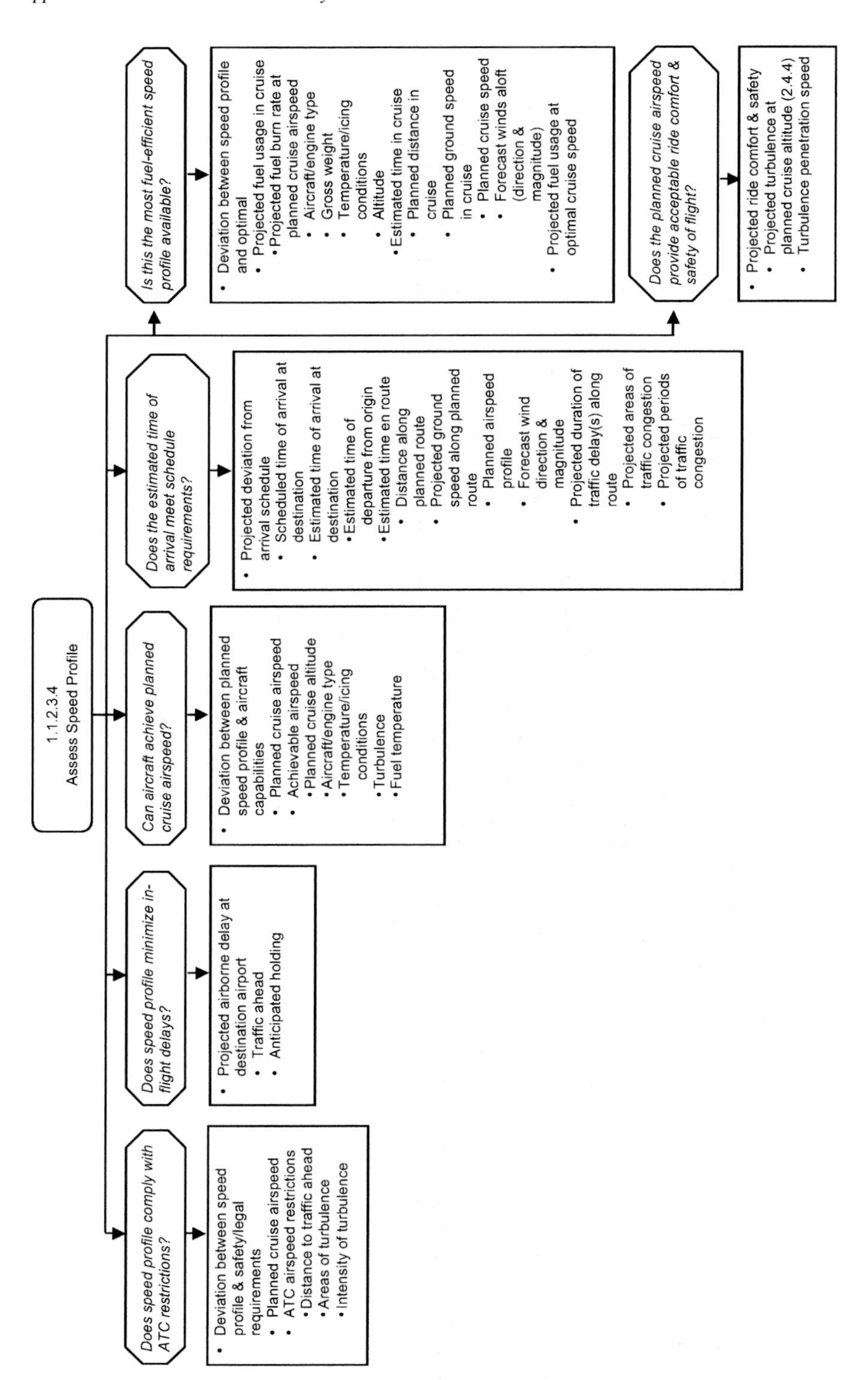
1.1.2.3.4
Assess Speed Profile
Does speed profile comply with ATC restrictions?
• Deviation between speed profile & safety/legal requirements
• Planned cruise airspeed
• ATC airspeed restrictions
• Distance to traffic ahead
• Areas of turbulence
• Intensity of turbulence
Does speed profile minimize in-flight delays?
• Projected airborne delay at destination airport
• Traffic ahead
• Anticipated holding
Can aircraft achieve planned cruise airspeed?
• Deviation between planned speed profile & aircraft capabilities
• Planned cruise airspeed
• Achievable airspeed
• Planned cruise altitude
• Aircraft/engine type
• Temperature/icing conditions
• Turbulence
• Fuel temperature
Does the estimated time of arrival meet schedule requirements?
• Projected deviation from arrival schedule
• Scheduled time of arrival at destination
• Estimated time of arrival at destination
• Estimated time of departure from origin
• Estimated time en route
• Distance along planned route
• Projected ground speed along planned route
• Planned airspeed profile
• Forecast wind direction & magnitude
• Projected duration of traffic delay(s) along route
• Projected areas of traffic congestion
• Projected periods of traffic congestion
Is this the most fuel-efficient speed profile available?
• Deviation between speed profile and optimal
• Projected fuel usage in cruise
• Projected fuel burn rate at planned cruise airspeed
• Aircraft/engine type
• Gross weight
• Temperature/icing conditions
• Altitude
• Estimated time in cruise
• Planned distance in cruise
• Planned ground speed in cruise
• Planned cruise speed
• Forecast winds aloft (direction & magnitude)
• Projected fuel usage at optimal cruise speed
Does the planned cruise airspeed provide acceptable ride comfort & safety of flight?
• Projected ride comfort & safety
• Projected turbulence at planned cruise altitude (2.4.4)
• Turbulence penetration speed

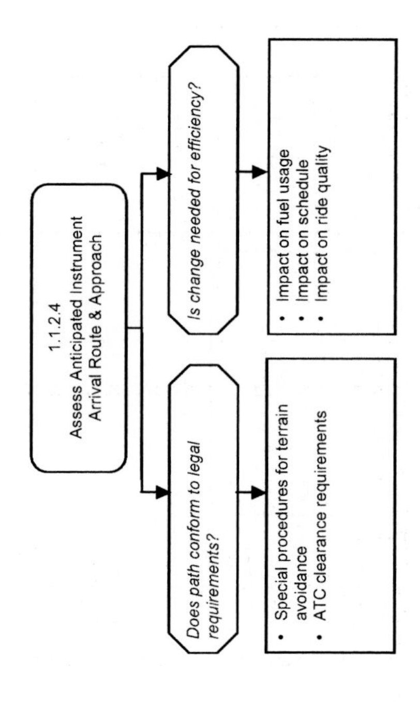
1.1.2.4
Assess Anticipated Instrument Arrival Route & Approach
Does path conform to legal requirements?
Is change needed for efficiency?
Special procedures for terrain avoidance
ATC clearance requirements
Impact on fuel usage
Impact on schedule
Impact on ride quality

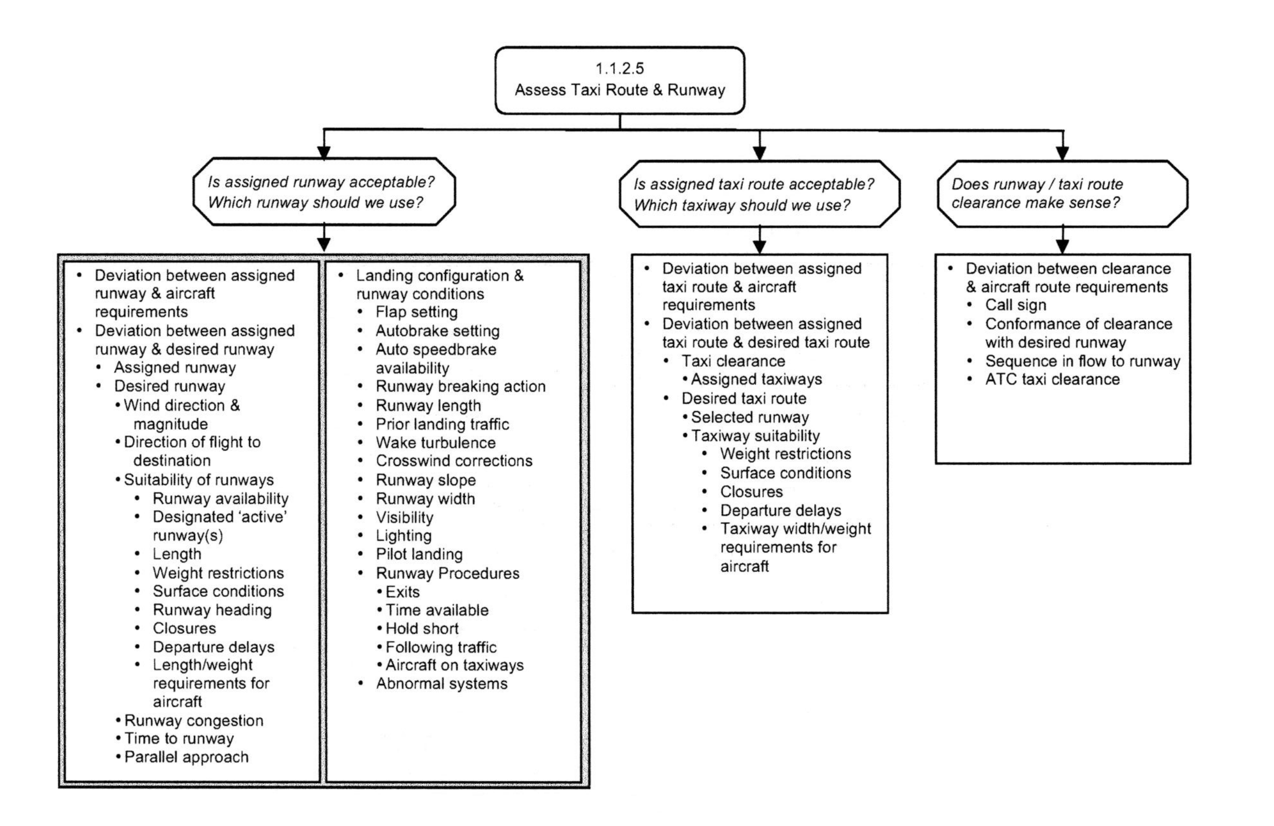
1.1.2.5
Assess Taxi Route & Runway
Is assigned runway acceptable? Which runway should we use?
• Deviation between assigned runway & aircraft requirements
• Deviation between assigned runway & desired runway
• Assigned runway
• Desired runway
• Wind direction & magnitude
• Direction of flight to destination
• Suitability of runways
• Runway availability
• Designated 'active' runway(s)
• Length
• Weight restrictions
• Surface conditions
• Runway heading
• Closures
• Departure delays
• Length/weight requirements for aircraft
• Runway congestion
• Time to runway
• Parallel approach
• Landing configuration & runway conditions
• Flap setting
• Autobrake setting
• Auto speedbrake availability
• Runway breaking action
• Runway length
• Prior landing traffic
• Wake turbulence
• Crosswind corrections
• Runway slope
• Runway width
• Visibility
• Lighting
• Pilot landing
• Runway Procedures
• Exits
• Time available
• Hold short
• Following traffic
• Aircraft on taxiways
• Abnormal systems
Is assigned taxi route acceptable? Which taxiway should we use?
• Deviation between assigned taxi route & aircraft requirements
• Deviation between assigned taxi route & desired taxi route
• Taxi clearance
• Assigned taxiways
• Desired taxi route
• Selected runway
• Taxiway suitability
• Weight restrictions
• Surface conditions
• Closures
• Departure delays
• Taxiway width/weight requirements for aircraft
Does runway / taxi route clearance make sense?
• Deviation between clearance & aircraft route requirements
• Call sign
• Conformance of clearance with desired runway
• Sequence in flow to runway
• ATC taxi clearance

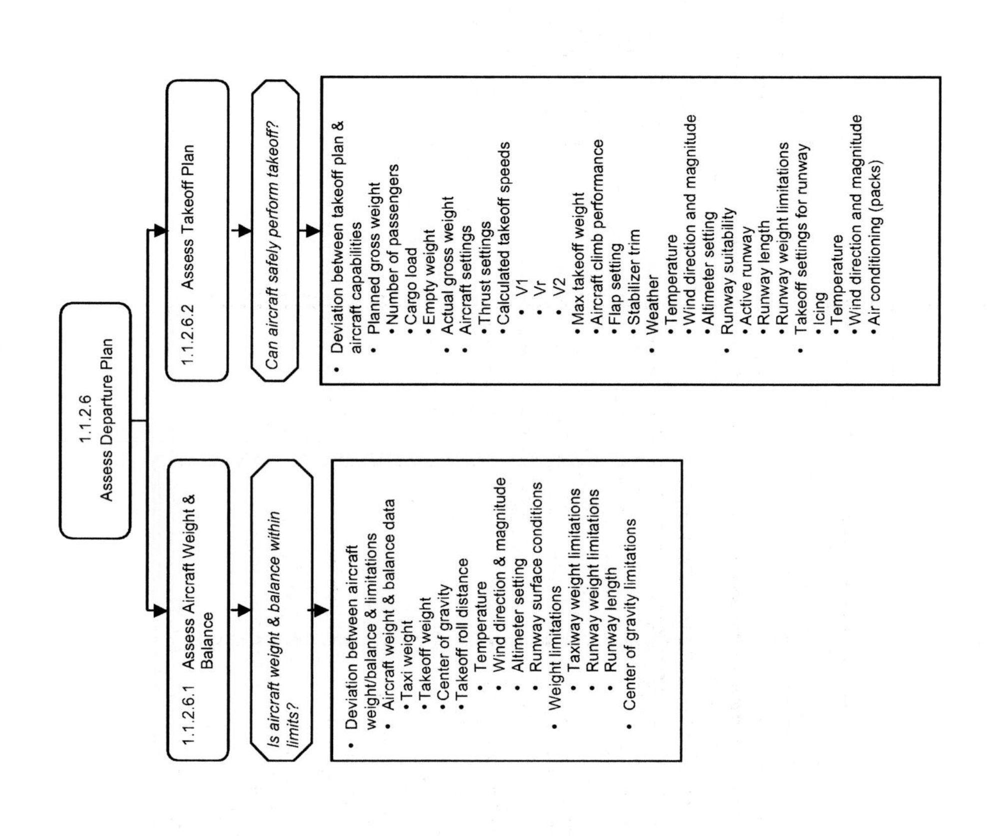
1.1.2.6
Assess Departure Plan
1.1.2.6.1 Assess Aircraft Weight & Balance
Is aircraft weight & balance within limits?
• Deviation between aircraft weight/balance & limitations
• Aircraft weight & balance data
• Taxi weight
• Takeoff weight
• Center of gravity
• Takeoff roll distance
• Temperature
• Wind direction & magnitude
• Altimeter setting
• Runway surface conditions
• Weight limitations
• Taxiway weight limitations
• Runway weight limitations
• Runway length
• Center of gravity limitations
1.1.2.6.2 Assess Takeoff Plan
Can aircraft safely perform takeoff?
• Deviation between takeoff plan & aircraft capabilities
• Planned gross weight
• Number of passengers
• Cargo load
• Empty weight
• Actual gross weight
• Aircraft settings
• Thrust settings
• Calculated takeoff speeds
• V1
• Vr
• V2
• Max takeoff weight
• Aircraft climb performance
• Flap setting
• Stabilizer trim
• Weather
• Temperature
• Wind direction and magnitude
• Altimeter setting
• Runway suitability
• Active runway
• Runway length
• Runway weight limitations
• Takeoff settings for runway
• Icing
• Temperature
• Wind direction and magnitude
• Air conditioning (packs)

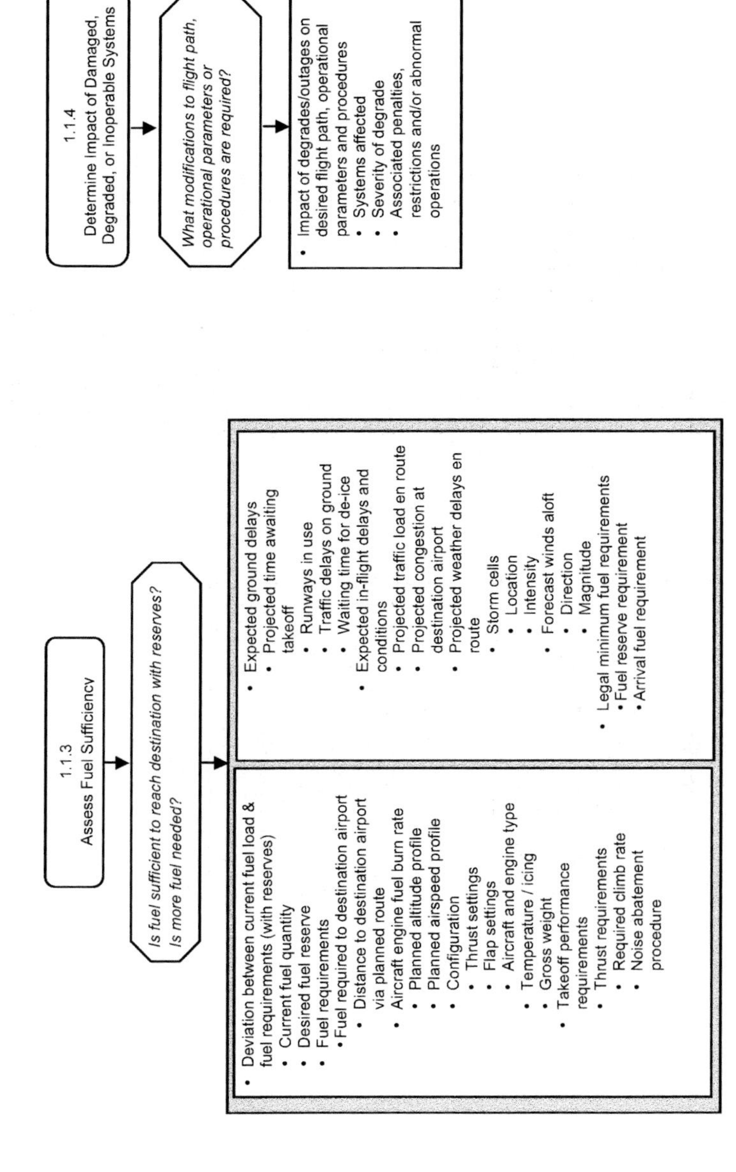
1.1.3
Assess Fuel Sufficiency
Is fuel sufficient to reach destination with reserves?
Is more fuel needed?
• Deviation between current fuel load & fuel requirements (with reserves)
• Current fuel quantity
• Desired fuel reserve
• Fuel requirements
• Fuel required to destination airport
• Distance to destination airport via planned route
• Aircraft engine fuel burn rate
• Planned altitude profile
• Planned airspeed profile
• Configuration
• Thrust settings
• Flap settings
• Aircraft and engine type
• Temperature / icing
• Gross weight
• Takeoff performance requirements
• Thrust requirements
• Required climb rate
• Noise abatement procedure
• Expected ground delays
• Projected time awaiting takeoff
• Runways in use
• Traffic delays on ground
• Waiting time for de-ice
• Expected in-flight delays and conditions
• Projected traffic load en route
• Projected congestion at destination airport
• Projected weather delays en route
• Storm cells
• Location
• Intensity
• Forecast winds aloft
• Direction
• Magnitude
• Legal minimum fuel requirements
• Fuel reserve requirement
• Arrival fuel requirement
1.1.4
Determine Impact of Damaged, Degraded, or Inoperable Systems
What modifications to flight path, operational parameters or procedures are required?
• Impact of degrades/outages on desired flight path, operational parameters and procedures
• Systems affected
• Severity of degrade
• Associated penalties, restrictions and/or abnormal operations

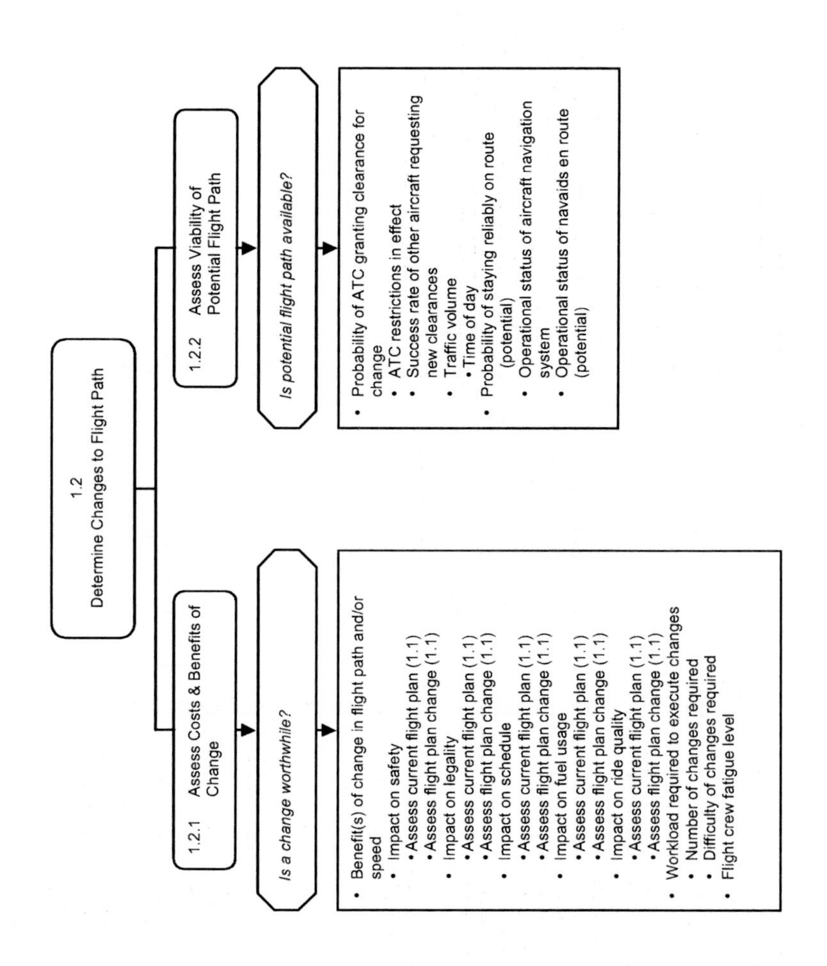
1.2
Determine Changes to Flight Path
1.2.1 Assess Costs & Benefits of Change
Is a change worthwhile?
• Benefit(s) of change in flight path and/or speed
• Impact on safety
• Assess current flight plan (1.1)
• Assess flight plan change (1.1)
• Impact on legality
• Assess current flight plan (1.1)
• Assess flight plan change (1.1)
• Impact on schedule
• Assess current flight plan (1.1)
• Assess flight plan change (1.1)
• Impact on fuel usage
• Assess current flight plan (1.1)
• Assess flight plan change (1.1)
• Impact on ride quality
• Assess current flight plan (1.1)
• Assess flight plan change (1.1)
• Workload required to execute changes
• Number of changes required
• Difficulty of changes required
• Flight crew fatigue level
1.2.2 Assess Viability of Potential Flight Path
Is potential flight path available?
• Probability of ATC granting clearance for change
• ATC restrictions in effect
• Success rate of other aircraft requesting new clearances
• Traffic volume
• Time of day
• Probability of staying reliably on route (potential)
• Operational status of aircraft navigation system
• Operational status of navaids en route (potential)

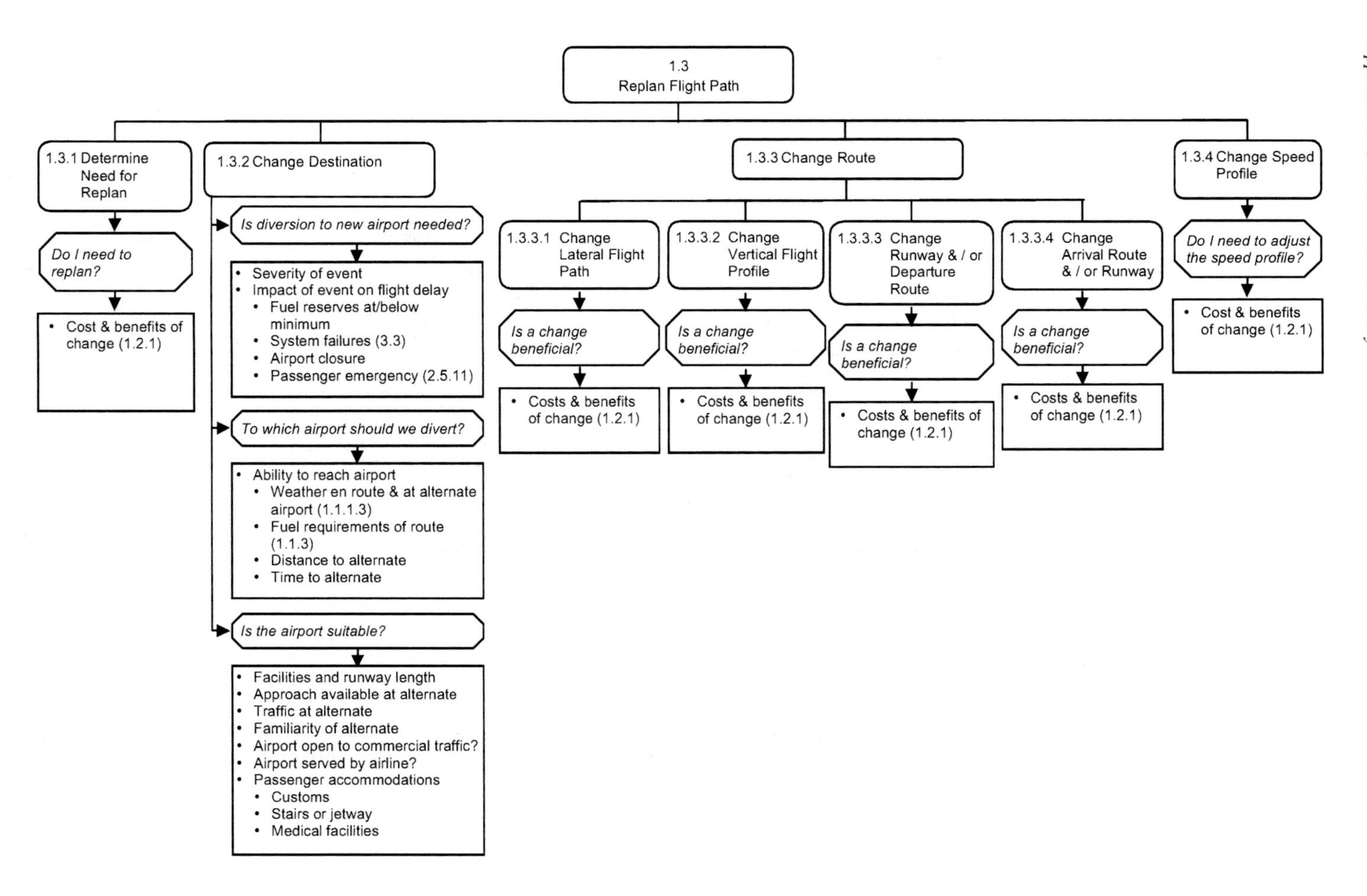
1.3
Replan Flight Path
1.3.1 Determine Need for Replan
Do I need to replan?
• Cost & benefits of change (1.2.1)
1.3.2 Change Destination
Is diversion to new airport needed?
• Severity of event
• Impact of event on flight delay
• Fuel reserves at/below minimum
• System failures (3.3)
• Airport closure
• Passenger emergency (2.5.11)
To which airport should we divert?
• Ability to reach airport
• Weather en route & at alternate airport (1.1.1.3)
• Fuel requirements of route (1.1.3)
• Distance to alternate
• Time to alternate
Is the airport suitable?
• Facilities and runway length
• Approach available at alternate
• Traffic at alternate
• Familiarity of alternate
• Airport open to commercial traffic?
• Airport served by airline?
• Passenger accommodations
• Customs
• Stairs or jetway
• Medical facilities
1.3.3 Change Route
1.3.3.1 Change Lateral Flight Path
Is a change beneficial?
• Costs & benefits of change (1.2.1)
1.3.3.2 Change Vertical Flight Profile
Is a change beneficial?
• Costs & benefits of change (1.2.1)
1.3.3.3 Change Runway & / or Departure Route
Is a change beneficial?
• Costs & benefits of change (1.2.1)
1.3.3.4 Change Arrival Route & / or Runway
Is a change beneficial?
• Costs & benefits of change (1.2.1)
1.3.4 Change Speed Profile
Do I need to adjust the speed profile?
• Cost & benefits of change (1.2.1)

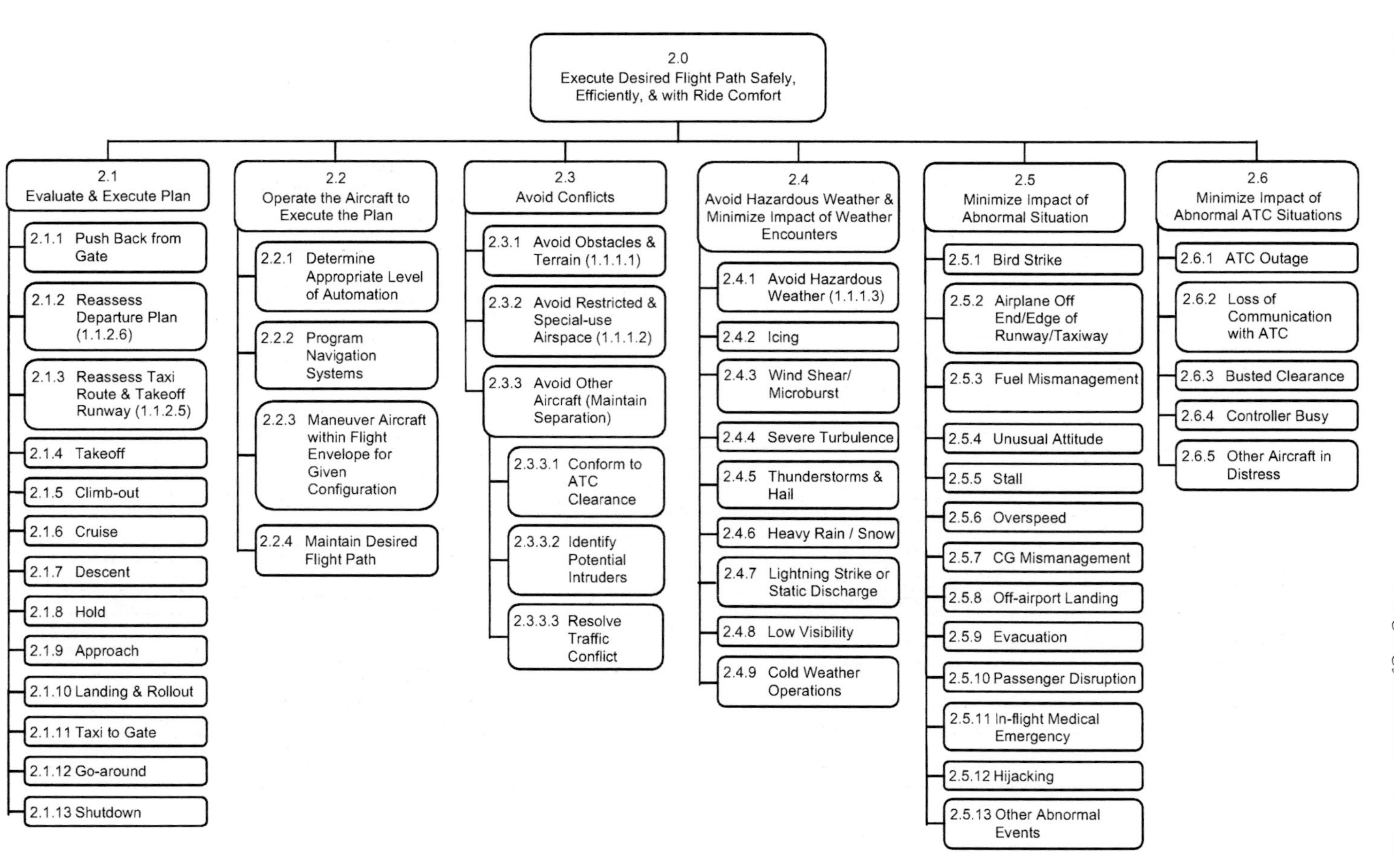
2.0 Execute Desired Flight Path Safely, Efficiently, & with Ride Comfort
2.1 Evaluate & Execute Plan
2.1.1 Push Back from Gate
2.1.2 Reassess Departure Plan (1.1.2.6)
2.1.3 Reassess Taxi Route & Takeoff Runway (1.1.2.5)
2.1.4 Takeoff
2.1.5 Climb-out
2.1.6 Cruise
2.1.7 Descent
2.1.8 Hold
2.1.9 Approach
2.1.10 Landing & Rollout
2.1.11 Taxi to Gate
2.1.12 Go-around
2.1.13 Shutdown
2.2 Operate the Aircraft to Execute the Plan
2.2.1 Determine Appropriate Level of Automation
2.2.2 Program Navigation Systems
2.2.3 Maneuver Aircraft within Flight Envelope for Given Configuration
2.2.4 Maintain Desired Flight Path
2.3 Avoid Conflicts
2.3.1 Avoid Obstacles & Terrain (1.1.1.1)
2.3.2 Avoid Restricted & Special-use Airspace (1.1.1.2)
2.3.3 Avoid Other Aircraft (Maintain Separation)
2.3.3.1 Conform to ATC Clearance
2.3.3.2 Identify Potential Intruders
2.3.3.3 Resolve Traffic Conflict
2.4 Avoid Hazardous Weather & Minimize Impact of Weather Encounters
2.4.1 Avoid Hazardous Weather (1.1.1.3)
2.4.2 Icing
2.4.3 Wind Shear/ Microburst
2.4.4 Severe Turbulence
2.4.5 Thunderstorms & Hail
2.4.6 Heavy Rain / Snow
2.4.7 Lightning Strike or Static Discharge
2.4.8 Low Visibility
2.4.9 Cold Weather Operations
2.5 Minimize Impact of Abnormal Situation
2.5.1 Bird Strike
2.5.2 Airplane Off End/Edge of Runway/Taxiway
2.5.3 Fuel Mismanagement
2.5.4 Unusual Attitude
2.5.5 Stall
2.5.6 Overspeed
2.5.7 CG Mismanagement
2.5.8 Off-airport Landing
2.5.9 Evacuation
2.5.10 Passenger Disruption
2.5.11 In-flight Medical Emergency
2.5.12 Hijacking
2.5.13 Other Abnormal Events
2.6 Minimize Impact of Abnormal ATC Situations
2.6.1 ATC Outage
2.6.2 Loss of Communication with ATC
2.6.3 Busted Clearance
2.6.4 Controller Busy
2.6.5 Other Aircraft in Distress

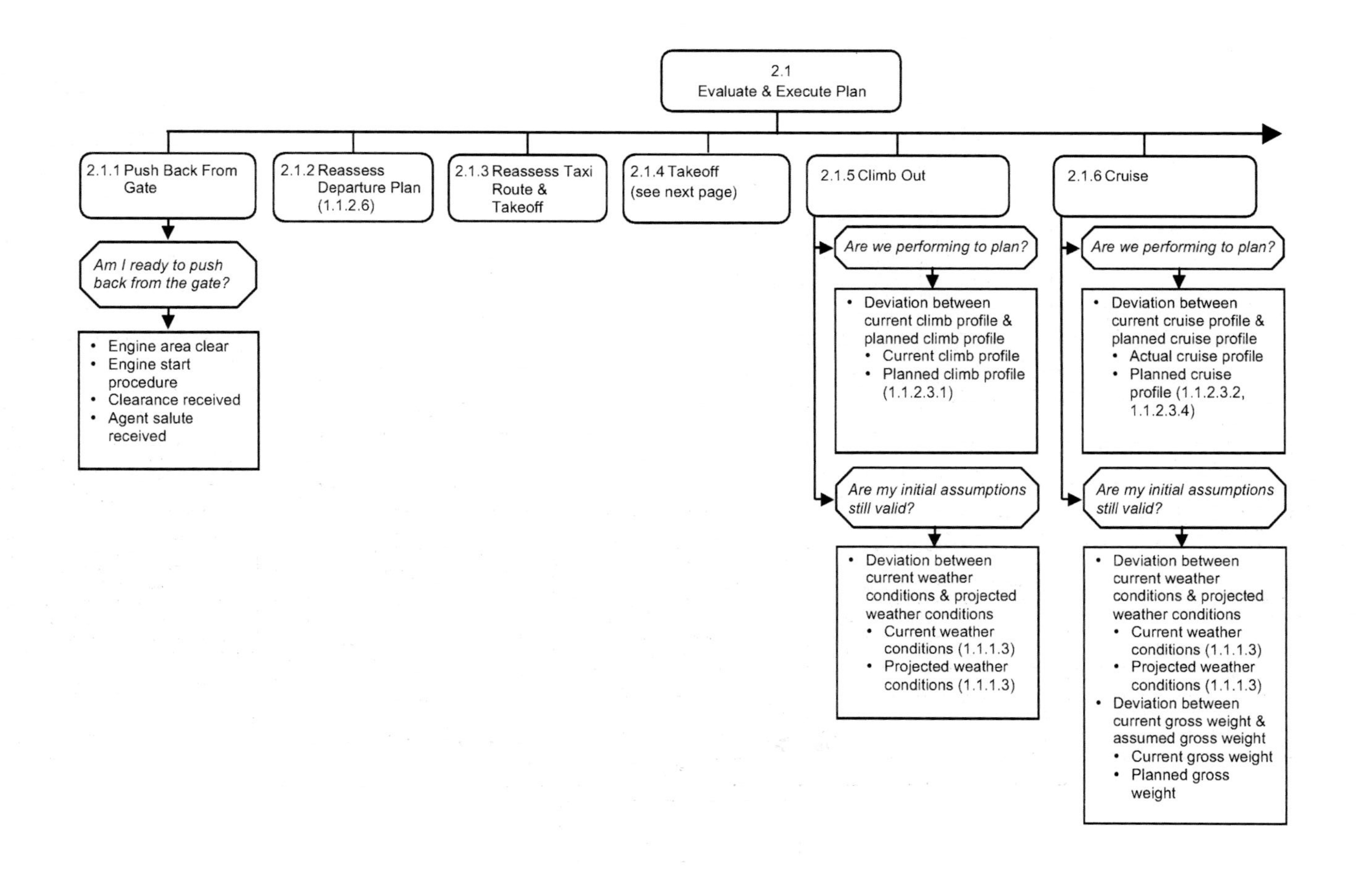
2.1
Evaluate & Execute Plan
2.1.1 Push Back From Gate
Am I ready to push back from the gate?
• Engine area clear
• Engine start procedure
• Clearance received
• Agent salute received
2.1.2 Reassess Departure Plan (1.1.2.6)
2.1.3 Reassess Taxi Route & Takeoff
2.1.4 Takeoff (see next page)
2.1.5 Climb Out
Are we performing to plan?
• Deviation between current climb profile & planned climb profile
• Current climb profile
• Planned climb profile (1.1.2.3.1)
Are my initial assumptions still valid?
• Deviation between current weather conditions & projected weather conditions
• Current weather conditions (1.1.1.3)
• Projected weather conditions (1.1.1.3)
2.1.6 Cruise
Are we performing to plan?
• Deviation between current cruise profile & planned cruise profile
• Actual cruise profile
• Planned cruise profile (1.1.2.3.2, 1.1.2.3.4)
Are my initial assumptions still valid?
• Deviation between current weather conditions & projected weather conditions
• Current weather conditions (1.1.1.3)
• Projected weather conditions (1.1.1.3)
• Deviation between current gross weight & assumed gross weight
• Current gross weight
• Planned gross weight

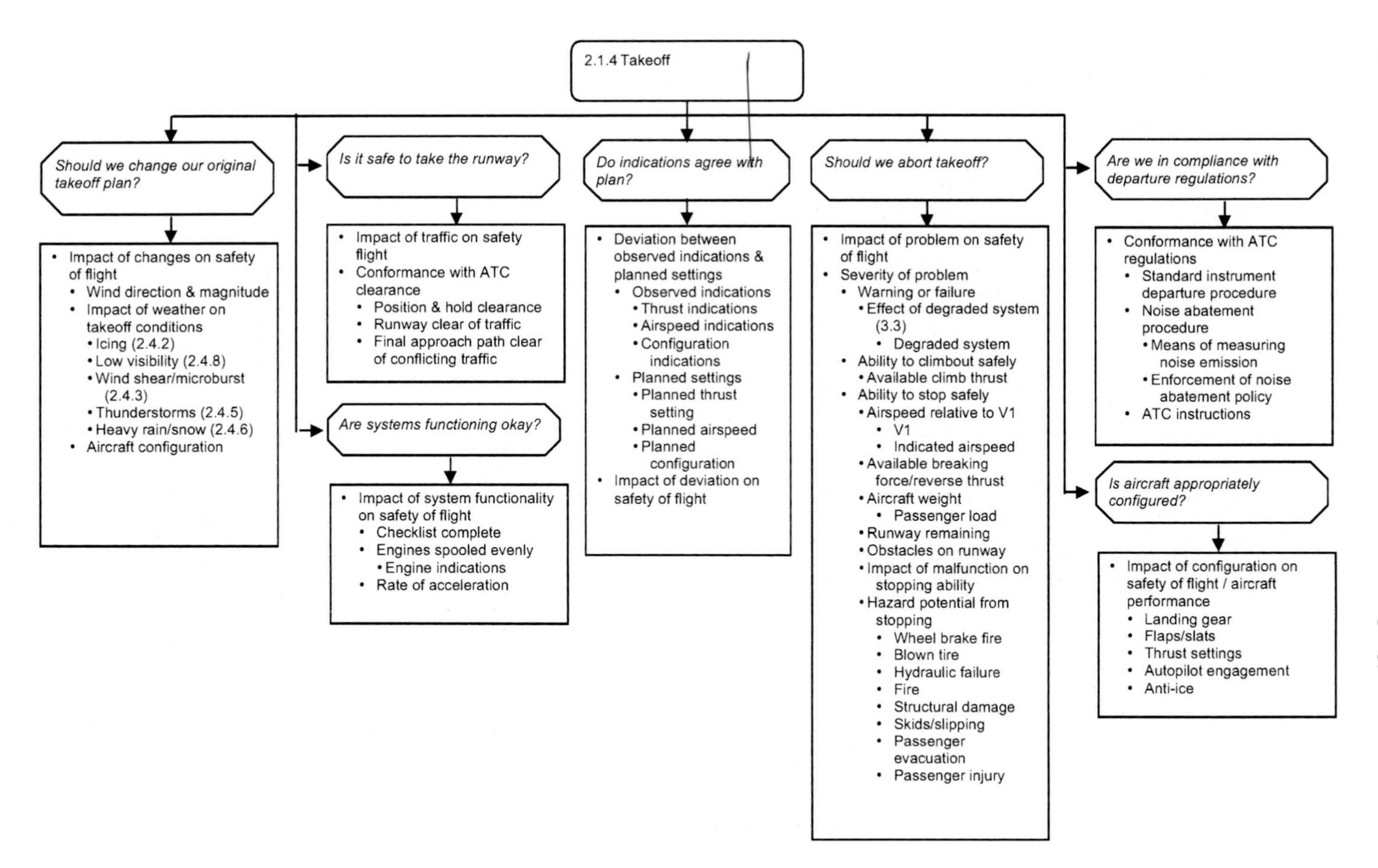
2.1.4 Takeoff
Should we change our original takeoff plan?
• Impact of changes on safety of flight
• Wind direction & magnitude
• Impact of weather on takeoff conditions
• Icing (2.4.2)
• Low visibility (2.4.8)
• Wind shear/microburst (2.4.3)
• Thunderstorms (2.4.5)
• Heavy rain/snow (2.4.6)
• Aircraft configuration
Is it safe to take the runway?
• Impact of traffic on safety flight
• Conformance with ATC clearance
• Position & hold clearance
• Runway clear of traffic
• Final approach path clear of conflicting traffic
Are systems functioning okay?
• Impact of system functionality on safety of flight
• Checklist complete
• Engines spooled evenly
• Engine indications
• Rate of acceleration
Do indications agree with plan?
• Deviation between observed indications & planned settings
• Observed indications
• Thrust indications
• Airspeed indications
• Configuration indications
• Planned settings
• Planned thrust setting
• Planned airspeed
• Planned configuration
• Impact of deviation on safety of flight
Should we abort takeoff?
• Impact of problem on safety of flight
• Severity of problem
• Warning or failure
• Effect of degraded system (3.3)
• Degraded system
• Ability to climbout safely
• Available climb thrust
• Ability to stop safely
• Airspeed relative to V1
• V1
• Indicated airspeed
• Available breaking force/reverse thrust
• Aircraft weight
• Passenger load
• Runway remaining
• Obstacles on runway
• Impact of malfunction on stopping ability
• Hazard potential from stopping
• Wheel brake fire
• Blown tire
• Hydraulic failure
• Fire
• Structural damage
• Skids/slipping
• Passenger evacuation
• Passenger injury
Are we in compliance with departure regulations?
• Conformance with ATC regulations
• Standard instrument departure procedure
• Noise abatement procedure
• Means of measuring noise emission
• Enforcement of noise abatement policy
• ATC instructions
Is aircraft appropriately configured?
• Impact of configuration on safety of flight / aircraft performance
• Landing gear
• Flaps/slats
• Thrust settings
• Autopilot engagement
• Anti-ice

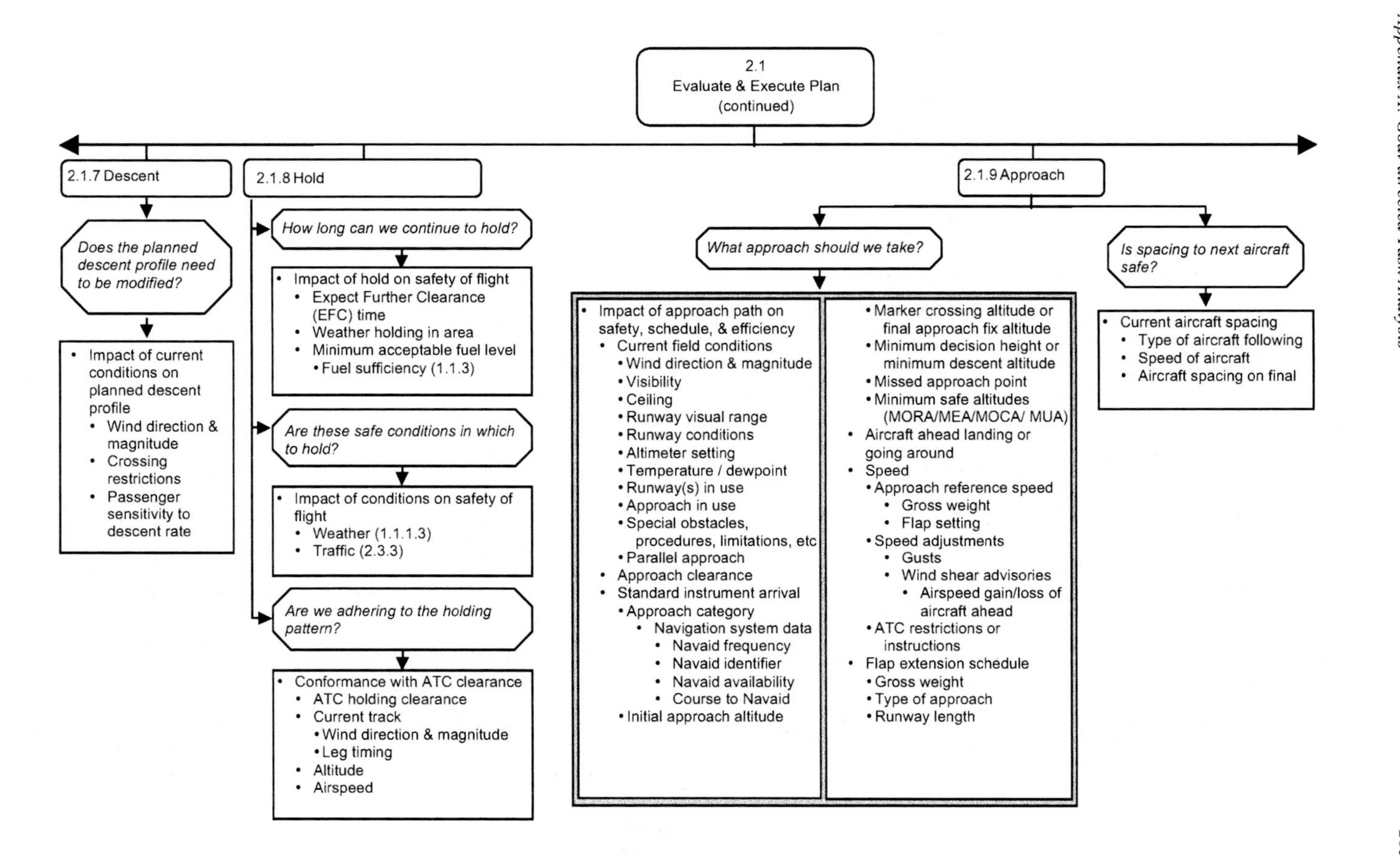
2.1
Evaluate & Execute Plan
(continued)
2.1.7 Descent
Does the planned descent profile need to be modified?
• Impact of current conditions on planned descent profile
• Wind direction & magnitude
• Crossing restrictions
• Passenger sensitivity to descent rate
2.1.8 Hold
How long can we continue to hold?
• Impact of hold on safety of flight
• Expect Further Clearance (EFC) time
• Weather holding in area
• Minimum acceptable fuel level
• Fuel sufficiency (1.1.3)
Are these safe conditions in which to hold?
• Impact of conditions on safety of flight
• Weather (1.1.1.3)
• Traffic (2.3.3)
Are we adhering to the holding pattern?
• Conformance with ATC clearance
• ATC holding clearance
• Current track
• Wind direction & magnitude
• Leg timing
• Altitude
• Airspeed
2.1.9 Approach
What approach should we take?
• Impact of approach path on safety, schedule, & efficiency
• Current field conditions
• Wind direction & magnitude
• Visibility
• Ceiling
• Runway visual range
• Runway conditions
• Altimeter setting
• Temperature / dewpoint
• Runway(s) in use
• Approach in use
• Special obstacles, procedures, limitations, etc
• Parallel approach
• Approach clearance
• Standard instrument arrival
• Approach category
• Navigation system data
• Navaid frequency
• Navaid identifier
• Navaid availability
• Course to Navaid
• Initial approach altitude
• Marker crossing altitude or final approach fix altitude
• Minimum decision height or minimum descent altitude
• Missed approach point
• Minimum safe altitudes (MORA/MEA/MOCA/ MUA)
• Aircraft ahead landing or going around
• Speed
• Approach reference speed
• Gross weight
• Flap setting
• Speed adjustments
• Gusts
• Wind shear advisories
• Airspeed gain/loss of aircraft ahead
• ATC restrictions or instructions
• Flap extension schedule
• Gross weight
• Type of approach
• Runway length
Is spacing to next aircraft safe?
• Current aircraft spacing
• Type of aircraft following
• Speed of aircraft
• Aircraft spacing on final

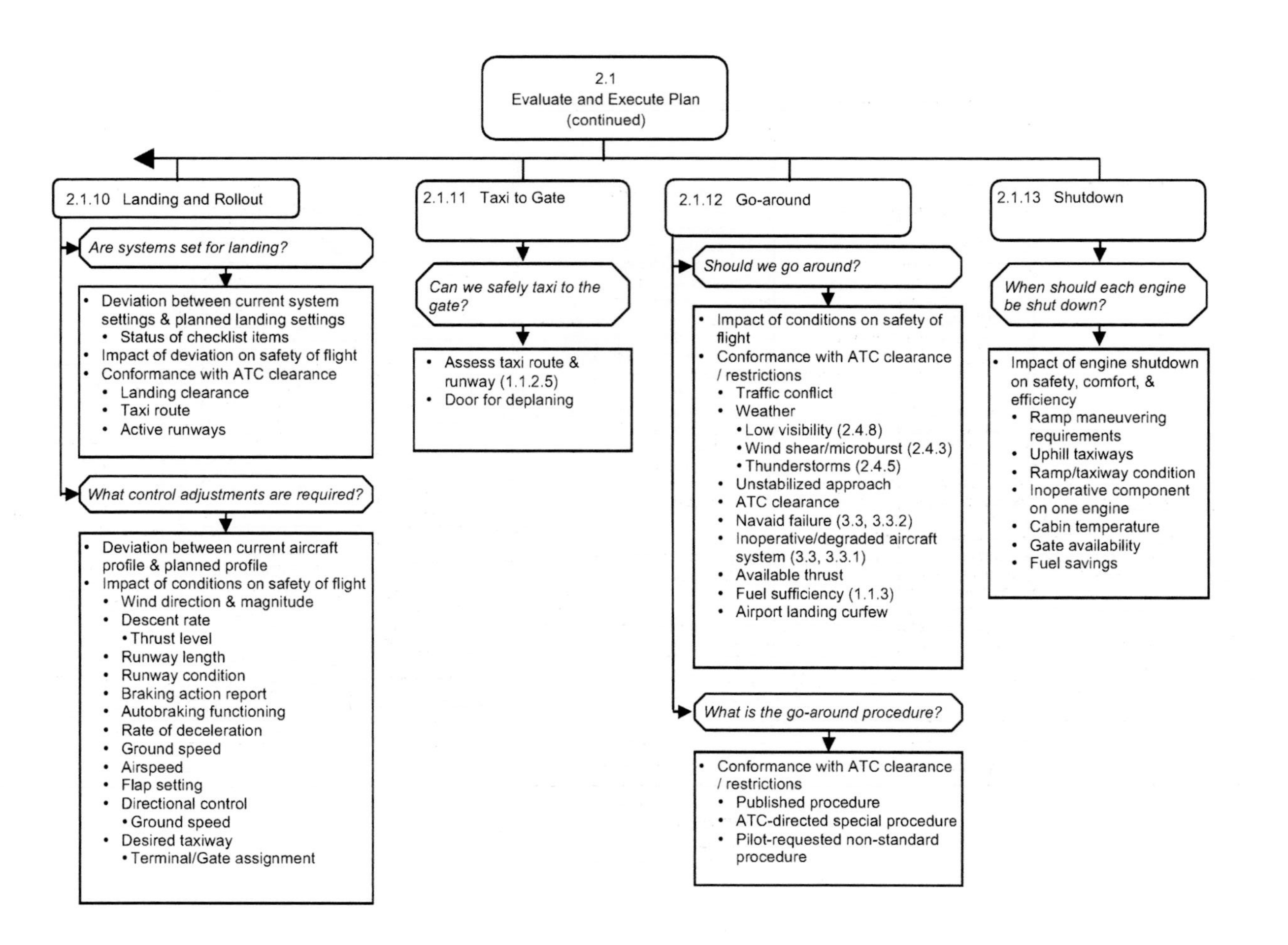
2.1
Evaluate and Execute Plan
(continued)
2.1.10 Landing and Rollout
Are systems set for landing?
• Deviation between current system settings & planned landing settings
• Status of checklist items
• Impact of deviation on safety of flight
• Conformance with ATC clearance
• Landing clearance
• Taxi route
• Active runways
What control adjustments are required?
• Deviation between current aircraft profile & planned profile
• Impact of conditions on safety of flight
• Wind direction & magnitude
• Descent rate
• Thrust level
• Runway length
• Runway condition
• Braking action report
• Autobraking functioning
• Rate of deceleration
• Ground speed
• Airspeed
• Flap setting
• Directional control
• Ground speed
• Desired taxiway
• Terminal/Gate assignment
2.1.11 Taxi to Gate
Can we safely taxi to the gate?
• Assess taxi route & runway (1.1.2.5)
• Door for deplaning
2.1.12 Go-around
Should we go around?
• Impact of conditions on safety of flight
• Conformance with ATC clearance / restrictions
• Traffic conflict
• Weather
• Low visibility (2.4.8)
• Wind shear/microburst (2.4.3)
• Thunderstorms (2.4.5)
• Unstabilized approach
• ATC clearance
• Navaid failure (3.3, 3.3.2)
• Inoperative/degraded aircraft system (3.3, 3.3.1)
• Available thrust
• Fuel sufficiency (1.1.3)
• Airport landing curfew
What is the go-around procedure?
• Conformance with ATC clearance / restrictions
• Published procedure
• ATC-directed special procedure
• Pilot-requested non-standard procedure
2.1.13 Shutdown
When should each engine be shut down?
• Impact of engine shutdown on safety, comfort, & efficiency
• Ramp maneuvering requirements
• Uphill taxiways
• Ramp/taxiway condition
• Inoperative component on one engine
• Cabin temperature
• Gate availability
• Fuel savings

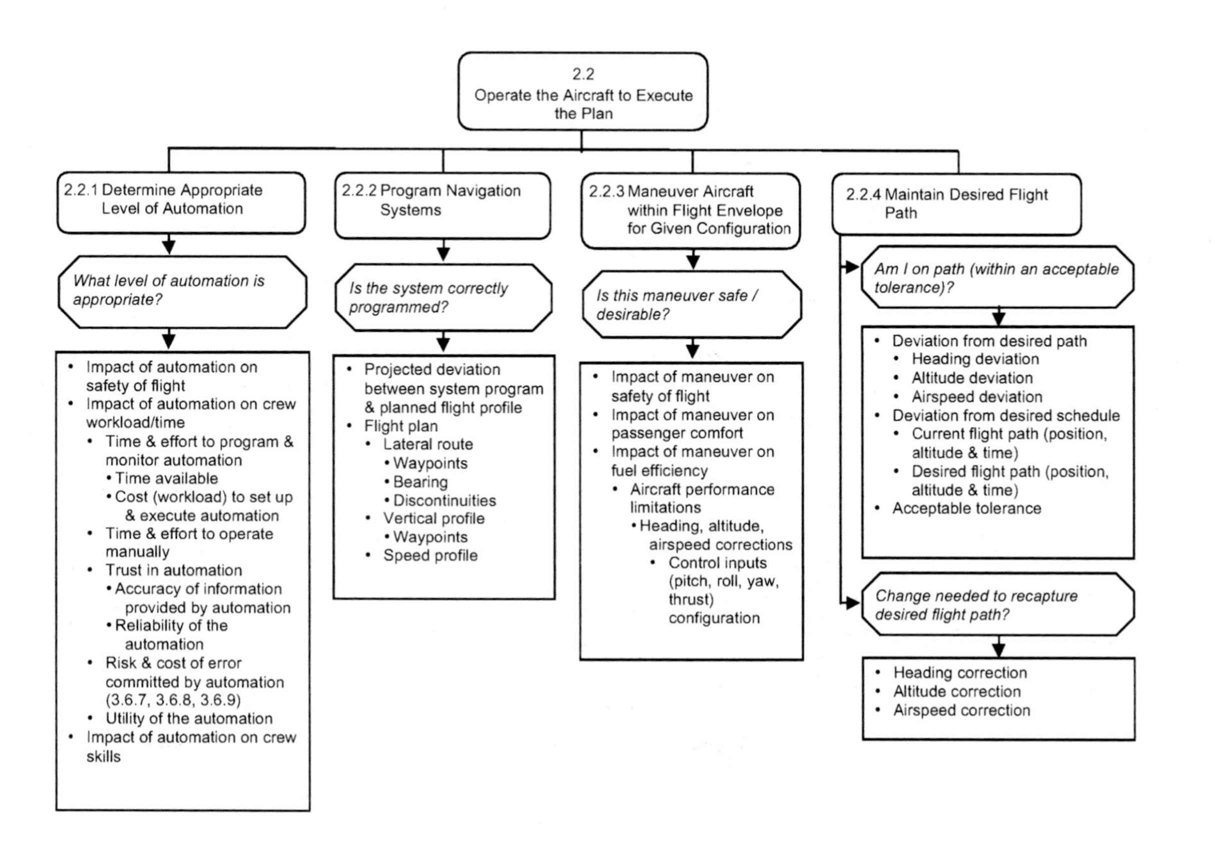
2.2
Operate the Aircraft to Execute the Plan
2.2.1 Determine Appropriate Level of Automation
What level of automation is appropriate?
• Impact of automation on safety of flight
• Impact of automation on crew workload/time
• Time & effort to program & monitor automation
• Time available
• Cost (workload) to set up & execute automation
• Time & effort to operate manually
• Trust in automation
• Accuracy of information provided by automation
• Reliability of the automation
• Risk & cost of error committed by automation (3.6.7, 3.6.8, 3.6.9)
• Utility of the automation
• Impact of automation on crew skills
2.2.2 Program Navigation Systems
Is the system correctly programmed?
• Projected deviation between system program & planned flight profile
• Flight plan
• Lateral route
• Waypoints
• Bearing
• Discontinuities
• Vertical profile
• Waypoints
• Speed profile
2.2.3 Maneuver Aircraft within Flight Envelope for Given Configuration
Is this maneuver safe / desirable?
• Impact of maneuver on safety of flight
• Impact of maneuver on passenger comfort
• Impact of maneuver on fuel efficiency
• Aircraft performance limitations
• Heading, altitude, airspeed corrections
• Control inputs (pitch, roll, yaw, thrust)
configuration
2.2.4 Maintain Desired Flight Path
Am I on path (within an acceptable tolerance)?
• Deviation from desired path
• Heading deviation
• Altitude deviation
• Airspeed deviation
• Deviation from desired schedule
• Current flight path (position, altitude & time)
• Desired flight path (position, altitude & time)
• Acceptable tolerance
Change needed to recapture desired flight path?
• Heading correction
• Altitude correction
• Airspeed correction

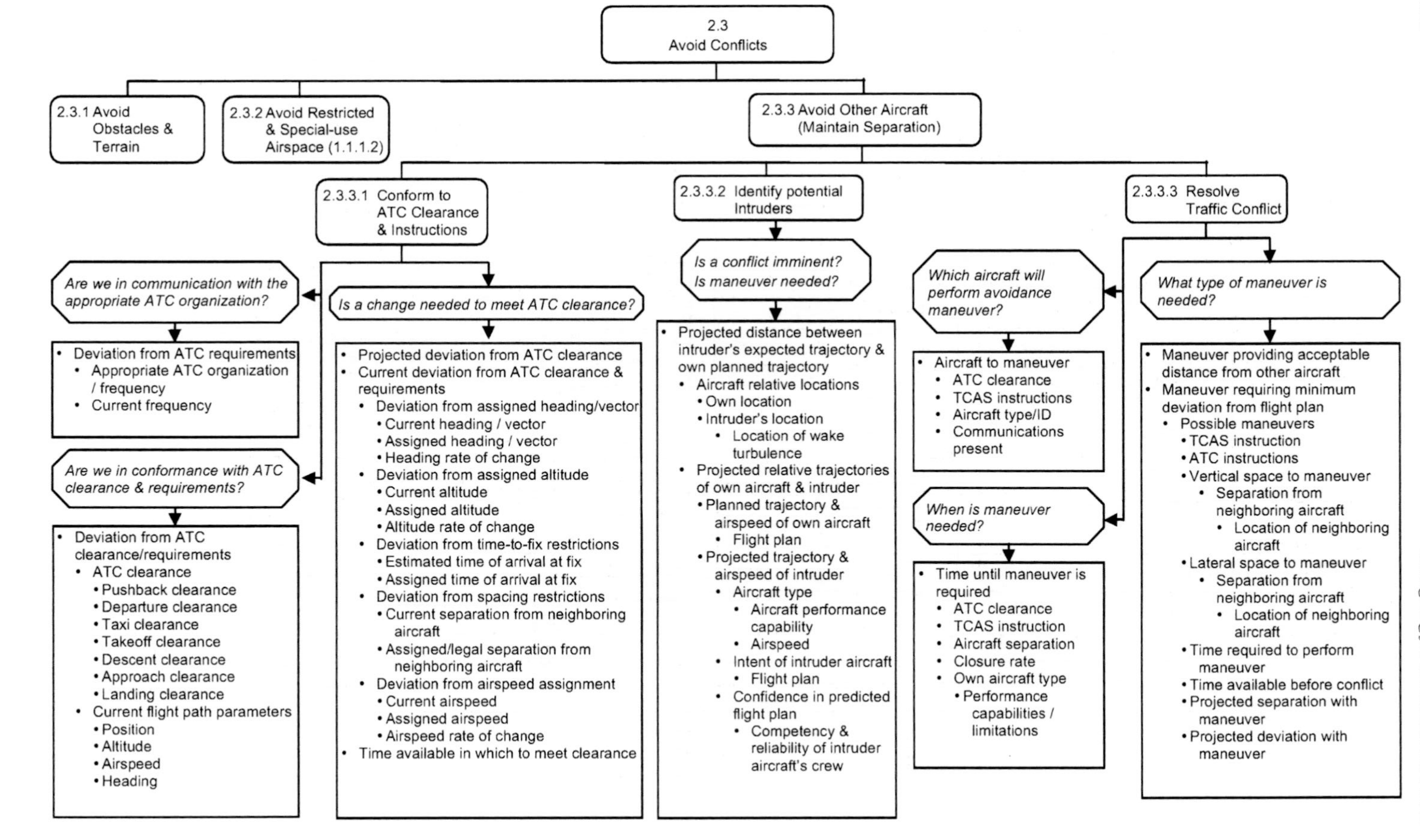
2.3
Avoid Conflicts
2.3.1 Avoid Obstacles & Terrain
2.3.2 Avoid Restricted & Special-use Airspace (1.1.1.2)
2.3.3 Avoid Other Aircraft (Maintain Separation)
2.3.3.1 Conform to ATC Clearance & Instructions
2.3.3.2 Identify potential Intruders
2.3.3.3 Resolve Traffic Conflict
Are we in communication with the appropriate ATC organization?
• Deviation from ATC requirements
• Appropriate ATC organization / frequency
• Current frequency
Are we in conformance with ATC clearance & requirements?
• Deviation from ATC clearance/requirements
• ATC clearance
• Pushback clearance
• Departure clearance
• Taxi clearance
• Takeoff clearance
• Descent clearance
• Approach clearance
• Landing clearance
• Current flight path parameters
• Position
• Altitude
• Airspeed
• Heading
Is a change needed to meet ATC clearance?
• Projected deviation from ATC clearance
• Current deviation from ATC clearance & requirements
• Deviation from assigned heading/vector
• Current heading / vector
• Assigned heading / vector
• Heading rate of change
• Deviation from assigned altitude
• Current altitude
• Assigned altitude
• Altitude rate of change
• Deviation from time-to-fix restrictions
• Estimated time of arrival at fix
• Assigned time of arrival at fix
• Deviation from spacing restrictions
• Current separation from neighboring aircraft
• Assigned/legal separation from neighboring aircraft
• Deviation from airspeed assignment
• Current airspeed
• Assigned airspeed
• Airspeed rate of change
• Time available in which to meet clearance
Is a conflict imminent? Is maneuver needed?
• Projected distance between intruder's expected trajectory & own planned trajectory
• Aircraft relative locations
• Own location
• Intruder's location
• Location of wake turbulence
• Projected relative trajectories of own aircraft & intruder
• Planned trajectory & airspeed of own aircraft
• Flight plan
• Projected trajectory & airspeed of intruder
• Aircraft type
• Aircraft performance capability
• Airspeed
• Intent of intruder aircraft
• Flight plan
• Confidence in predicted flight plan
• Competency & reliability of intruder aircraft's crew
Which aircraft will perform avoidance maneuver?
• Aircraft to maneuver
• ATC clearance
• TCAS instructions
• Aircraft type/ID
• Communications present
When is maneuver needed?
• Time until maneuver is required
• ATC clearance
• TCAS instruction
• Aircraft separation
• Closure rate
• Own aircraft type
• Performance capabilities / limitations
What type of maneuver is needed?
• Maneuver providing acceptable distance from other aircraft
• Maneuver requiring minimum deviation from flight plan
• Possible maneuvers
• TCAS instruction
• ATC instructions
• Vertical space to maneuver
• Separation from neighboring aircraft
• Location of neighboring aircraft
• Lateral space to maneuver
• Separation from neighboring aircraft
• Location of neighboring aircraft
• Time required to perform maneuver
• Time available before conflict
• Projected separation with maneuver
• Projected deviation with maneuver

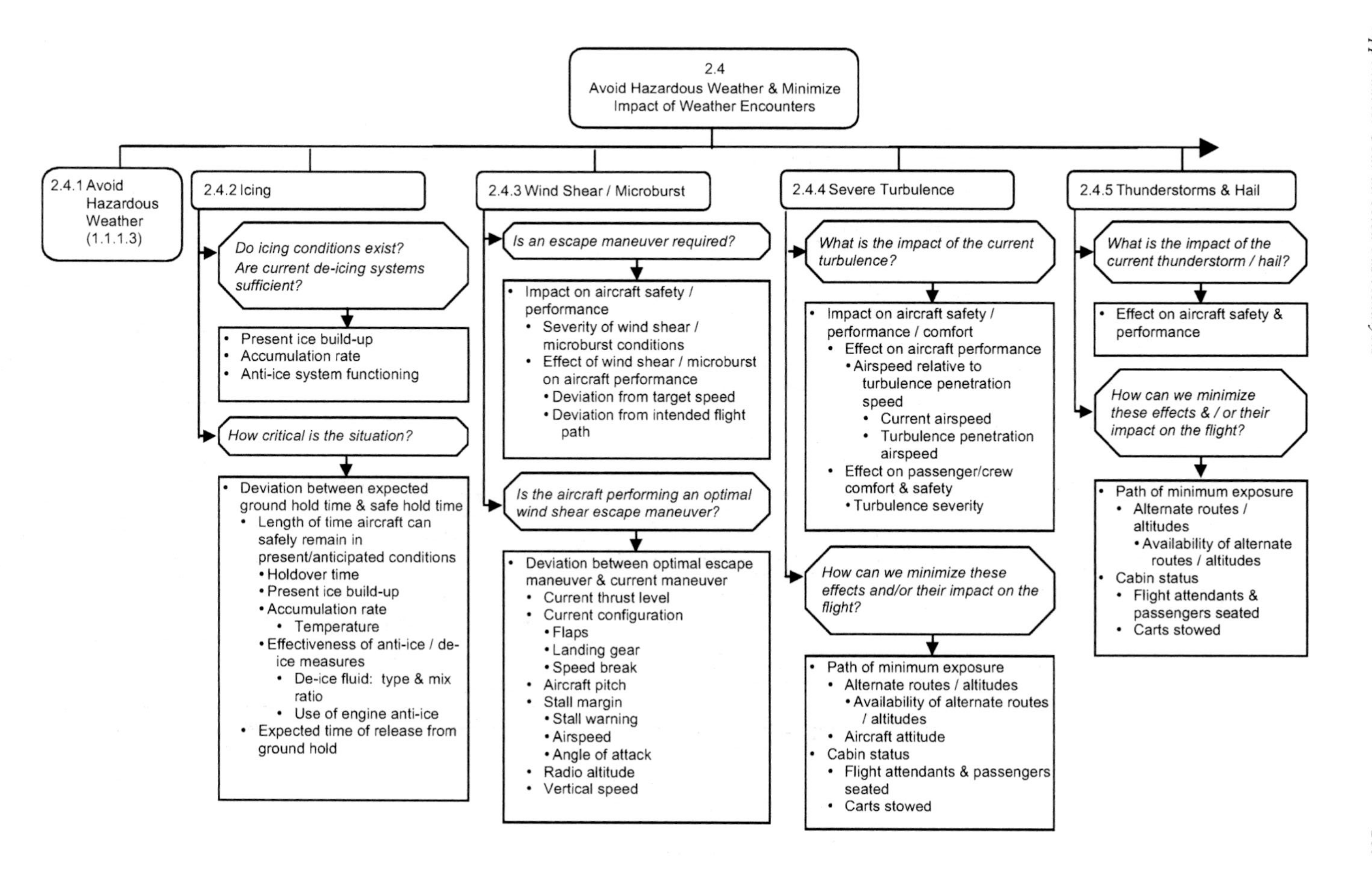

2.4
Avoid Hazardous Weather & Minimize Impact of Weather Encounters
2.4.1 Avoid Hazardous Weather (1.1.1.3)
2.4.2 Icing
Do icing conditions exist? Are current de-icing systems sufficient?
• Present ice build-up
• Accumulation rate
• Anti-ice system functioning
How critical is the situation?
• Deviation between expected ground hold time & safe hold time
• Length of time aircraft can safely remain in present/anticipated conditions
• Holdover time
• Present ice build-up
• Accumulation rate
• Temperature
• Effectiveness of anti-ice / de-ice measures
• De-ice fluid: type & mix ratio
• Use of engine anti-ice
• Expected time of release from ground hold
2.4.3 Wind Shear / Microburst
Is an escape maneuver required?
• Impact on aircraft safety / performance
• Severity of wind shear / microburst conditions
• Effect of wind shear / microburst on aircraft performance
• Deviation from target speed
• Deviation from intended flight path
Is the aircraft performing an optimal wind shear escape maneuver?
• Deviation between optimal escape maneuver & current maneuver
• Current thrust level
• Current configuration
• Flaps
• Landing gear
• Speed break
• Aircraft pitch
• Stall margin
• Stall warning
• Airspeed
• Angle of attack
• Radio altitude
• Vertical speed
2.4.4 Severe Turbulence
What is the impact of the current turbulence?
• Impact on aircraft safety / performance / comfort
• Effect on aircraft performance
• Airspeed relative to turbulence penetration speed
• Current airspeed
• Turbulence penetration airspeed
• Effect on passenger/crew comfort & safety
• Turbulence severity
How can we minimize these effects and/or their impact on the flight?
• Path of minimum exposure
• Alternate routes / altitudes
• Availability of alternate routes / altitudes
• Aircraft attitude
• Cabin status
• Flight attendants & passengers seated
• Carts stowed
2.4.5 Thunderstorms & Hail
What is the impact of the current thunderstorm / hail?
• Effect on aircraft safety & performance
How can we minimize these effects & / or their impact on the flight?
• Path of minimum exposure
• Alternate routes / altitudes
• Availability of alternate routes / altitudes
• Cabin status
• Flight attendants & passengers seated
• Carts stowed

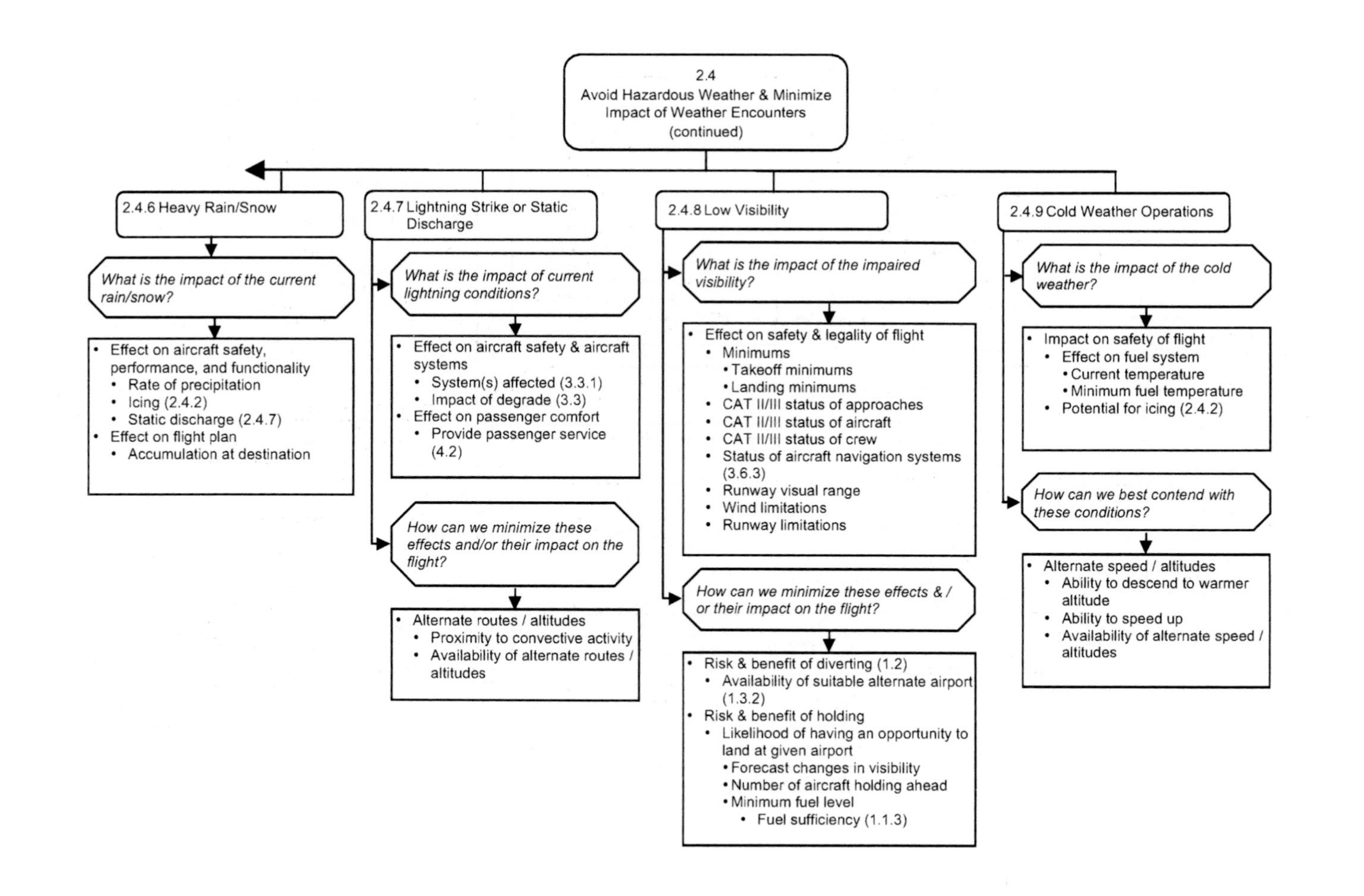
2.4
Avoid Hazardous Weather & Minimize Impact of Weather Encounters
(continued)
2.4.6 Heavy Rain/Snow
What is the impact of the current rain/snow?
• Effect on aircraft safety, performance, and functionality
• Rate of precipitation
• Icing (2.4.2)
• Static discharge (2.4.7)
• Effect on flight plan
• Accumulation at destination
2.4.7 Lightning Strike or Static Discharge
What is the impact of current lightning conditions?
• Effect on aircraft safety & aircraft systems
• System(s) affected (3.3.1)
• Impact of degrade (3.3)
• Effect on passenger comfort
• Provide passenger service (4.2)
How can we minimize these effects and/or their impact on the flight?
• Alternate routes / altitudes
• Proximity to convective activity
• Availability of alternate routes / altitudes
2.4.8 Low Visibility
What is the impact of the impaired visibility?
• Effect on safety & legality of flight
• Minimums
• Takeoff minimums
• Landing minimums
• CAT II/III status of approaches
• CAT II/III status of aircraft
• CAT II/III status of crew
• Status of aircraft navigation systems (3.6.3)
• Runway visual range
• Wind limitations
• Runway limitations
How can we minimize these effects & / or their impact on the flight?
• Risk & benefit of diverting (1.2)
• Availability of suitable alternate airport (1.3.2)
• Risk & benefit of holding
• Likelihood of having an opportunity to land at given airport
• Forecast changes in visibility
• Number of aircraft holding ahead
• Minimum fuel level
• Fuel sufficiency (1.1.3)
2.4.9 Cold Weather Operations
What is the impact of the cold weather?
• Impact on safety of flight
• Effect on fuel system
• Current temperature
• Minimum fuel temperature
• Potential for icing (2.4.2)
How can we best contend with these conditions?
• Alternate speed / altitudes
• Ability to descend to warmer altitude
• Ability to speed up
• Availability of alternate speed / altitudes

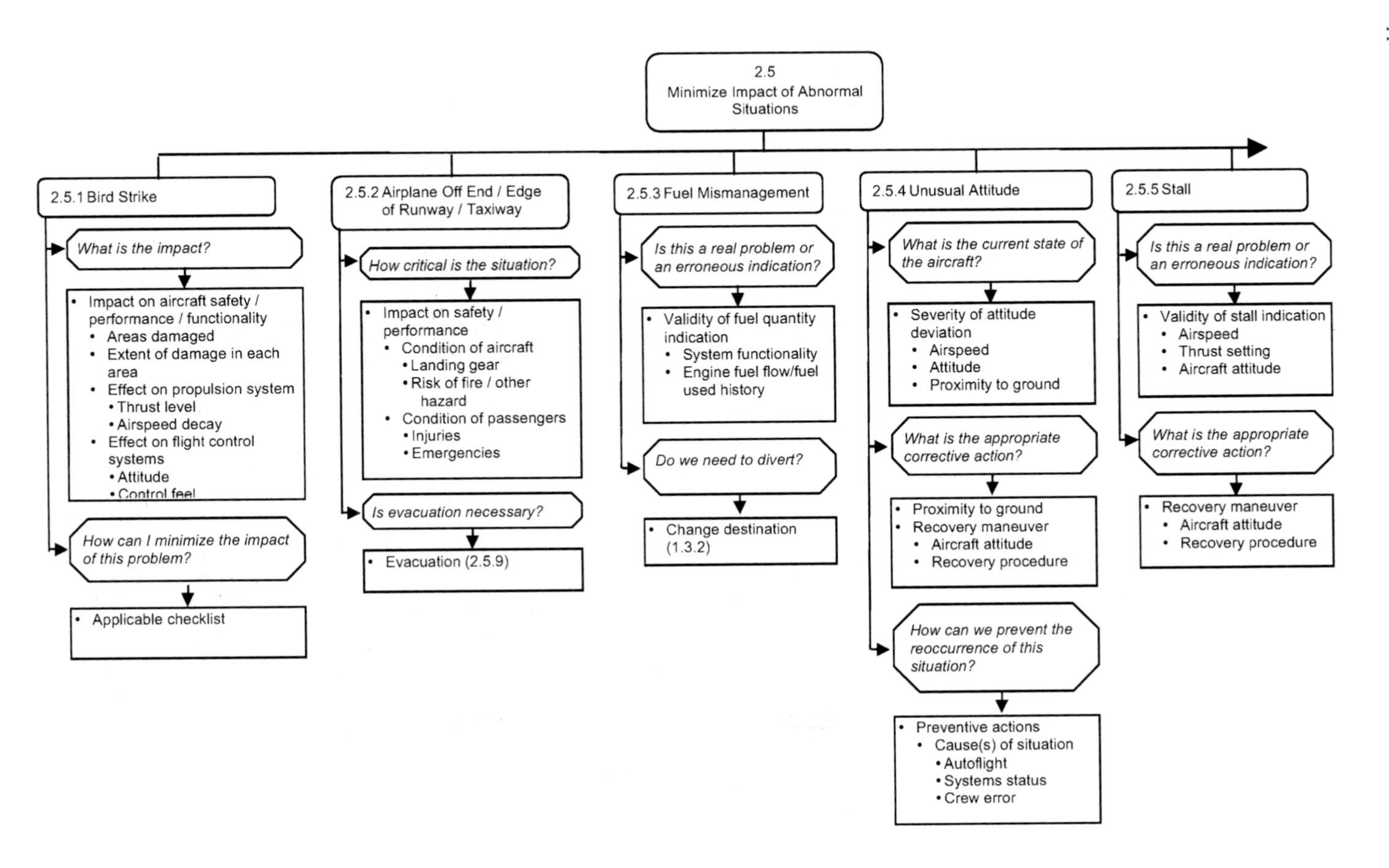
2.5
Minimize Impact of Abnormal Situations
2.5.1 Bird Strike
What is the impact?
Impact on aircraft safety / performance / functionality
Areas damaged
Extent of damage in each area
Effect on propulsion system
Thrust level
Airspeed decay
Effect on flight control systems
Attitude
Control feel
How can I minimize the impact of this problem?
Applicable checklist
2.5.2 Airplane Off End / Edge of Runway / Taxiway
How critical is the situation?
Impact on safety / performance
Condition of aircraft
Landing gear
Risk of fire / other hazard
Condition of passengers
Injuries
Emergencies
Is evacuation necessary?
Evacuation (2.5.9)
2.5.3 Fuel Mismanagement
Is this a real problem or an erroneous indication?
Validity of fuel quantity indication
System functionality
Engine fuel flow/fuel used history
Do we need to divert?
Change destination (1.3.2)
2.5.4 Unusual Attitude
What is the current state of the aircraft?
Severity of attitude deviation
Airspeed
Attitude
Proximity to ground
What is the appropriate corrective action?
Proximity to ground
Recovery maneuver
Aircraft attitude
Recovery procedure
How can we prevent the reoccurrence of this situation?
Preventive actions
Cause(s) of situation
Autoflight
Systems status
Crew error
2.5.5 Stall
Is this a real problem or an erroneous indication?
Validity of stall indication
Airspeed
Thrust setting
Aircraft attitude
What is the appropriate corrective action?
Recovery maneuver
Aircraft attitude
Recovery procedure

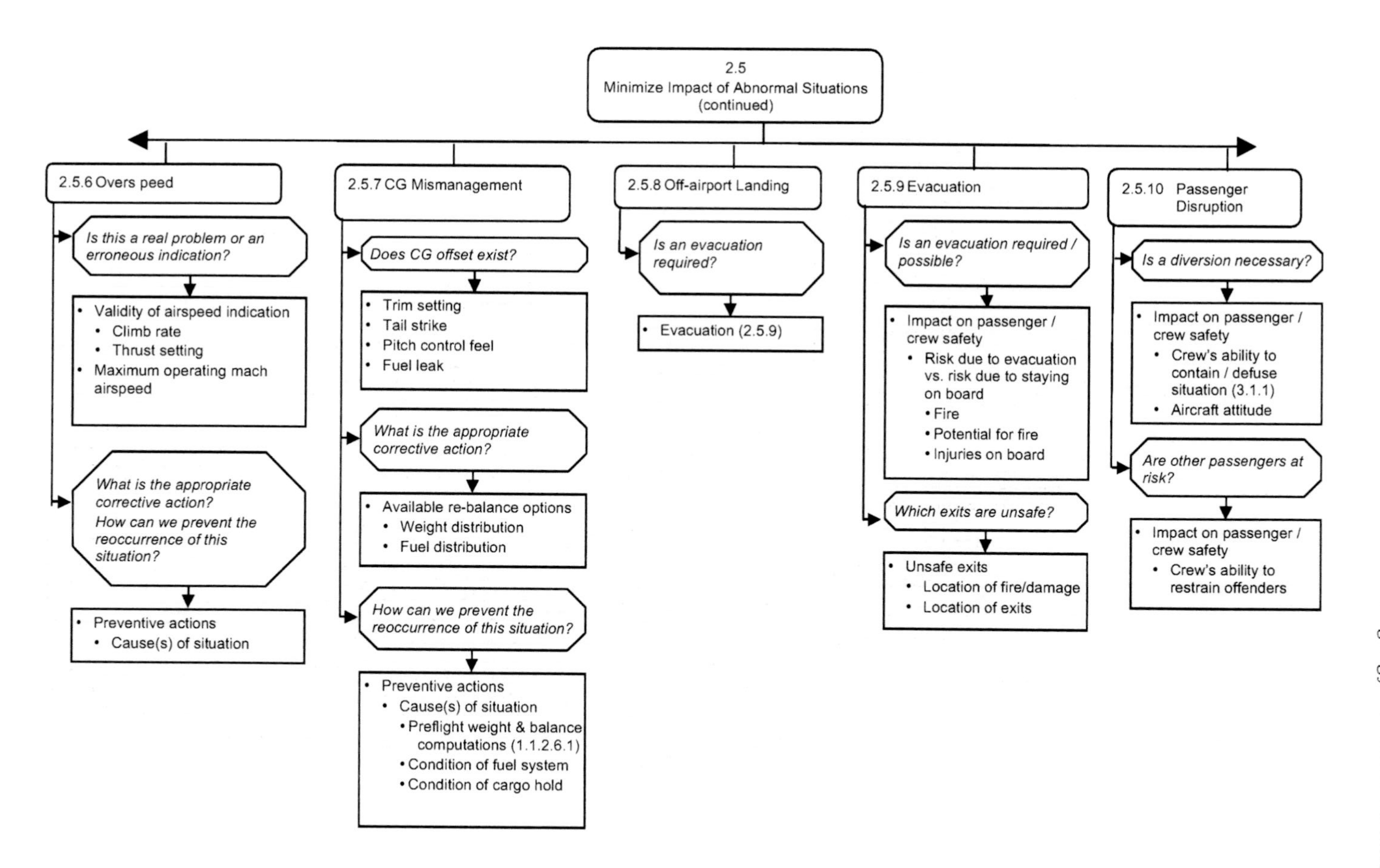
2.5
Minimize Impact of Abnormal Situations
(continued)
2.5.6 Overs peed
Is this a real problem or an erroneous indication?
• Validity of airspeed indication
• Climb rate
• Thrust setting
• Maximum operating mach airspeed
What is the appropriate corrective action?
How can we prevent the reoccurrence of this situation?
• Preventive actions
• Cause(s) of situation
2.5.7 CG Mismanagement
Does CG offset exist?
• Trim setting
• Tail strike
• Pitch control feel
• Fuel leak
What is the appropriate corrective action?
• Available re-balance options
• Weight distribution
• Fuel distribution
How can we prevent the reoccurrence of this situation?
• Preventive actions
• Cause(s) of situation
• Preflight weight & balance computations (1.1.2.6.1)
• Condition of fuel system
• Condition of cargo hold
2.5.8 Off-airport Landing
Is an evacuation required?
• Evacuation (2.5.9)
2.5.9 Evacuation
Is an evacuation required / possible?
• Impact on passenger / crew safety
• Risk due to evacuation vs. risk due to staying on board
• Fire
• Potential for fire
• Injuries on board
Which exits are unsafe?
• Unsafe exits
• Location of fire/damage
• Location of exits
2.5.10 Passenger Disruption
Is a diversion necessary?
• Impact on passenger / crew safety
• Crew's ability to contain / defuse situation (3.1.1)
• Aircraft attitude
Are other passengers at risk?
• Impact on passenger / crew safety
• Crew's ability to restrain offenders

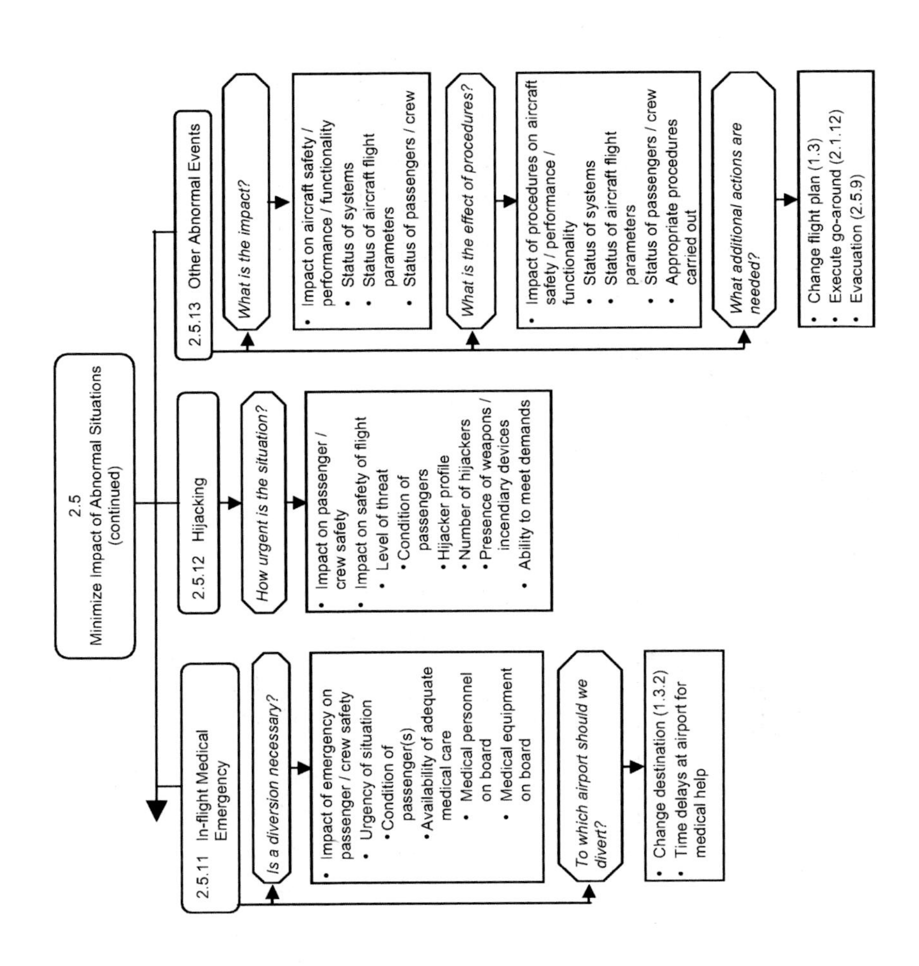
2.5
Minimize Impact of Abnormal Situations
(continued)
2.5.11 In-flight Medical Emergency
Is a diversion necessary?
• Impact of emergency on passenger / crew safety
• Urgency of situation
• Condition of passenger(s)
• Availability of adequate medical care
• Medical personnel on board
• Medical equipment on board
To which airport should we divert?
• Change destination (1.3.2)
• Time delays at airport for medical help
2.5.12 Hijacking
How urgent is the situation?
• Impact on passenger / crew safety
• Impact on safety of flight
• Level of threat
• Condition of passengers
• Hijacker profile
• Number of hijackers
• Presence of weapons / incendiary devices
• Ability to meet demands
2.5.13 Other Abnormal Events
What is the impact?
• Impact on aircraft safety / performance / functionality
• Status of systems
• Status of aircraft flight parameters
• Status of passengers / crew
What is the effect of procedures?
• Impact of procedures on aircraft safety / performance / functionality
• Status of systems
• Status of aircraft flight parameters
• Status of passengers / crew
• Appropriate procedures carried out
What additional actions are needed?
• Change flight plan (1.3)
• Execute go-around (2.1.12)
• Evacuation (2.5.9)

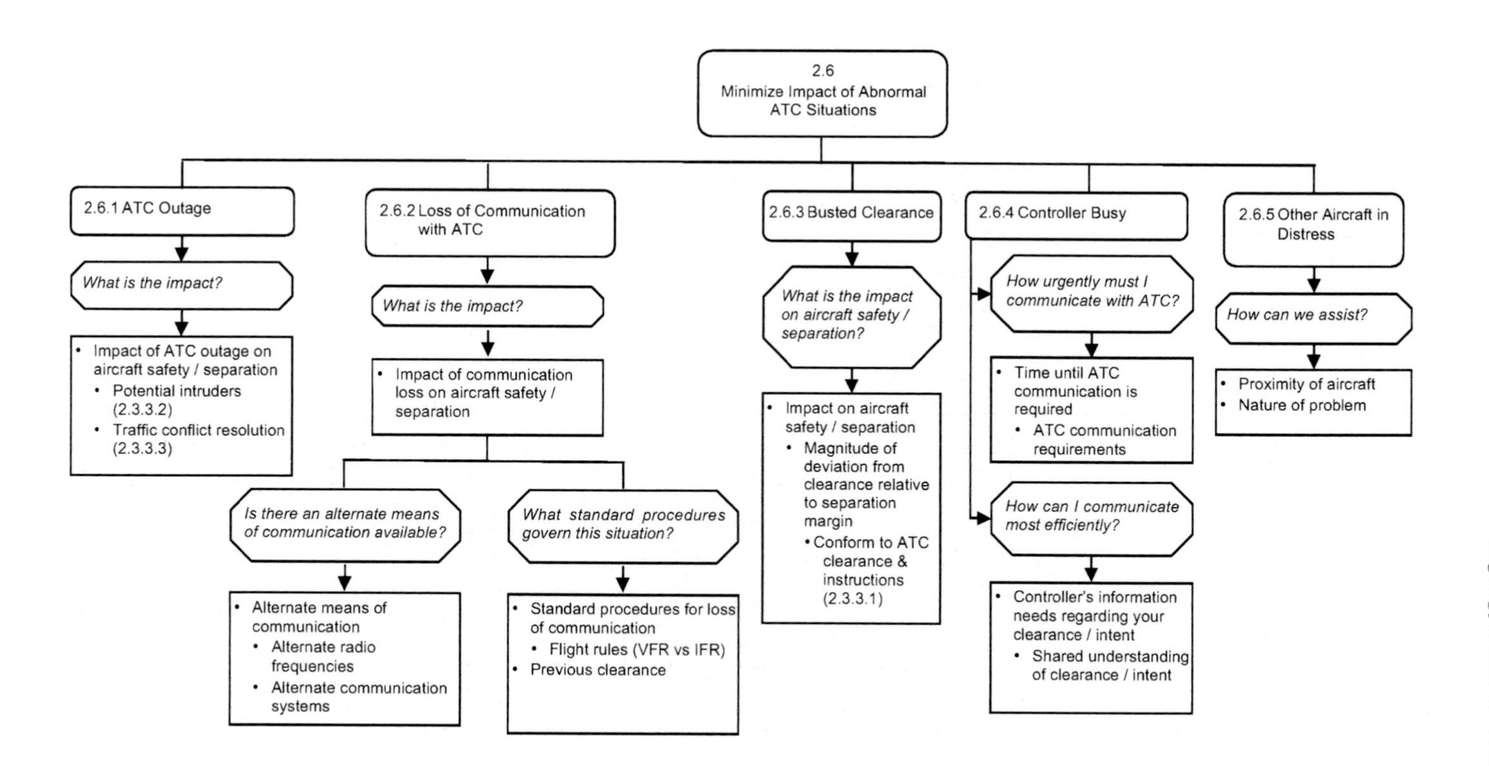
2.6
Minimize Impact of Abnormal ATC Situations
2.6.1 ATC Outage
What is the impact?
• Impact of ATC outage on aircraft safety / separation
• Potential intruders (2.3.3.2)
• Traffic conflict resolution (2.3.3.3)
2.6.2 Loss of Communication with ATC
What is the impact?
• Impact of communication loss on aircraft safety / separation
Is there an alternate means of communication available?
• Alternate means of communication
• Alternate radio frequencies
• Alternate communication systems
What standard procedures govern this situation?
• Standard procedures for loss of communication
• Flight rules (VFR vs IFR)
• Previous clearance
2.6.3 Busted Clearance
What is the impact on aircraft safety / separation?
• Impact on aircraft safety / separation
• Magnitude of deviation from clearance relative to separation margin
• Conform to ATC clearance & instructions (2.3.3.1)
2.6.4 Controller Busy
How urgently must I communicate with ATC?
• Time until ATC communication is required
• ATC communication requirements
How can I communicate most efficiently?
• Controller's information needs regarding your clearance / intent
• Shared understanding of clearance / intent
2.6.5 Other Aircraft in Distress
How can we assist?
• Proximity of aircraft
• Nature of problem

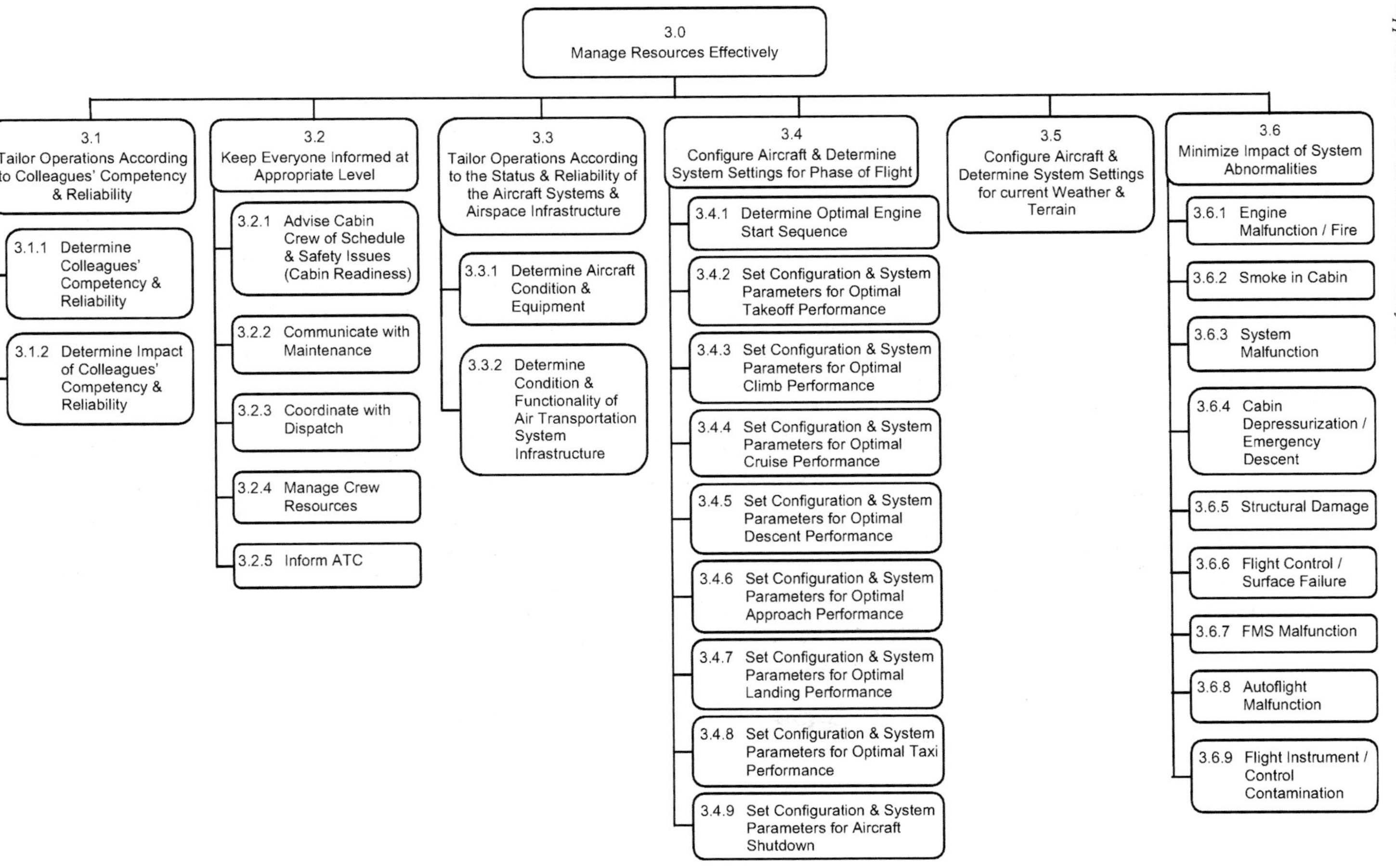
3.0
Manage Resources Effectively
3.1
Tailor Operations According to Colleagues' Competency & Reliability
3.1.1 Determine Colleagues' Competency & Reliability
3.1.2 Determine Impact of Colleagues' Competency & Reliability
3.2
Keep Everyone Informed at Appropriate Level
3.2.1 Advise Cabin Crew of Schedule & Safety Issues (Cabin Readiness)
3.2.2 Communicate with Maintenance
3.2.3 Coordinate with Dispatch
3.2.4 Manage Crew Resources
3.2.5 Inform ATC
3.3
Tailor Operations According to the Status & Reliability of the Aircraft Systems & Airspace Infrastructure
3.3.1 Determine Aircraft Condition & Equipment
3.3.2 Determine Condition & Functionality of Air Transportation System Infrastructure
3.4
Configure Aircraft & Determine System Settings for Phase of Flight
3.4.1 Determine Optimal Engine Start Sequence
3.4.2 Set Configuration & System Parameters for Optimal Takeoff Performance
3.4.3 Set Configuration & System Parameters for Optimal Climb Performance
3.4.4 Set Configuration & System Parameters for Optimal Cruise Performance
3.4.5 Set Configuration & System Parameters for Optimal Descent Performance
3.4.6 Set Configuration & System Parameters for Optimal Approach Performance
3.4.7 Set Configuration & System Parameters for Optimal Landing Performance
3.4.8 Set Configuration & System Parameters for Optimal Taxi Performance
3.4.9 Set Configuration & System Parameters for Aircraft Shutdown
3.5
Configure Aircraft & Determine System Settings for current Weather & Terrain
3.6
Minimize Impact of System Abnormalities
3.6.1 Engine Malfunction / Fire
3.6.2 Smoke in Cabin
3.6.3 System Malfunction
3.6.4 Cabin Depressurization / Emergency Descent
3.6.5 Structural Damage
3.6.6 Flight Control / Surface Failure
3.6.7 FMS Malfunction
3.6.8 Autoflight Malfunction
3.6.9 Flight Instrument / Control Contamination

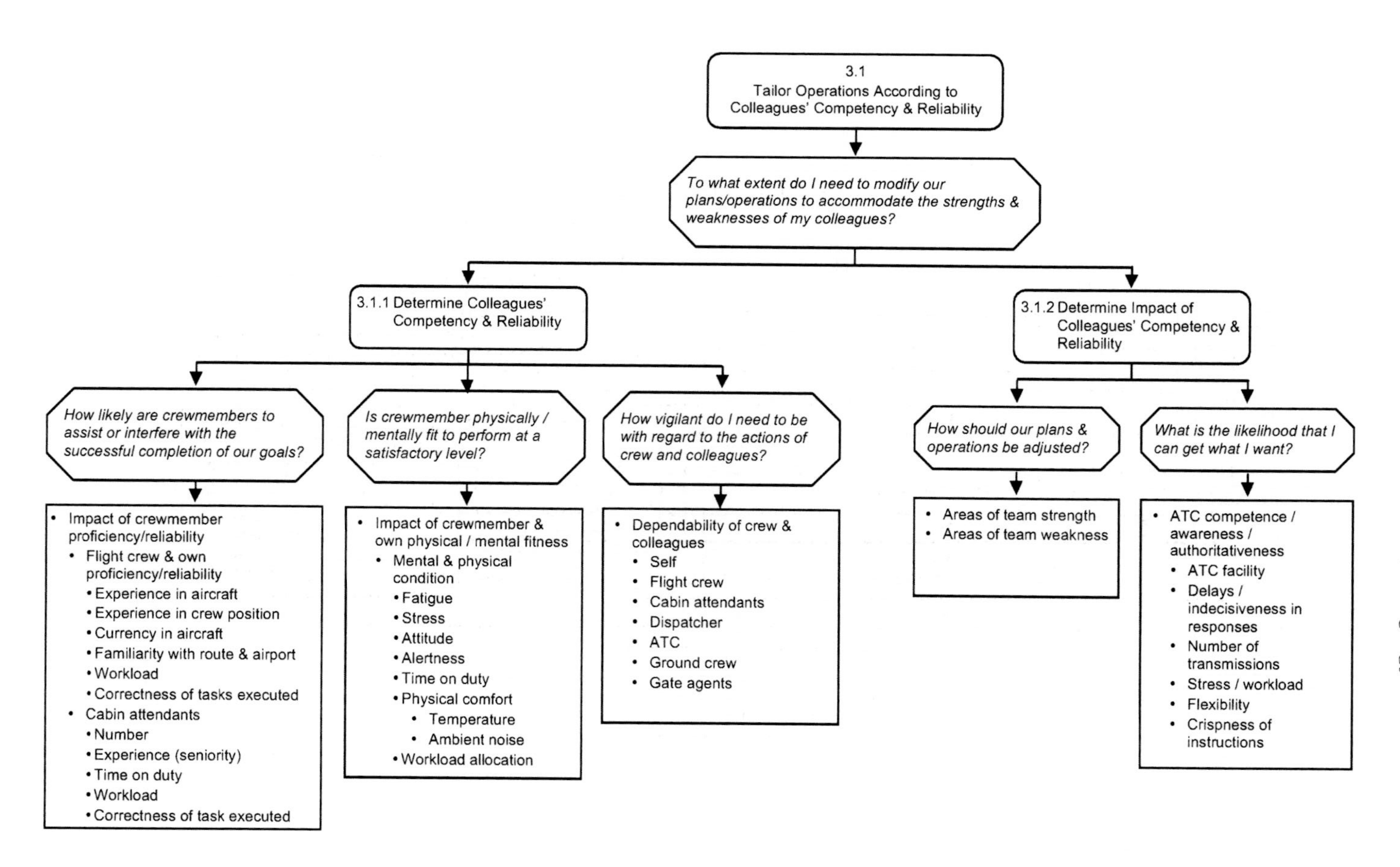
3.1
Tailor Operations According to Colleagues' Competency & Reliability
To what extent do I need to modify our plans/operations to accommodate the strengths & weaknesses of my colleagues?
3.1.1 Determine Colleagues' Competency & Reliability
3.1.2 Determine Impact of Colleagues' Competency & Reliability
How likely are crewmembers to assist or interfere with the successful completion of our goals?
Is crewmember physically / mentally fit to perform at a satisfactory level?
How vigilant do I need to be with regard to the actions of crew and colleagues?
How should our plans & operations be adjusted?
What is the likelihood that I can get what I want?
Impact of crewmember proficiency/reliability
Flight crew & own proficiency/reliability
Experience in aircraft
Experience in crew position
Currency in aircraft
Familiarity with route & airport
Workload
Correctness of tasks executed
Cabin attendants
Number
Experience (seniority)
Time on duty
Workload
Correctness of task executed
Impact of crewmember & own physical / mental fitness
Mental & physical condition
Fatigue
Stress
Attitude
Alertness
Time on duty
Physical comfort
Temperature
Ambient noise
Workload allocation
Dependability of crew & colleagues
Self
Flight crew
Cabin attendants
Dispatcher
ATC
Ground crew
Gate agents
Areas of team strength
Areas of team weakness
ATC competence / awareness / authoritativeness
ATC facility
Delays / indecisiveness in responses
Number of transmissions
Stress / workload
Flexibility
Crispness of instructions

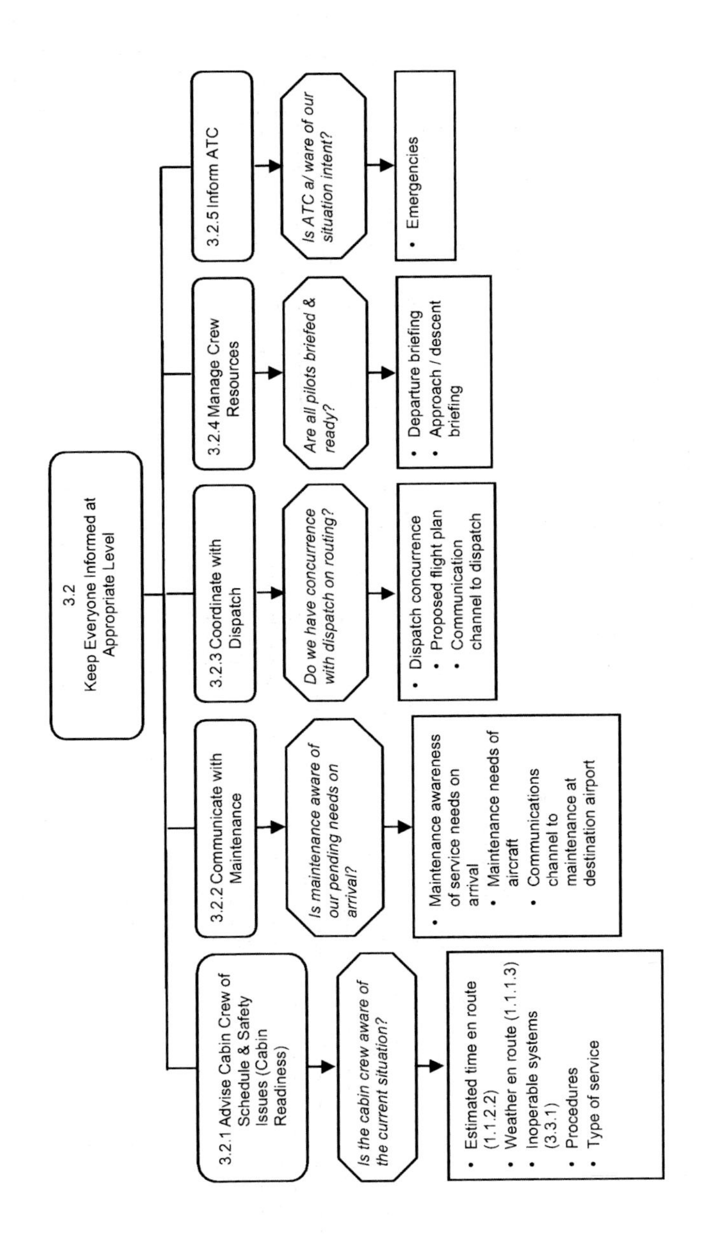
3.2
Keep Everyone Informed at Appropriate Level
3.2.1 Advise Cabin Crew of Schedule & Safety Issues (Cabin Readiness)
Is the cabin crew aware of the current situation?
• Estimated time en route (1.1.2.2)
• Weather en route (1.1.1.3)
• Inoperable systems (3.3.1)
• Procedures
• Type of service
3.2.2 Communicate with Maintenance
Is maintenance aware of our pending needs on arrival?
• Maintenance awareness of service needs on arrival
• Maintenance needs of aircraft
• Communications channel to maintenance at destination airport
3.2.3 Coordinate with Dispatch
Do we have concurrence with dispatch on routing?
• Dispatch concurrence
• Proposed flight plan
• Communication channel to dispatch
3.2.4 Manage Crew Resources
Are all pilots briefed & ready?
• Departure briefing
• Approach / descent briefing
3.2.5 Inform ATC
Is ATC a/ ware of our situation intent?
• Emergencies

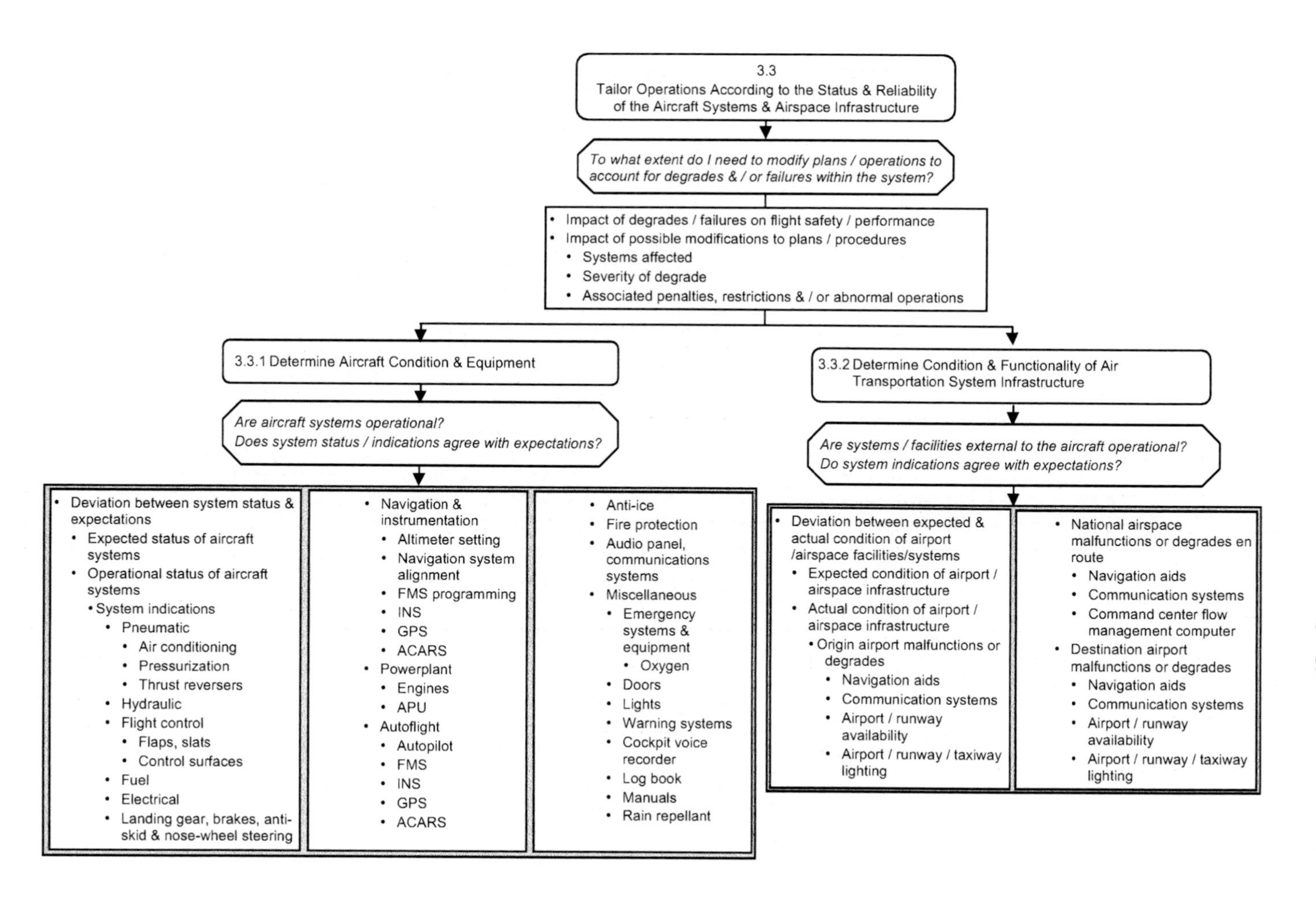
3.3
Tailor Operations According to the Status & Reliability of the Aircraft Systems & Airspace Infrastructure
To what extent do I need to modify plans / operations to account for degrades & / or failures within the system?
• Impact of degrades / failures on flight safety / performance
• Impact of possible modifications to plans / procedures
• Systems affected
• Severity of degrade
• Associated penalties, restrictions & / or abnormal operations
3.3.1 Determine Aircraft Condition & Equipment
Are aircraft systems operational?
Does system status / indications agree with expectations?
• Deviation between system status & expectations
• Expected status of aircraft systems
• Operational status of aircraft systems
• System indications
• Pneumatic
• Air conditioning
• Pressurization
• Thrust reversers
• Hydraulic
• Flight control
• Flaps, slats
• Control surfaces
• Fuel
• Electrical
• Landing gear, brakes, anti-skid & nose-wheel steering
• Navigation & instrumentation
• Altimeter setting
• Navigation system alignment
• FMS programming
• INS
• GPS
• ACARS
• Powerplant
• Engines
• APU
• Autoflight
• Autopilot
• FMS
• INS
• GPS
• ACARS
• Anti-ice
• Fire protection
• Audio panel, communications systems
• Miscellaneous
• Emergency systems & equipment
• Oxygen
• Doors
• Lights
• Warning systems
• Cockpit voice recorder
• Log book
• Manuals
• Rain repellant
3.3.2 Determine Condition & Functionality of Air Transportation System Infrastructure
Are systems / facilities external to the aircraft operational?
Do system indications agree with expectations?
• Deviation between expected & actual condition of airport /airspace facilities/systems
• Expected condition of airport / airspace infrastructure
• Actual condition of airport / airspace infrastructure
• Origin airport malfunctions or degrades
• Navigation aids
• Communication systems
• Airport / runway availability
• Airport / runway / taxiway lighting
• National airspace malfunctions or degrades en route
• Navigation aids
• Communication systems
• Command center flow management computer
• Destination airport malfunctions or degrades
• Navigation aids
• Communication systems
• Airport / runway availability
• Airport / runway / taxiway lighting

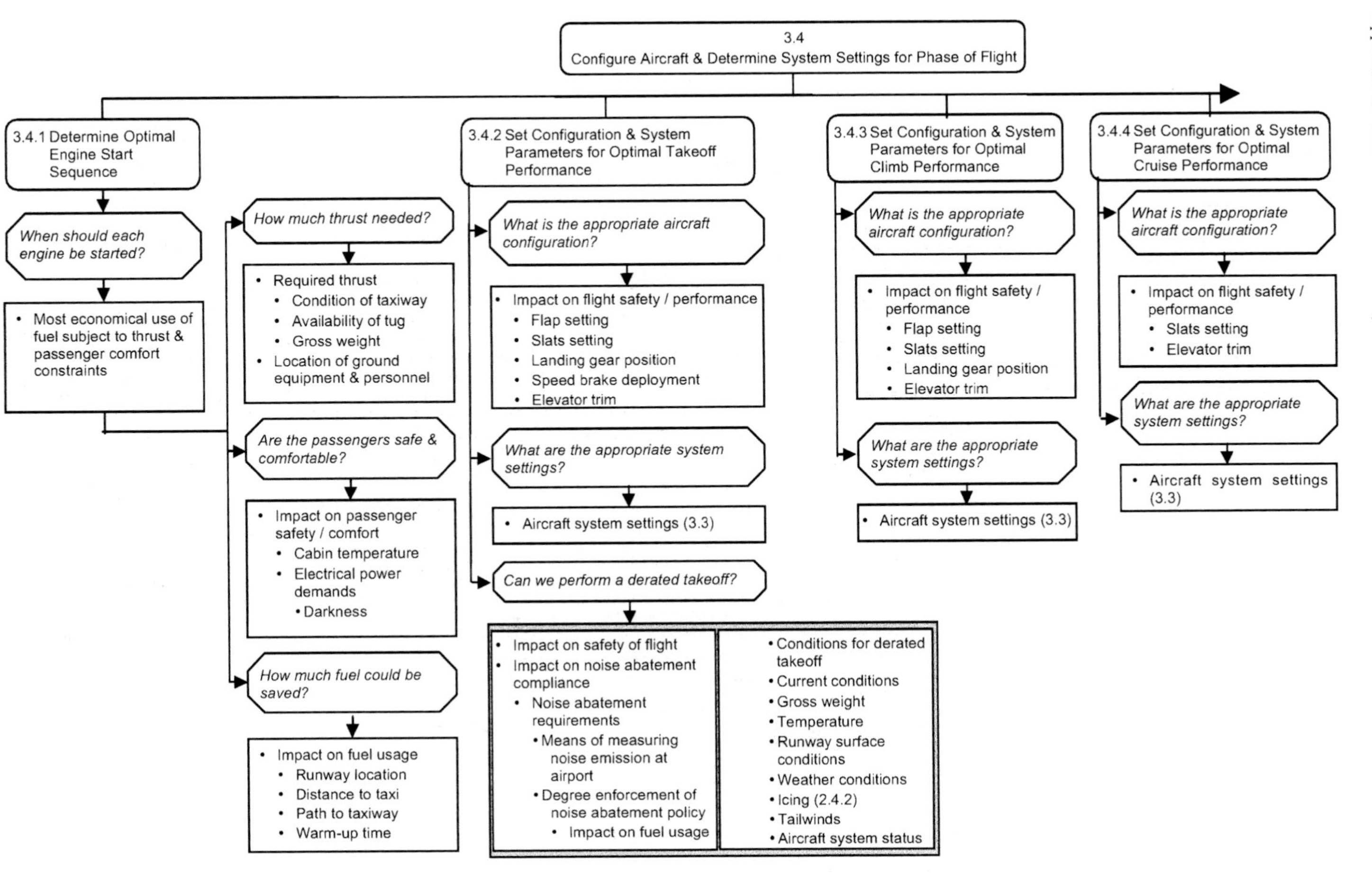

3.4
Configure Aircraft & Determine System Settings for Phase of Flight
3.4.1 Determine Optimal Engine Start Sequence
When should each engine be started?
• Most economical use of fuel subject to thrust & passenger comfort constraints
How much thrust needed?
• Required thrust
• Condition of taxiway
• Availability of tug
• Gross weight
• Location of ground equipment & personnel
Are the passengers safe & comfortable?
• Impact on passenger safety / comfort
• Cabin temperature
• Electrical power demands
• Darkness
How much fuel could be saved?
• Impact on fuel usage
• Runway location
• Distance to taxi
• Path to taxiway
• Warm-up time
3.4.2 Set Configuration & System Parameters for Optimal Takeoff Performance
What is the appropriate aircraft configuration?
• Impact on flight safety / performance
• Flap setting
• Slats setting
• Landing gear position
• Speed brake deployment
• Elevator trim
What are the appropriate system settings?
• Aircraft system settings (3.3)
Can we perform a derated takeoff?
• Impact on safety of flight
• Impact on noise abatement compliance
• Noise abatement requirements
• Means of measuring noise emission at airport
• Degree enforcement of noise abatement policy
• Impact on fuel usage
• Conditions for derated takeoff
• Current conditions
• Gross weight
• Temperature
• Runway surface conditions
• Weather conditions
• Icing (2.4.2)
• Tailwinds
• Aircraft system status
3.4.3 Set Configuration & System Parameters for Optimal Climb Performance
What is the appropriate aircraft configuration?
• Impact on flight safety / performance
• Flap setting
• Slats setting
• Landing gear position
• Elevator trim
What are the appropriate system settings?
• Aircraft system settings (3.3)
3.4.4 Set Configuration & System Parameters for Optimal Cruise Performance
What is the appropriate aircraft configuration?
• Impact on flight safety / performance
• Slats setting
• Elevator trim
What are the appropriate system settings?
• Aircraft system settings (3.3)

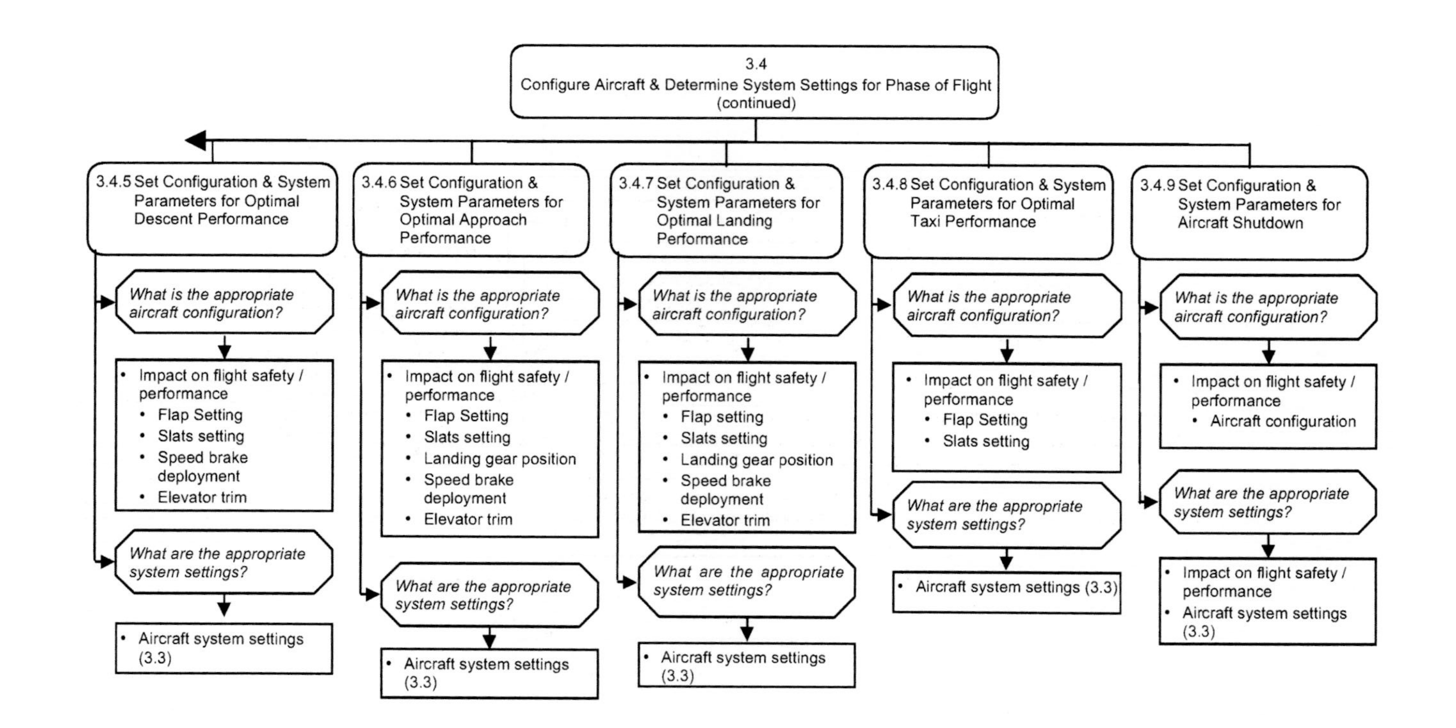
3.4
Configure Aircraft & Determine System Settings for Phase of Flight
(continued)
3.4.5 Set Configuration & System Parameters for Optimal Descent Performance
What is the appropriate aircraft configuration?
• Impact on flight safety / performance
• Flap Setting
• Slats setting
• Speed brake deployment
• Elevator trim
What are the appropriate system settings?
• Aircraft system settings (3.3)
3.4.6 Set Configuration & System Parameters for Optimal Approach Performance
What is the appropriate aircraft configuration?
• Impact on flight safety / performance
• Flap Setting
• Slats setting
• Landing gear position
• Speed brake deployment
• Elevator trim
What are the appropriate system settings?
• Aircraft system settings (3.3)
3.4.7 Set Configuration & System Parameters for Optimal Landing Performance
What is the appropriate aircraft configuration?
• Impact on flight safety / performance
• Flap setting
• Slats setting
• Landing gear position
• Speed brake deployment
• Elevator trim
What are the appropriate system settings?
• Aircraft system settings (3.3)
3.4.8 Set Configuration & System Parameters for Optimal Taxi Performance
What is the appropriate aircraft configuration?
• Impact on flight safety / performance
• Flap Setting
• Slats setting
What are the appropriate system settings?
• Aircraft system settings (3.3)
3.4.9 Set Configuration & System Parameters for Aircraft Shutdown
What is the appropriate aircraft configuration?
• Impact on flight safety / performance
• Aircraft configuration
What are the appropriate system settings?
• Impact on flight safety / performance
• Aircraft system settings (3.3)

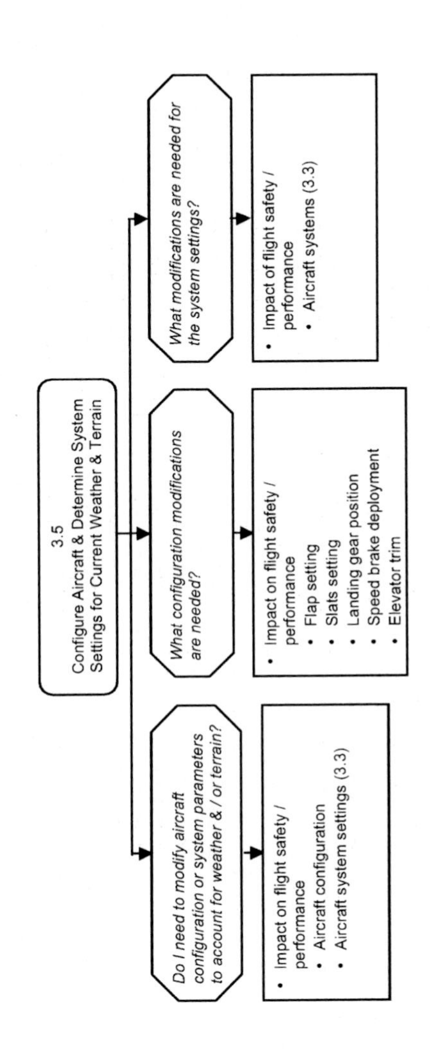
3.5
Configure Aircraft & Determine System Settings for Current Weather & Terrain
Do I need to modify aircraft configuration or system parameters to account for weather & / or terrain?
• Impact on flight safety / performance
• Aircraft configuration
• Aircraft system settings (3.3)
What configuration modifications are needed?
• Impact on flight safety / performance
• Flap setting
• Slats setting
• Landing gear position
• Speed brake deployment
• Elevator trim
What modifications are needed for the system settings?
• Impact of flight safety / performance
• Aircraft systems (3.3)

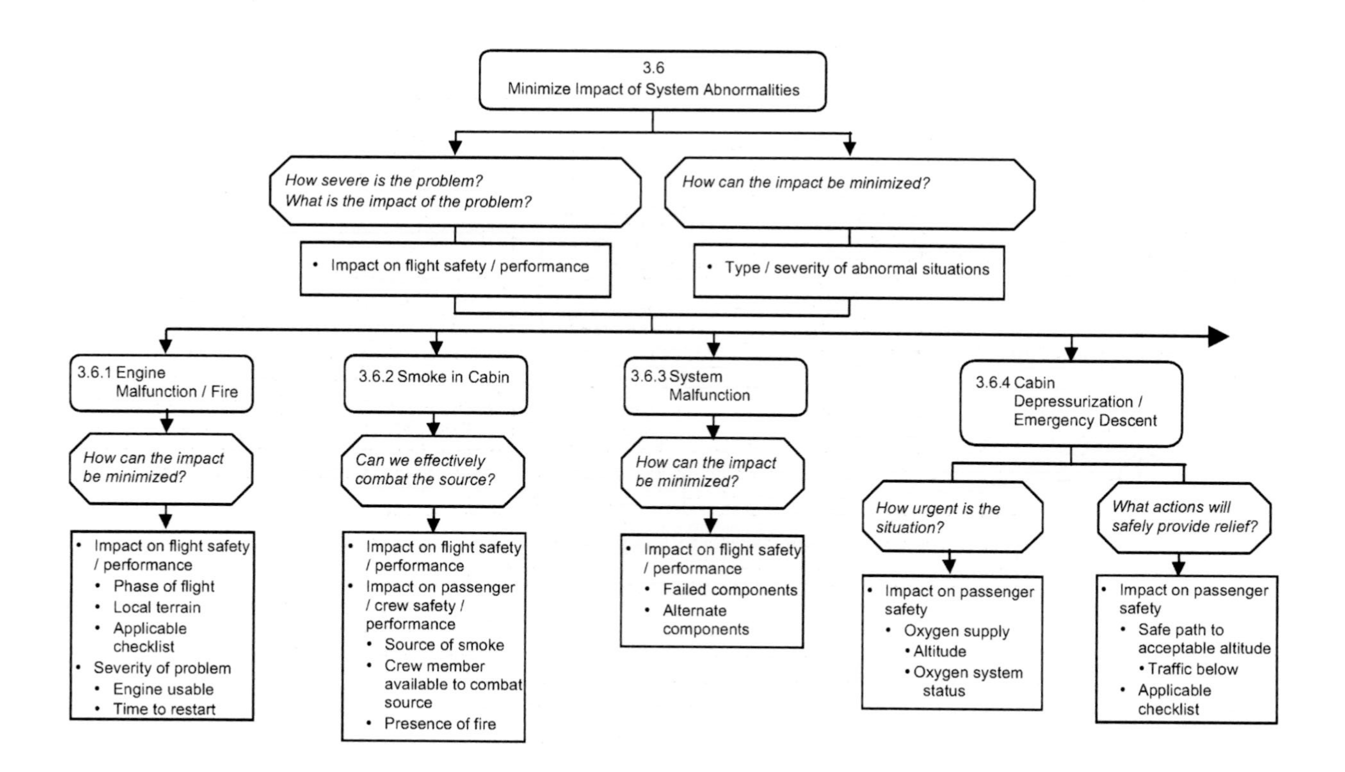
3.6
Minimize Impact of System Abnormalities
How severe is the problem?
What is the impact of the problem?
• Impact on flight safety / performance
How can the impact be minimized?
• Type / severity of abnormal situations
3.6.1 Engine Malfunction / Fire
How can the impact be minimized?
• Impact on flight safety / performance
• Phase of flight
• Local terrain
• Applicable checklist
• Severity of problem
• Engine usable
• Time to restart
3.6.2 Smoke in Cabin
Can we effectively combat the source?
• Impact on flight safety / performance
• Impact on passenger / crew safety / performance
• Source of smoke
• Crew member available to combat source
• Presence of fire
3.6.3 System Malfunction
How can the impact be minimized?
• Impact on flight safety / performance
• Failed components
• Alternate components
3.6.4 Cabin Depressurization / Emergency Descent
How urgent is the situation?
• Impact on passenger safety
• Oxygen supply
• Altitude
• Oxygen system status
What actions will safely provide relief?
• Impact on passenger safety
• Safe path to acceptable altitude
• Traffic below
• Applicable checklist

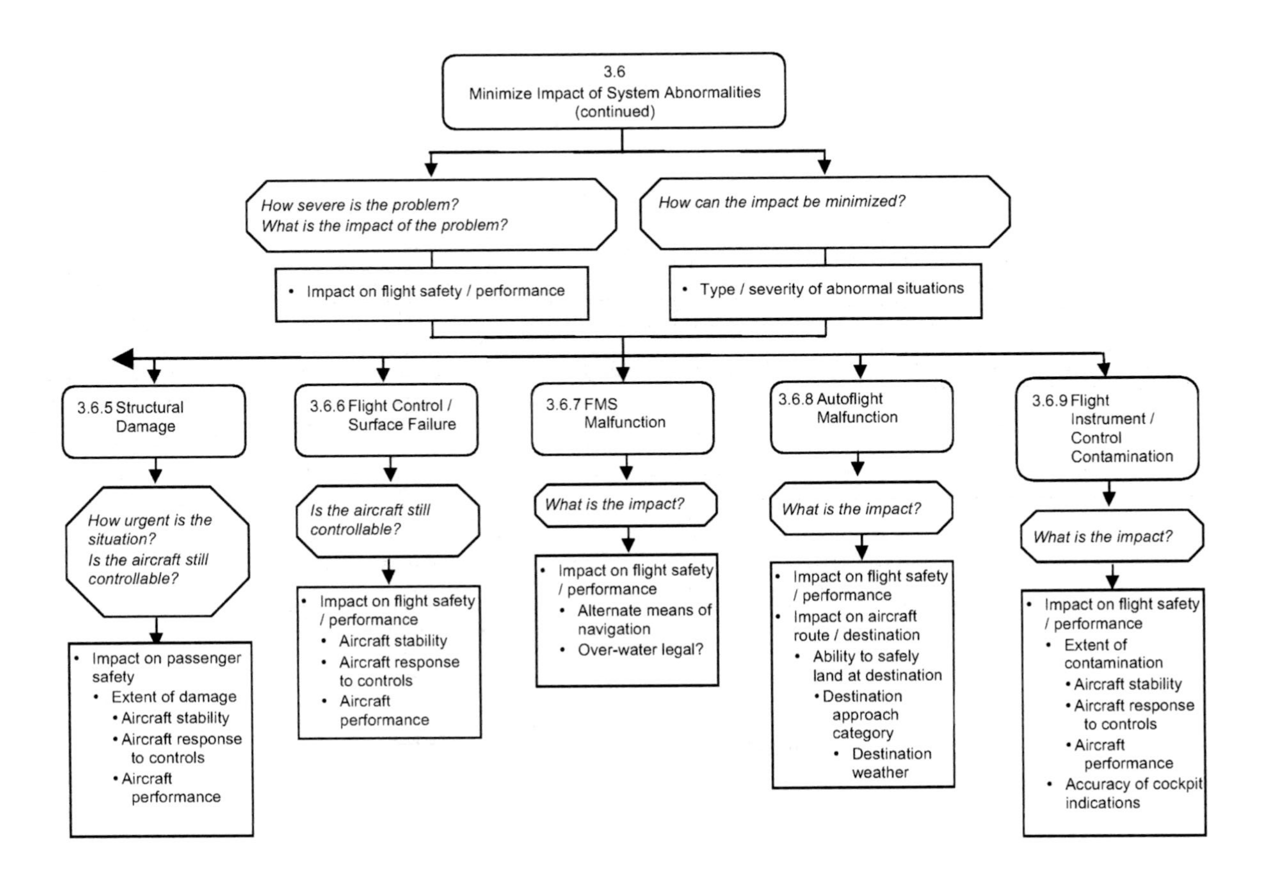

3.6
Minimize Impact of System Abnormalities
(continued)
How severe is the problem?
What is the impact of the problem?
How can the impact be minimized?
• Impact on flight safety / performance
• Type / severity of abnormal situations
3.6.5 Structural Damage
How urgent is the situation?
Is the aircraft still controllable?
• Impact on passenger safety
• Extent of damage
• Aircraft stability
• Aircraft response to controls
• Aircraft performance
3.6.6 Flight Control / Surface Failure
Is the aircraft still controllable?
• Impact on flight safety / performance
• Aircraft stability
• Aircraft response to controls
• Aircraft performance
3.6.7 FMS Malfunction
What is the impact?
• Impact on flight safety / performance
• Alternate means of navigation
• Over-water legal?
3.6.8 Autoflight Malfunction
What is the impact?
• Impact on flight safety / performance
• Impact on aircraft route / destination
• Ability to safely land at destination
• Destination approach category
• Destination weather
3.6.9 Flight Instrument / Control Contamination
What is the impact?
• Impact on flight safety / performance
• Extent of contamination
• Aircraft stability
• Aircraft response to controls
• Aircraft performance
• Accuracy of cockpit indications

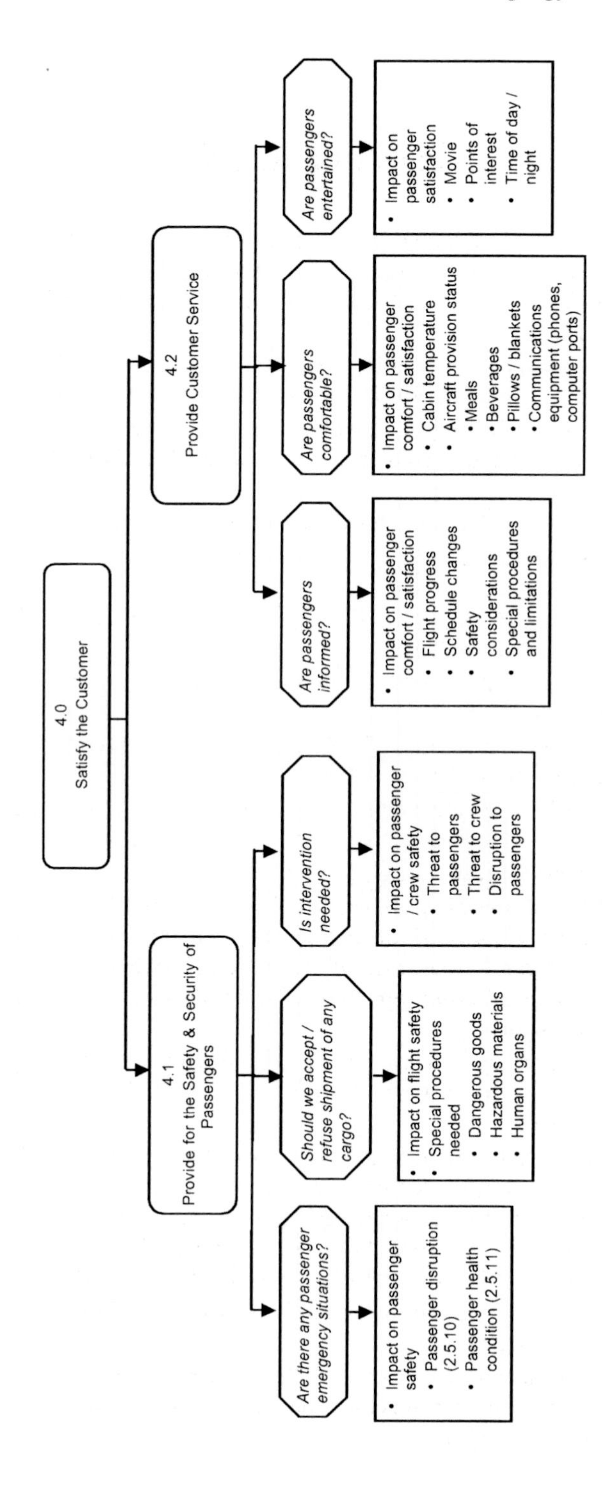
4.0
Satisfy the Customer
4.1
Provide for the Safety & Security of Passengers
4.2
Provide Customer Service
Are there any passenger emergency situations?
• Impact on passenger safety
• Passenger disruption (2.5.10)
• Passenger health condition (2.5.11)
Should we accept / refuse shipment of any cargo?
• Impact on flight safety
• Special procedures needed
• Dangerous goods
• Hazardous materials
• Human organs
Is intervention needed?
• Impact on passenger / crew safety
• Threat to passengers
• Threat to crew
• Disruption to passengers
Are passengers informed?
• Impact on passenger comfort / satisfaction
• Flight progress
• Schedule changes
• Safety considerations
• Special procedures and limitations
Are passengers comfortable?
• Impact on passenger comfort / satisfaction
• Cabin temperature
• Aircraft provision status
• Meals
• Beverages
• Pillows / blankets
• Communications equipment (phones, computer ports)
Are passengers entertained?
• Impact on passenger satisfaction
• Movie
• Points of interest
• Time of day / night

References

Aeronautica Civil of the Republic of Colombia. (1996). *Aircraft accident report: Controlled flight into terrain, American Airlines Flight 965, Boeing 757-223, N651AA, near Cali, Colombia, December 20, 1995*. Santafe de Bogota, D.C., Colombia: Author.

Air Transportation Office of the Philippines. (2002). *Aircraft Accident Investigation Report: N 581 FE MD-11 Aircraft*. Manila, Philippines: Author.

Allender, L., Kelley, T., Archer, S., & Adkins, R. (1997). IMPRINT: the transition and further development of a soldier-system analysis tool. *MANPRINT Quarterly, 5*(1), 1-7.

Anderson, J. R. (1993). *Rules of the mind*. Hillsdale, NJ: Lawrence Erlbaum.

Anderson, J. R., & Lebiere, C. (1998). *Atomic components of thought*. Hillsdale, NJ: Lawrence Erlbaum.

Andre, A. D., & Cutler, H. A. (1998). Displaying uncertainty in advanced navigation systems. *Proceedings of the Human Factors and Ergonomics Society 42nd Annual Meeting,* (pp 31-35). Santa Monica, CA: Human Factors and Ergonomics Society.

Associated Press. (1996, December 17). Safety system may have contributed to LA ship crash. *The Boston Globe,* pp. A33.

Baddeley, A. D. (1972). Selective attention and performance in dangerous environments. *British Journal of Psychology, 63*, 537-546.

Baddeley, A. D. (1986). *Human memory*. Oxford: Clarendon Press.

Banbury, S., Selcon, S., Endsley, M., Gorton, T., & Tatlock, K. (1998). Being certain about uncertainty: How the representation of system reliability affects pilot decision making. *Proceedings of the Human Factors and Ergonomics Society 42nd Annual Meeting* (pp. 36-41). Santa Monica, CA: Human Factors and Ergonomics Society.

Banbury, S. P., Macken, W. J., Tremblay, S., & Jones, D. M. (2001). Auditory distraction and short-term memory: Phenomena and practical implications. *Human Factors, 43*(1), 12-29.

Banda, C., Bushnell, D., Chen, S., Chiu, A., Constantine, B., Murray, J., Neukom, C., Prevost, M., Shankar, R., & Staveland, L. (1991). *Army-NASA aircrew/aircraft integration program: Phase IV A3I man-machine integration design and analysis system (MIDAS) software detailed design document* (NASA Contractor Report 177593). Moffett Field, CA: NASA Ames Research Center.

Bartlett, F. C. (1943). Fatigue following highly skilled work. *Proceedings of the Royal Society (B), 131*, 147-257.

Begault, D. R., & Wenzel, E. M. (1992). Techniques and applications for binaural sound manipulation in human-machine interfaces. *International Journal of Aviation Psychology, 2*(1), 1-22.

Bell, H. H., & Lyon, D. R. (2000). Using observer ratings to assess situation awareness. In M. R. Endsley & D. J. Garland (Eds), *Situation awareness analysis and measurement* (pp. 129-146). Mahwah, NJ: Lawrence Erlbaum.

Bell, H. H., & Waag, W. L. (1995). Using observer ratings to assess situational awareness in tactical air environments. In D. J. Garland & M. R. Endsley (Eds), *Experimental analysis and measurement of situation awareness* (pp. 93-99). Daytona Beach, FL: Embry-Riddle Aeronautical University Press.

Belz, S. B., Robinson, G. S., & Casali, J. G. (1999). A new class of auditory warning signals for complex systems: Auditory icons. *Human Factors, 41*(4), 608-618.

Bereiter, S., & Miller, S. (1989). A field-based study of troubleshooting in computer controlled manufacturing systems. *IEEE Transactions on Systems, Man, and Cybernetics, 19*, 205-219.

Berg, B. G. (1990). Observer efficiency and weights in a multiple observation task. *Journal of the Acoustical Society of America, 88*, 149-158.

Billings, C. E. (1991). *Human-centered aircraft automation: A concept and guidelines* (NASA Technical Memorandum 103885). Moffet Field, CA: NASA Ames Research Center.

Billings, C. E. (1997). *Aviation automation: The search for a human-centered approach*. Mahwah, NJ: Lawrence Erlbaum.

Bliss, J. P. (1993). *The cry-wolf phenomenon and its effect on alarm responses.* Unpublished doctoral dissertation, University of Central Florida, Orlando, Florida.

Bliss, J. P., Dunn, M., & Fuller, B. S. (1995). Reversal of the cry-wolf effect: An investigation of two methods to increase alarm response rates. *Perceptual and Motor Skills, 80*, 1231-1242.

Bliss, J. P., Gilson, R. D., & Deaton, J. E. (1995). Human probability matching behavior in response to alarms of varying reliability. *Ergonomics, 38*, 2300-2312.

Bolstad, C. (2001). Situation awareness: Does it change with age? *Proceedings of the Human Factors and Ergonomics Society 45th Annual Meeting* (pp. 272-276). Santa Monica, CA: Human Factors and Ergonomics Society.

Bolstad, C. A., & Endsley, M. R. (1998). *Information dissonance, shared mental models and shared displays: An empirical evaluation of information dominance techniques* (AFRL-HE-WP-TR-1999-0213). Wright Patterson AFB, OH: Air Force Research Laboratory.

Bolstad, C. A., & Endsley, M. R. (1999a). *The effect of taskload and shared displays on team situation awareness* (AFRL-HE-WP-TR-1999-0243). Wright Patterson AFB, OH: Air Force Research Laboratory.

Bolstad, C. A., & Endsley, M. R. (1999b). Shared mental models and shared displays: An empirical evaluation of team performance. *Proceedings of the 43rd Annual Meeting of the Human Factors and Ergonomics Society* (pp. 213-217). Santa Monica, CA: Human Factors and Ergonomics Society.

Bolstad, C. A., & Endsley, M. R. (2000). The effect of task load and shared displays on team situation awareness. *Proceedings of the 14th Triennial Congress of the International Ergonomics Association and the 44th Annual*

Meeting of the Human Factors and Ergonomics Society (pp. 189-192). Santa Monica, CA: Human Factors and Ergonomics Society.

Bolstad, C. A., & Endsley, M. R. (2002). *A taxonomy of collaboration: An analysis of tools for supporting team SA* (SATech 02-06). Marietta, GA: SA Technologies.

Breznitz, S. (1983). *Cry wolf: The psychology of false alarms.* Hillsdale, NJ: Lawrence Erlbaum.

Bringelson, L. S., & Pettitt, M. A. (1995). Applying airline crew resource management in emergency medicine. *Proceedings of the Human Factors and Ergonomics Society 39th Annual Meeting* (pp. 728-732). Santa Monica, CA: Human Factors and Ergonomics Society.

Broadbent, D. E. (1954). Some effects of noise on visual performance. *Quarterly Journal of Experimental Psychology, 6*, 1-5.

Burrough, B. (1998). *Dragonfly: NASA and the crisis aboard Mir.* New York: Harper Collins.

Busquets, A. M., Parrish, R. V., Williams, S. P., & Nold, D. E. (1994). Comparison of pilots' acceptance and spatial awareness when using EFIS vs. pictorial display formats for complex, curved landing approaches. In R. D. Gilson & D. J. Garland & J. M. Koonce (Eds), *Situational awareness in complex systems* (pp. 139-167). Daytona Beach, FL: Embry-Riddle Aeronautical University Press.

Cale, M. L., Paglione, M. M., Ryan, H. F., Oaks, R. D., & Summerill, J. S. (1998, December). Application of generic metrics to assess the accuracy of strategic conflict probes. *Proceedings of the 2nd USA/Europe Air Traffic Management R&D Seminar.* Orlando, FL: FAA.

Calhoun, G. L., Janson, W. P., & Valencia, G. (1988). Effectiveness of three-dimensional auditory directional cues. *Proceedings of the Human Factors Society 32nd Annual Meeting* (pp. 68-72). Santa Monica, CA: Human Factors Society.

Calhoun, G. L., Valencia, G., & Furness, T. A. (1987). Three-dimensional auditory cue simulation for crew station design/evaluation. *Proceedings of the Human Factors Society 31st Annual Meeting* (pp. 1398-1402). Santa Monica, CA: Human Factors Society.

Campbell, D. J. (1988). Task complexity: A review and analysis. *Academy of Management Review, 13*, 40-52.

Carmino, A., Idee, E., Larchier Boulanger, J., & Morlat, G. (1988). Representational errors: Why some may be termed diabolical. In L. P. Goodstein & H. B. Anderson & S. E. Olsen (Eds), *Tasks, errors and mental models* (pp. 240-250). London: Taylor & Francis.

Casey, S. (1993). *Set phasers on stun.* Santa Barbara, CA: Aegean.

Casson, R. W. (1983). Schema in cognitive anthropology. *Annual Review of Anthropology, 12*, 429-462.

Chapanis, A. (1996). *Human factors in systems engineering.* New York: John Wiley & Sons.

Chechile, R. A. (1992). A review of Anets: A model of cognitive complexity of displays. *Proceedings of the Human Factors Society 36th Annual Meeting* (pp. 1176-1180). Santa Monica, CA: Human Factors Society.

Chechile, R., Eggleston, R., Fleischman, R., & Sasseville, A. (1989). Modeling the cognitive content of displays. *Human Factors, 31*, 31-43.

Chechile, R., Peretz, N., & O'Hearn, B. (1989). *A cognitive and perceptual model of display complexity* (AAMRL/HEA). Wright Patterson AFB, OH: Armstrong Aeromedical Research Laboratory.

Childs, J. A., Ross, K. G., & Ross, W. A. (2000). Identifying breakdowns in team situation awareness in a complex synthetic environment to support multi-focused research and development. *Proceedings of the Human Performance, Situation Awareness and Automation: User-Centered Design for the New Millennium Conference* (pp. 305-310). Atlanta, GA: SA Technologies, Inc.

Christ, R. E., McKeever, K. J., & Huff, J. W. (1994). Collective training of multiple team organizational units: Influence of intra- and inter-team processes. In G. E. Bradley & H. W. Hendrick (Eds), *Human Factors in Organizational Design and Management - IV* (pp. 323-326). Amsterdam: Elsevier.

Citera, M., McNeese, M. D., Brown, C. E., Selvaraj, J. A., Zaff, B. S., & Whitaker, R. D. (1995). Fitting information systems to collaborating design teams. *Journal of the American Society for Information Science, 46*(7), 551-559.

Clark, H. H., & Schaefer, E. F. (1989). Contributing to discourse. *Cognitive Science, 13*, 259-294.

Combs, R. B., & Aghazadeh, F. (1988). Ergonomically designed chemical plant control room. In F. Aghazedeh (Ed.), *Trends in ergonomics and human factors* (pp. 357-364). Amsterdam: North-Holland.

Cook, N. J., Salas, E., Cannon-Bowers, J. A., & Stout, R. J. (2000). Measuring team knowledge. *Human Factors, 42*(1), 151-173.

Cooper, G., & Harper, R. (1969). *The use of pilot ratings in the evaluation of aircraft handling qualities* (NASA TN-D-5153). Moffett Field, CA: NASA Ames Research Center.

Cowan, N. (1988). Evolving conceptions of memory storage, selective attention, and their mutual constraints within the human information processing system. *Psychological Bulletin, 104*(2), 163-191.

Davies, D. R., & Parasuraman, R. (1980). *The psychology of vigilance*. London: Academic Press.

Deaton, J. E., & Glenn, F. A. (1999). The development of specifications for an automated monitoring system interface associated with aircraft condition. *International Journal of Aviation Psychology, 9*(2), 175-187.

DeCelles, J. L. (1991). *The delayed GPWS response syndrome*. Herndon, VA: Aviation Research and Education Foundation.

deGroot, A. (1965). *Thought and choice in chess*. The Hague, The Netherlands: Mouton.

Deutsch, S. E., & Adams, M. (1995). The operator-model architecture and its psychological framework. *Proceedings of the Sixth IFAC Symposium on Man-Machine Systems* (pp. 41-46). Cambridge, MA: Massachusetts Institute of Technology.

Deutsch, S. E., Macmillan, J., Cramer, N. L., & Chopra, S. (1997). *Operator model architecture (OMAR) final report* (Report Number 8179). Cambridge, MA: BBN Corporation.

Dreyfus, S. E. (1981). *Formal models vs. human situational understanding: Inherent limitations on the modeling of business expertise* (ORC 81-3). Berkeley: Operations Research Center, University of California.

Duffy, L. R. (1993). Team decision making biases: An information processing perspective. In G. A. Klein & J. Orasanu & R. Calderwood & C. E. Zsambok (Eds), *Decision making in action: Models and methods* (pp. 346-359). Norwood, NJ: Ablex.

Duley, J. A., Westerman, S., Molloy, R., & Parasuraman, R. (1997). Effects of display superimposition on monitoring of automation. *Proceedings of the Ninth International Symposium on Aviation Psychology* (pp. 322-328). Columbus, OH: The Ohio State University.

Durso, F. T., Hackworth, C. A., Truitt, T. R., Crutchfield, J., Nikolic, D., & Manning, C. A. (1998). Situation awareness as a predictor of performance for en route air traffic controllers. *Air Traffic Control Quarterly, 6*(1), 1-20.

Edworthy, J. (1997). Cognitive compatibility and warning design. *International Journal of Cognitive Ergonomics, 1*(3), 193-209.

Edworthy, J., Loxley, S., & Dennis, I. (1991). Improving auditory warning design: Relationship between warning sound parameters and perceived urgency. *Human Factors, 33*(2), 205-231.

Eggleston, R. G., Chechile, R. A., Fleischman, R. N., & Sasseville, A. (1986). Modeling the cognitive complexity of visual displays. *Proceedings of the Human Factors Society 30th Annual Meeting* (pp. 675-678). Santa Monica, CA: Human Factors Society.

Endsley, M. R. (1988). Design and evaluation for situation awareness enhancement. *Proceedings of the Human Factors Society 32nd Annual Meeting* (pp. 97-101). Santa Monica, CA: Human Factors Society.

Endsley, M. R. (1989). *Final report: Situation awareness in an advanced strategic mission* (NOR DOC 89-32). Hawthorne, CA: Northrop Corporation.

Endsley, M. R. (1990a). A methodology for the objective measurement of situation awareness. *Situational Awareness in Aerospace Operations (AGARD-CP-478)* (pp. 1/1 - 1/9). Neuilly Sur Seine, France: NATO - AGARD.

Endsley, M. R. (1990b). Predictive utility of an objective measure of situation awareness. *Proceedings of the Human Factors Society 34th Annual Meeting* (pp. 41-45). Santa Monica, CA: Human Factors Society.

Endsley, M. R. (1993a). Situation awareness and workload: Flip sides of the same coin. *Proceedings of the Seventh International Symposium on Aviation Psychology* (pp. 906-911). Columbus, OH: Department of Aviation, The Ohio State University.

Endsley, M. R. (1993b). A survey of situation awareness requirements in air-to-air combat fighters. *International Journal of Aviation Psychology, 3*(2), 157-168.

Endsley, M. R. (1995a). Measurement of situation awareness in dynamic systems. *Human Factors, 37*(1), 65-84.

Endsley, M. R. (1995b). A taxonomy of situation awareness errors. In R. Fuller & N. Johnston & N. McDonald (Eds.), *Human factors in aviation operations* (pp. 287-292). Aldershot, England: Avebury Aviation, Ashgate Publishing Ltd.

Endsley, M. R. (1995c). Toward a theory of situation awareness in dynamic systems. *Human Factors, 37*(1), 32-64.

Endsley, M. R. (1995d). Towards a new paradigm for automation: Designing for situation awareness. *Proceedings of the 6th IFAC/IFIP/IFOR/IEA Symposium on Analysis, Design and Evaluation of Man-Machine Systems* (pp. 421-426). Cambridge, MA: MIT.

Endsley, M. R. (1996a). Automation and situation awareness. In R. Parasuraman & M. Mouloua (Eds.), *Automation and human performance: Theory and applications* (pp. 163-181). Mahwah, NJ: Lawrence Erlbaum.

Endsley, M. R. (1996b). Situation awareness measurement in test and evaluation. In T. G. O'Brien & S. G. Charlton (Eds), *Handbook of human factors testing & evaluation* (pp. 159-180). Mahwah, NJ: Lawrence Erlbaum.

Endsley, M. R. (1997). Level of automation: Integrating humans and automated systems. *Proceedings of the Human Factors and Ergonomics Society 41st Annual Meeting* (pp. 200-204). Santa Monica, CA: Human Factors and Ergonomics Society.

Endsley, M. R. (2000). Direct measurement of situation awareness: Validity and use of SAGAT. In M. R. Endsley & D. J. Garland (Eds), *Situation awareness analysis and measurement* (pp. 147-174). Mahwah, NJ: LEA.

Endsley, M. R. (2000). Theoretical underpinnings of situation awareness: A critical review. In M. R. Endsley & D. J. Garland (Eds), *Situation awareness analysis and measurement* (pp. 3-32). Mahwah, NJ: LEA.

Endsley, M. R., & Bolstad, C. A. (1993). Human capabilities and limitations in situation awareness. *Combat Automation for Airborne Weapon Systems: Man/Machine Interface Trends and Technologies (AGARD-CP-520)* (pp. 19/11-19/10). Neuilly Sur Seine, France: NATO - AGARD.

Endsley, M. R., & Bolstad, C. A. (1994). Individual differences in pilot situation awareness. *International Journal of Aviation Psychology, 4*(3), 241-264.

Endsley, M. R., Farley, T. C., Jones, W. M., Midkiff, A. H., & Hansman, R. J. (1998). *Situation awareness information requirements for commercial airline pilots* (ICAT-98-1). Cambridge, MA: Massachusetts Institute of Technology International Center for Air Transportation.

Endsley, M. R., & Garland, D. J. (Eds). (2000). *Situation awareness analysis and measurement.* Mahwah, NJ: Lawrence Erlbaum.

Endsley, M. R., Hansman, R. J., & Farley, T. C. (1998). Shared situation awareness in the flight deck - ATC system. *Proceedings of the AIAA/IEEE/SAE 17th Digital Avionics Systems Conference.* Bellevue, WA: IEEE.

Endsley, M. R., & Jones, W. M. (1997). *Situation awareness, information dominance, and information warfare* (AL/CF-TR-1997-0156). Wright-Patterson AFB, OH: United States Air Force Armstrong Laboratory.

Endsley, M. R., & Jones, W. M. (2001). A model of inter- and intrateam situation awareness: Implications for design, training and measurement. In M. McNeese, E. Salas & M. Endsley (Eds), *New trends in cooperative activities:*

Understanding system dynamics in complex environments (pp. 46-67). Santa Monica, CA: Human Factors and Ergonomics Society.

Endsley, M. R., & Kaber, D. B. (1997). The use of level of automation as a means of alleviating out-of-the-loop performance problems: A taxonomy and empirical analysis. In P. Seppala & T. Luopajarvi & C. H. Nygard & M. Mattila (Eds), *13th Triennial Congress of the International Ergonomics Association* (Vol. 1, pp. 168-170). Helsinki: Finnish Institute of Occupational Health.

Endsley, M. R., & Kaber, D. B. (1999). Level of automation effects on performance, situation awareness and workload in a dynamic control task. *Ergonomics, 42*(3), 462-492.

Endsley, M. R., & Kiris, E. O. (1994a). Information presentation for expert systems in future fighter aircraft. *International Journal of Aviation Psychology, 4*(4), 333-348.

Endsley, M. R., & Kiris, E. O. (1994b). The out-of-the-loop performance problem: Impact of level of control and situation awareness. In M. Mouloua & R. Parasuraman (Eds), *Human performance in automated systems: Current research and trends* (pp. 50-56). Hillsdale, NJ: Lawrence Erlbaum.

Endsley, M. R., & Kiris, E. O. (1994c). *Situation awareness in FAA airway facilities maintenance control centers (MCC): Final report.* Lubbock, TX: Texas Tech University.

Endsley, M. R., & Kiris, E. O. (1995a). The out-of-the-loop performance problem and level of control in automation. *Human Factors, 37*(2), 381-394.

Endsley, M. R., & Kiris, E. O. (1995b). *Situation awareness global assessment technique (SAGAT) TRACON air traffic control version user guide.* Lubbock, TX: Texas Tech University.

Endsley, M. R., Mogford, R., Allendoerfer, K., Snyder, M. D., & Stein, E. S. (1997). *Effect of free flight conditions on controller performance, workload and situation awareness: A preliminary investigation of changes in locus of control using existing technology* (DOT/FAA/CT-TN 97/12). Atlantic City, NJ: Federal Aviation Administration William J. Hughes Technical Center.

Endsley, M. R., Mogford, R. H., & Stein, E. S. (1997). Controller situation awareness in free flight. *Proceedings of the Human Factors and Ergonomics Society 41st Annual Meeting* (pp. 4-8). Santa Monica, CA: Human Factors and Ergonomics Society.

Endsley, M. R., Onal, E., & Kaber, D. B. (1997). The impact of intermediate levels of automation on situation awareness and performance in dynamic control systems. In D. I. Gertman & D. L. Schurman & H. S. Blackman (Eds.), *Proceedings of the 1997 IEEE Sixth Conference on Human Factors and Power Plants.* (pp. 7-7/7-12). New York: IEEE.

Endsley, M. R., & Robertson, M. M. (1996a). *Team situation awareness in aircraft maintenance.* Lubbock, TX: Texas Tech University.

Endsley, M. R., & Robertson, M. M. (1996b). Team situation awareness in aviation maintenance. *Proceedings of the 40th Annual Meeting of the Human Factors and Ergonomics Society* (pp. 1077-1081). Santa Monica, CA: Human Factors and Ergonomics Society.

Endsley, M. R., & Robertson, M. M. (2000). Situation awareness in aircraft maintenance teams. *International Journal of Industrial Ergonomics, 26*, 301-325.

Endsley, M. R., & Rodgers, M. D. (1994). Situation awareness information requirements for en route air traffic control. *Proceedings of the Human Factors and Ergonomics Society 38th Annual Meeting* (pp. 71-75). Santa Monica, CA: Human Factors and Ergonomics Society.

Endsley, M. R., & Rodgers, M. D. (1998). Distribution of attention, situation awareness, and workload in a passive air traffic control task: Implications for operational errors and automation. *Air Traffic Control Quarterly, 6*(1), 21-44.

Endsley, M. R., & Rosiles, S. A. (1995). Vertical auditory localization for spatial orientation. *Proceedings of the Human Factors and Ergonomics Society 39th Annual Meeting* (pp. 55-59). Santa Monica, CA: Human Factors and Ergonomics Society.

Endsley, M. R., & Selcon, S. J. (1997). Designing to aid decisions through situation awareness enhancement. *Proceedings of the 2nd Symposium on Situation Awareness in Tactical Aircraft* (pp. 107-112). Patuxent River, MD: Naval Air Warfare Center.

Endsley, M. R., Sollenberger, R., Nakata, A., Hough, D., & Stein, E. (1999). *Situation awareness in air traffic control: Enhanced displays for advanced operations*. Atlantic City, NJ: Federal Aviation Administration William J. Hughes Technical Center.

Endsley, M. R., Sollenberger, R., Nakata, A., & Stein, E. (2000). *Situation awareness in air traffic control: Enhanced displays for advanced operations* (DOT/FAA/CT-TN00/01). Atlantic City, NJ: Federal Aviation Administration William J. Hughes Technical Center.

Endsley, M. R., Sollenberger, R., & Stein, E. (1999). The use of predictive displays for aiding controller situation awareness. *Proceedings of the 43rd Annual Meeting of the Human Factors and Ergonomics Society* (pp. 51-55). Santa Monica, CA: Human Factors and Ergonomics Society.

Endsley, M. R., & Strauch, B. (1997). Automation and situation awareness: The accident at Cali, Colombia. *Proceedings of the Ninth International Symposium on Aviation Psychology* (pp. 877-881). Columbus, OH: Ohio State University.

Ephrath, A. R., & Young, L. R. (1981). Monitoring vs. man-in-the-loop detection of aircraft control failures. In J. Rasmussen & W. B. Rouse (Eds), *Human detection and diagnosis of system failures*. New York: Plenum Press.

Evans, S. (1978). *Updated user's guide for the COMBIMAN* (AMRL-TR-78-31). Wright-Patterson AFB, OH: Aerospace Medical Research Laboratory.

Farley, T. C., Hansman, R. J., Amonlirdviman, K., & Endsley, M. R. (2000). Shared information between pilots and controllers in tactical air traffic control. *Journal of Guidance, Control and Dynamics, 23*(5), 826-836.

Finger, R., & Bisantz, A. M. (2000). Utilizing graphical formats to convey uncertainty in a decision making task. *Proceedings of the IEA 2000/HFES 2000 Congress* (pp. 1:13-11:16). Santa Monica, CA: Human Factors and Ergonomics Society.

Forbes, T. W. (1946). Auditory signals for instrument flying. *Journal of Aeronautical Science, 13*, 255-258.

Fracker, M. L. (1987). *Situation awareness: A decision model.* Unpublished manuscript, Dayton, OH: Armstrong Research Laboratory, U.S. Air Force.

Fracker, M. L., & Vidulich, M. A. (1991). Measurement of situation awareness: A brief review. In Y. Queinnec & F. Danniellou (Eds), *Designing for Everyone: Proceedings of the 11th Congress of the International Ergonomics Association* (pp. 795-797). London: Taylor & Francis.

Gilliland, K., & Schlegel, R. E. (1994). Tactile stimulation of the human head for information display. *Human Factors, 36*(4), 700-717.

Gilson, R. D., Mouloua, M., Graft, A. S., & McDonald, D. P. (2001). Behavioral influences of proximal alarms. *Human Factors, 4*(4), 595-610.

Graeber, R. C. (1996). Integrating human factors and safety into airplane design and operations. In B. J. Hayward & A. R. Lowe (Eds), *Applied aviation psychology: Achievement, change and challenge* (pp. 27-38). Aldershot, UK: Avebury Aviation.

Green, P., & Wei-Haas, L. (1985). The rapid development of user interfaces: Experience with the wizard of oz method. *Proceedings of the Human Factors Society 29th Annual Meeting* (pp. 470-474). Santa Monica, CA: Human Factors Society.

Guerlain, S. A., Smith, P., Heinz Obradovich, J., Rudman, S., Strohm, P., Smith, J. W., Svirbely, J., & Sachs, L. (1999). Interactive critiquing as a form of decision support: An empirical evaluation. *Human Factors, 41*(1), 72-89.

Hahn, E. C., & Hansman, R. J. (1992). Experimental studies on the effect of automation on pilot situational awareness in the datalink ATC environment. *Proceedings of the SAE AEROTECH Conference and Exposition*. Anaheim, CA: SAE.

Hancock, P., & Meshkati, N. (Eds.). (1988). *Human mental workload.* Amsterdam: North-Holland.

Hart, S. G., & Staveland, L. E. (1988). Development of NASA-TLX (Task Load Index): Results of empirical and theoretical research. In P. A. Hancock & N. Meshkati (Eds), *Human mental workload* (pp. 139-183). Amsterdam: North-Holland.

Hellier, E., Edworthy, J., & Dennis, I. (1993). Improving auditory warning design: Quantifying and predicting the effects of different warning parameters on perceived urgency. *Human Factors, 35*(4), 693-706.

Helmreich, R. L., Foushee, H. C., Benson, R., & Russini, R. (1986). Cockpit management attitudes: Exploring the attitude-performance linkage. *Aviation, Space and Environmental Medicine, 57*, 1198-1200.

Hilburn, B., Jorna, P. G., Bryne, E. A., & Parasuraman, R. (1997). The effect of adaptive air traffic control decision aiding on controller mental workload. In M. Mouloua & J. M. Koonce (Eds), *Human automation interaction: Research and practice* (pp. 84-91). Mahwah, NJ: LEA.

Hoffberg, L. I. (1991). Designing user interface guidelines for time-shift programming on a video cassette recorder. *Proceedings of the Human Factors*

Society 35th Annual Meeting (pp. 501-504). Santa Monica, CA: Human Factors Society.

Hoffer, J. A., George, J. F., & Valacich, J. S. (2002). *Modern systems analysis and design*. Upper Saddle River, NJ: Prentice-Hall.

Hogg, D. N., Torralba, B., & Volden, F. S. (1993). *A situation awareness methodology for the evaluation of process control systems: Studies of feasibility and the implication of use* (1993-03-05). Storefjell, Norway: OECD Halden Reactor Project.

Hollnagel, E., & Woods, D. D. (1983). Cognitive systems engineering: New wine in new bottles. *International Journal of Man-Machine Studies, 18*, 583-591.

Jentsch, F., Barnett, J., & Bowers, C. (1997). Loss of aircrew situation awareness: a cross-validation. *Proceedings of the Human Factors and Ergonomics Society 41st Annual Meeting* (p. 1379). Santa Monica, CA: Human Factors and Ergonomics Society.

Johnson, E. N., & Pritchett, A. R. (1995). Experimental study of vertical flight path mode awareness. *Proceedings of the 6th IFAC/IFIP/IFORS/IEA Symposium on Analysis, Design and Evaluation of Man-Machine Systems* (pp. 185-190). Cambridge, MA: MIT.

Jones, D. G., & Endsley, M. R. (1996). Sources of situation awareness errors in aviation. *Aviation, Space and Environmental Medicine, 67*(6), 507-512.

Jones, D. G., & Endsley, M. R. (2000a). Can real-time probes provide a valid measure of situation awareness? *Proceedings of the Human Performance, Situation Awareness and Automation: User-Centered Design for the New Millennium Conference* (pp. 245-250). Atlanta, GA: SA Technologies, Inc.

Jones, D. G., & Endsley, M. R. (2000b). Overcoming representational errors in complex environments. *Human Factors, 42*(3), 367-378.

Jones, D. G., & Endsley, M. R. (2002). *Measurement of shared SA in teams: In initial investigation* (SATech-02-05). Marietta, GA: SA Technologies.

Jones, W. M. (1997). Enhancing team situation awareness: Aiding pilots in forming initial mental models of team members. *Proceedings of the Ninth International Symposium on Aviation Psychology* (pp. 1436-1441). Columbus, OH: The Ohio State University.

Jones, W. M. (2000). Pilot use of TCAS. Personal communication.

Kaber, D. B. (1996). *The effect of level of automation and adaptive automation on performance in dynamic control environments.* Unpublished Doctoral Dissertation, Texas Tech University, Lubbock, TX.

Kaber, D. B., & Riley, J. (1999). Adaptive automation of a dynamic control task based on secondary task workload measurement. *International Journal of Cognitive Ergonomics, 3*(3), 169-187.

Kahneman, D., Slovic, P., & Tversky, A. (1982). *Judgment under uncertainty: Heuristics and biases*. Cambridge, UK: Cambridge University Press.

Kantowitz, B. H., & Sorkin, R. D. (1983). *Human factors: Understanding people-system relationships*. New York: John Wiley & Sons.

Kaplan, C. A., & Simon, H. A. (1990). In search of insight. *Cognitive Psychology, 22*, 374-419.

Kelley, J. F. (1983). An empirical methodology for writing user-friendly natural language computer applications. *Proceedings of the CHI '83 Conference on Human Factors in Computing Systems* (pp. 193-196). New York: Association for Computing Machinery.

Kerstholt, J. H., Passenier, P. O., Houttuin, K., & Schuffel, H. (1996). The effect of a priori probability and complexity on decision making in a supervisory control task. *Human Factors, 38*(1), 65-78.

Kessel, C. J., & Wickens, C. D. (1982). The transfer of failure-detection skills between monitoring and controlling dynamic systems. *Human Factors, 24*(1), 49-60.

Kesting, I. G., Miller, B. T., & Lockhart, C. H. (1988). Auditory alarms during anesthesia monitoring. *Anesthesiology, 69*, 106-107.

Kibbe, M. (1988). Information transfer from intelligent EW displays. *Proceedings of the Human Factors Society 32nd Annual Meeting* (pp. 107-110). Santa Monica, CA: Human Factors Society.

Kibbe, M., & McDowell, E. D. (1995). Operator decision making: Information on demand. In R. Fuller & N. Johnston & N. McDonald (Eds), *Human factors in aviation operations* (Vol. 3, pp. 43-48). Aldershot, UK: Avebury.

Kieras, D. E., & Polson, P. G. (1985). An approach to the formal analysis of user complexity. *International Journal of Man-Machine Studies, 22*, 365-394.

Kirschenbaum, S. S., & Arruda, J. E. (1994). Effects of graphic and verbal probability information on command decision making. *Human Factors, 36*(3), 406-418.

Klein, G. A. (1989). Recognition-primed decisions. In W. B. Rouse (Ed.), *Advances in man-machine systems research* (Vol. 5, pp. 47-92). Greenwich, Conn: JAI Press, Inc.

Klein, G. A. (1993). Sources of error in naturalistic decision making tasks. *Proceedings of the Human Factors and Ergonomics Society 37th Annual Meeting* (pp. 368-371). Santa Monica, CA: Human Factors and Ergonomics Society.

Klein, G. A., Calderwood, R., & Clinton-Cirocco, A. (1986). Rapid decision making on the fire ground. *Proceedings of the Human Factors Society 30th Annual Meeting* (pp. 576-580). Santa Monica, CA: Human Factors Society.

Klein, G. A., Orasanu, J., Calderwood, R., & Zsambock, C. E. (Eds). (1993). *Decision making in action: models and methods*. Norwood, NJ: Ablex.

Klein, G. A., Zsambok, C. E., & Thordsen, M. L. (1993). Team decision training: Five myths and a model. *Military Review, April*, 36-42.

Kohn, L. T., Corrigan, J. M., & Donaldson, M. S. (Eds). (1999). *To err is human: Building a safer health system*. Washington, D.C.: National Academy Press.

Kragt, H., & Bonten, J. (1983). Evaluation of a conventional process alarm system in a fertilizer plant. *IEEE Transactions on Systems, Man and Cybernetics, 13*(4), 589-600.

Kuchar, J. K., & Hansman, R. J. (1993). An exploratory study of plan-view terrain displays for air carrier operations. *International Journal of Aviation Psychology, 3*(1), 39-54.

Kuhn, T. (1970). *The structure of scientific revolutions* (2nd ed). Chicago: University of Chicago Press.

Kuipers, A., Kappers, A., van Holten, C. R., van Bergen, J. H. W., & Oosterveld, W. J. (1990). Spatial disorientation incidents in the R.N.L.A.F. F16 and F5 aircraft and suggestions for prevention, *Situational awareness in aerospace operations (AGARD-CP-478)* (pp. OV/E/1 - OV/E/16). Neuilly Sur Seine, France: NATO - AGARD.

Landler, M. (2002, July 20). 2 pilots in fatal crash in Europe saw each other as they collided. *New York Times*.

Laughery, K. R., & Corker, K. M. (1997). Computer modeling and simulation of human/system performance. In G. Salvendy (Ed), *Handbook of human factors* (2nd ed). New York: Wiley & Sons.

Layton, C., Smith, P. J., & McCoy, C. E. (1994). Design of a cooperative problem-solving system for en route flight planning: An empirical evaluation. *Human Factors, 36*(1), 94-119.

Lee, J., & Moray, N. (1992). Trust, control strategies and allocation of function in human-machine systems. *Ergonomics, 35*(10), 1243-1270.

Lehto, M. R., Papastavrou, J. D., & Giffen, W. J. (1998). An empirical study of adaptive warnings: Human versus computer-adjusted warning thresholds. *International Journal of Cognitive Ergonomics, 2*(1-2), 19-33.

Lipshitz, R. (1987). *Decision making in the real world: Developing descriptions and prescriptions from decision maker's retrospective accounts*. Boston, MA: Boston University Center for Applied Sciences.

Loeb, V. (2002, March 24). Friendly fire deaths traced to dead battery: Taliban targeted but U.S. forces killed. *Washington Post,* pp. 21.

Logan, G. D. (1988). Automaticity, resources and memory: Theoretical controversies and practical implications. *Human Factors, 30*(5), 583-598.

Macworth, N. H. (1948). The breakdown of vigilance during prolonged visual search. *Quarterly Journal of Experimental Psychology, 1*, 5-61.

Macworth, N. H. (1970). *Vigilance and attention*. Baltimore: Penguin.

Matthews, M. D., Pleban, R. J., Endsley, M. R., & Strater, L. G. (2000). Measures of infantry situation awareness for a virtual MOUT environment. *Proceedings of the Human Performance, Situation Awareness and Automation: User-Centered Design for the New Millennium Conference* (pp. 262-267). Savannah, GA: SA Technologies, Inc.

McCabe, T. (1976). A complexity measure. *IEEE Transactions on Software Engineering, 2*(4), 308-320.

McCloskey, M. J. (1996). An analysis of uncertainty in the marine corps. *Proceedings of the Human Factors and Ergonomics Society 40th Annual Meeting* (pp. 194-198). Santa Monica, CA: Human Factors and Ergonomics Society.

McClumpha, A., & James, M. (1994). Understanding automated aircraft. In M. Mouloua & R. Parasuraman (Eds), *Human performance in automated systems: Current research and trends* (pp. 183-190). Hillsdale, NJ: LEA.

McKinley, R. L., Erickson, M. A., & D'Angelo, W. R. (1994). 3-Dimensional auditory displays: Development, applications, and performance. *Aviation, Space, and Environmental Medicine, 65*, A31-A38.

Meister, D., & Enderwick, T. P. (2002). *Human factors in system design, development and testing*. Mahwah, NJ: Lawrence Erlbaum.

Metzger, U., & Parasuraman, R. (2001a). Automation-related "complacency": Theory, empirical data and design implications. *Proceedings of the Human Factors and Ergonomics Society 45th Annual Meeting* (pp. 463-467). Santa Monica, CA: Human Factors and Ergonomics Society.

Metzger, U., & Parasuraman, R. (2001b). The role of the air traffic controller in future air traffic management: An empirical study of active control vs. passive monitoring. *Human Factors, 43*(4), 519-528.

Midkiff, A. H., & Hansman, R. J. (1992). Identification of important party-line information elements and implications for situational awareness in the datalink environment. *Proceedings of the SAE Aerotech Conference and Exposition*. Anaheim, CA: SAE.

Miller, C. (2000). From the Microsoft paperclip to the rotorcraft pilot's associate: Lessons learned from fielding adaptive automation systems. *Proceedings of the Human Performance, Situation Awareness and Automation: User-Centered Design for the New Millennium Conference* (p. 4). Savannah, GA: SA Technologies, Inc.

Miller, C., Hannen, M., & Guerlain, S. (1999). The rotocraft pilot's associate cockpit information manager: Acceptable behavior from a new crew member. *Proceedings of the American Helicopter Society's FORUM 55*. Montreal, Quebec: American Helicopter Society.

Miller, C. A., Pelican, M., & Goldman, R. (1999). Tasking interfaces for flexible interaction with automation: Keeping the operator in control. *Proceedings of the International Conference on Intelligent User Interfaces*. Redondo Beach, CA.

Miller, G. A. (1956). The magical number seven plus or minus two: Some limits on our capacity for processing information. *Psychological Review, 63*, 81-97.

Mintzburg, H. (1973). *The nature of managerial work*. New York: Harper and Row.

Momtahan, K., Hétu, R., & Tansley, B. (1993). Audibility and identification of auditory alarms in the operating room and intensive care unit. *Ergonomics, 36*(10), 1159-1176.

Montgomery, D. A., & Sorkin, R. D. (1996). Observer sensitivity to element reliability in a multielement visual display. *Human Factors, 38*(3), 484-494.

Moray, N. (1986). Monitoring behavior and supervisory control. In K. Boff (Ed.), *Handbook of perception and human performance* (Vol. II, pp. 40/41-40/51). New York: Wiley.

Moray, N. (2000). Are observers ever really complacent when monitoring automated systems? *Proceedings of the IES 2000/HFES 2000 Congress* (pp. 1/592-595). Santa Monica CA: Human Factors and Ergonomics Society.

Moray, N., & Rotenberg, I. (1989). Fault management in process control: Eye movements and action. *Ergonomics, 32*, 1319-1342.

Moshansky, V. P. (1992). *Commission of inquiry into Air Ontario accident at Dryden, Ontario* (Final Report Vol 1-4). Ottawa, ON: Minister of Supply and Services, Canada.

Mosier, K. L., & Chidester, T. R. (1991). Situation assessment and situation awareness in a team setting. In Y. Queinnec & F. Daniellou (Eds), *Designing for Everyone* (pp. 798-800). London: Taylor & Francis.

Mosier, K. L., Skitka, L. J., & Korte, K. J. (1994). Cognitive and social psychology issues in flight crew/automation interaction. In M. Mouloua & R. Parasuraman (Eds), *Human performance in automated systems: Current research and trends* (pp. 191-197). Hillsdale, NJ: Lawrence Erlbaum.

Nagel, D. C. (1988). Human error in aviation operations. In E. L. Weiner & D. C. Nagel (Eds), *Human factors in aviation* (pp. 263-303). San Diego: Academic Press.

National Transportation Safety Board. (1973). *Aircraft accidents report: Eastern Airlines 401/L-1011, Miami, Florida, December 29, 1972.* Washington, D.C.: Author.

National Transportation Safety Board. (1986). *China Airlines B-747-SP, 300 NM northwest of San Francisco, CA, 2/19/85* (NTSB Report No AAR-86/03). Washington, DC: Author.

National Transportation Safety Board. (1988). *Aircraft accidents report: Northwest Airlines, Inc., McDonnell-Douglas DC-9-82, N312RC, Detroit Metropolitan Wayne County Airport, August, 16, 1987* (NTSB/AAR-99-05). Washington, D.C.: Author.

National Transportation Safety Board. (1990). *Aircraft accidents report: USAIR, Inc., Boeing 737-400, LaGuardia Airport, Flushing, New York, September 20, 1989* (NTSB/AAR-90-03). Washington, D.C.: Author.

National Transportation Safety Board. (1991). *Aircraft accidents report: US Air, Boeing 737-300 and Skywest, Fairchild SA-227-AC at Los Angeles International Airport* (NTSB/AAR-91/08). Washington, D.C.: Author.

National Transportation Safety Board. (1994). *A review of flightcrews involved in major accidents of U.S. air carriers 1978 - 1990.* Washington, DC: Author.

National Transportation Safety Board. (1997). *Wheels-up landing: Continental Airlines Flight 1943, Douglas DC-9 N10556, Houston, Texas, February 19, 1996* (PB97-910401). Washington, DC: Author.

Noble, D., Boehm-Davis, D., & Grosz, C. (1987). *Rules, schema and decision making.* (NR 649-005). Vienna, VA: Engineering Research Associates.

Norman, D. A. (1981). *Steps towards a cognitive engineering* (Tech. Report). San Diego: University of California, Program in Cognitive Science.

Norman, D. A. (1986). Cognitive engineering. In D. A. Norman & S. W. Draper (Eds), *User centered system design*. Hillsdale, NJ: Lawrence Erlbaum.

Norman, D. A. (1989). *The problem of automation: Inappropriate feedback and interaction not overautomation* (ICS Report 8904). La Jolla, CA: Institute for Cognitive Science, U. C. San Diego.

Noyes, J. M., & Starr, A. F. (2000). Civil aircraft warning systems: Future directions in information management and presentation. *International Journal of Aviation Psychology, 10*(2), 169-188.

O'Brien, T. G., & Charlton, S. G. (1996). *Handbook of human factors testing and evaluation*. Mahwah, NJ: Lawrence Erlbaum.

Olson, W. A., & Sarter, N. B. (1999). Supporting informed consent in human machine collaboration: The role of conflict type, time pressure, and display design. *Proceedings of the Human Factors and Ergonomics Society 42nd Annual Meeting* (pp. 189-193). Santa Monica, CA: Human Factors and Ergonomics Society.

Orasanu, J. (1990). Shared mental models and crew decision making. *Proceedings of the 12th Annual Conference of the Cognitive Science Society*. Cambridge, MA.

Orasanu, J., & Fischer, U. (1997). Finding decisions in natural environments: the view from the cockpit. In C. E. Zsambok & G. Klein (Eds), *Naturalistic decision making* (pp. 343-357). Mahwah, NJ: Lawrence Erlbaum.

Orasanu, J., & Salas, E. (1993). Team decision making in complex environments. In G. A. Klein & J. Orasanu & R. Calderwood & C. E. Zsambok (Eds), *Decision making in action: Models and methods* (pp. 327-345). Norwood, NJ: Ablex.

Paielli, R. A. (1998). Empirical test of conflict probability estimation. *Proceedings of the 2nd USA/Europe Air Traffic Management R&D Seminar*. Orlando, FL.

Parasuraman, R. (1987). Human-computer monitoring. *Human Factors, 29*(6), 695-706.

Parasuraman, R. (1993). Effects of adaptive function allocation on human performance. In D. J. Garland & J. A. Wise (Eds), *Human factors and advanced aviation technologies* (pp. 147-158). Daytona Beach, FL: Embry-Riddle Aeronautical University Press.

Parasuraman, R., Molloy, R., Mouloua, M., & Hilburn, B. (1996). Monitoring of automated systems. In R. Parasuraman & M. Mouloua (Eds), *Automation and human performance: Theory and applications* (pp. 91-115). Mahwah, NJ: Lawrence Erlbaum Associates.

Parasuraman, R., Molloy, R., & Singh, I. L. (1993). Performance consequences of automation-induced complacency. *International Journal of Aviation Psychology, 3*(1), 1-23.

Parasuraman, R., & Mouloua, M. (Eds). (1996). *Automation and human performance: Theory and applications*. Mahwah, NJ: Lawrence Erlbaum.

Parasuraman, R., Mouloua, M., & Molloy, R. (1994). Monitoring automation failures in human-machine systems. In M. Mouloua & R. Parasuraman (Eds), *Human performance in automated systems: Current research and trends* (pp. 45-49). Hillsdale, NJ: LEA.

Parasuraman, R., Mouloua, M., & Molloy, R. (1996). Effects of adaptive task allocation on monitoring of automated systems. *Human Factors, 38*(4), 665-679.

Parasuraman, R., & Riley, V. (1997). Humans and automation: Use, misuse, disuse and abuse. *Human Factors, 39*(2), 230-253.

Parasuraman, R., Sheridan, T. B., & Wickens, C. D. (2000). A model of types and levels of human interaction with automation. *IEEE Transactions on Systems, Man and Cybernetics, 30*(3), 286-297.

Pew, R. W. (1995). The state of situation awareness measurement: Circa 1995. In M. R. Endsley & D. J. Garland (Eds), *Experimental analysis and measurement of situation awareness* (pp. 7-16). Daytona Beach, FL: Embry-Riddle Aeronautical University.

Pew, R. W., & Mavor, A. S. (Eds). (1998). *Modeling human and organizational behavior*. Washington, D.C.: National Academy Press.

Pope, A. T., Comstock, R. J., Bartolome, D. S., Bogart, E. H., & Burdette, D. W. (1994). Biocybernetic system validates index of operator engagement in automated task. In M. Mouloua & R. Parasuraman (Eds), *Human performance in automated systems: Current research and trends* (pp. 300-306). Hillsdale, NJ: Lawrence Erlbaum.

Prince, C., & Salas, E. (2000). Team situation awareness, errors, and crew resource management: Research integration for training guidance. In M. R. Endsley & D. J. Garland (Eds), *Situation awareness analysis and measurement* (pp. 325-347). Mahwah, NJ: Lawrence Erlbaum.

Prinzel, L. J., & Pope, A. T. (2000). The double-edged sword of self-efficacy: Implications for automation-induced complacency. *Proceedings of the IEA 2000/HFES 2000 Congress* (pp. 3-107). Santa Monica, CA: Human Factors and Ergonomics Society.

Pritchett, A. R. (1997). *Pilot non-conformance to alerting system commands during closely spaced parallel approaches.* Unpublished Doctoral dissertation. Cambridge, MA: Massachusetts Institute of Technology.

Pritchett, A. R., & Hansman, R. J. (1997). Pilot non-conformance to alerting system commands. *Proceedings of the Ninth International Symposium on Aviation Psychology* (pp. 274-279). Columbus, OH: Ohio State University.

Pritchett, A. R., Hansman, R. J., & Johnson, E. N. (1995). Use of testable responses for performance-based measurement of situation awareness. In D. J. Garland & M. R. Endsley (Eds), *Experimental analysis and measurement of situation awareness* (pp. 75-81). Daytona Beach, FL: Embry-Riddle University Press.

Rasmussen, J. (1980). What can be learned from human error reports? In K. D. Duncan & M. Gruneberg & D. Wallis (Eds), *Changes in working life* (pp. 97-113). New York: Wiley.

Redelmeier, D. A., & Tibshirani, R. J. (1997). Association between cellular-telephone calls and motor vehicle collisions. *The New England Journal of Medicine, 336*(7), 453-458.

Reid, G. B. (1987). *Subjective workload assessment technique (SWAT): A user's guide*. Wright-Patterson AFB, OH: Harry G. Armstrong Aerospace Medical Research Laboratory.

Reising, D. V. (1993). Diagnosing multiple simultaneous faults. *Proceedings of the Human Factors and Ergonomics Society 37th Annual Meeting* (pp. 524-528). Santa Monica, CA: Human Factors and Ergonomics Society.

Reising, D. V., & Sanderson, P. M. (1995). Mapping the domain of electronic repair shops: A field study in fault diagnosis. *Proceedings of the Human Factors and Ergonomics Society 39th Annual Meeting* (pp. 464-468). Santa Monica, CA: Human Factors and Ergonomics Society.

Riley, J. M., & Kaber, D. B. (2001). Utility of situation awareness and attention for describing telepresence experiences in a virtual teloperation task. *Proceedings of the International Conference on Computer-Aided Ergonomics and Safety*. Maui, HI: IEA.

Riley, J. M., Kaber, D. B., & Draper, J. V. (in press). Situation awareness and attention allocation measures for quantifying telepresence experiences in teleoperation. *Human Factors and Ergonomics in Manufacturing.*

Robinson, D. (2000). The development of flight crew situation awareness in commercial transport aircraft. *Proceedings of the Human Performance, Situation Awareness and Automation: User-Centered Design for a New Millennium Conference* (pp. 88-93). Marietta, GA: SA Technologies, Inc.

Rochlis, J. L., & Newman, D. J. (2000). A tactile display for international space station (ISS) extravehicular activity (EVA). *Aviation, Space and Environmental Medicine, 71*(6), 571-578.

Rouse, W. (1986). Measuring the complexity of monitoring and controlling large scale systems. *IEEE Transactions on Systems, Man and Cybernetics, 16*(2), 193-207.

Rouse, W. B. (1977). Human-computer interaction in multi-task situations. *IEEE Transactions on Systems, Man and Cybernetics, SMC-7*, 384-392.

Rouse, W. B. (1988). Adaptive aiding for human/computer control. *Human Factors, 30*(4), 431-438.

Rouse, W. B., & Morris, N. M. (1985). *On looking into the black box: Prospects and limits in the search for mental models* (DTIC #AD-A159080). Atlanta, GA: Center for Man-Machine Systems Research, Georgia Institute of Technology.

RTCA. (1995). *Report of the RTCA Board of Directors select committee on free flight*. Washington, D.C.: Author.

Salas, E., Dickinson, T. L., Converse, S., & Tannenbaum, S. I. (1992). Toward an understanding of team performance and training. In R. W. Swezey & E. Salas (Eds), *Teams: their training and performance* (pp. 3-29). Norwood, NJ: Ablex.

Salas, E., Prince, C., Baker, D. P., & Shrestha, L. (1995). Situation awareness in team performance: Implications for measurement and training. *Human Factors, 37*(1), 123-136.

SAMMIE CAD Ltd. (1990). *The SAMMIE system*. Loughborough, England: Author.

Sanders, M. S., & McCormick, E. J. (1992). *Human factors in engineering and design* (Seventh ed). New York: McGraw-Hill.

Sanderson, P. M., Reising, D. V., & Augustiniak, M. J. (1995). Diagnosis of multiple faults in systems: Effects of fault difficulty, expectancy, and prior exposure. *Proceedings of the Human Factors and Ergonomics Society 39th Annual Meeting* (pp. 459-463). Santa Monica, CA: Human Factors and Ergonomics Society.

Sarter, N. B., & Schroeder, B. (2001). Supporting decision making and action selection under time pressure and uncertainty: The case of in-flight icing. *Human Factors, 43*(4), 573-583.

Sarter, N. B., & Woods, D. D. (1991). Situation awareness: A critical but ill-defined phenomenon. *The International Journal of Aviation Psychology, 1*(1), 45-57.

Sarter, N. B., & Woods, D. D. (1994a). 'How in the world did I ever get into that mode': Mode error and awareness in supervisory control. In R. D. Gilson & D. J. Garland & J. M. Koonce (Eds), *Situational awareness in complex systems* (pp. 111-124). Daytona Beach, FL: Embry-Riddle Aeronautical University Press.

Sarter, N. B., & Woods, D. D. (1994b). Pilot interaction with cockpit automation II: An experimental study of pilots' model and awareness of flight management systems. *International Journal of Aviation Psychology, 4*(1), 1-28.

Scerbo, M. W. (1996). Theoretical perspectives on adaptive automation. In R. Parasuraman & M. Mouloua (Eds), *Automation and human performance: Theory and application* (pp. 37-63). Mahwah, NJ: Lawrence Erlbaum.

Schraagen, J. M., Chipman, S. F., & Shalin, V. L. (2000). *Cognitive task analysis*. Mahwah, NJ: Lawrence Erlbaum.

Seagull, F. J., & Sanderson, P. M. (2001). Anesthesia alarms in context: An observational study. *Human Factors, 43*(1), 66-78.

Segal, L. D. (1994). Actions speak louder than words. *Proceedings of the Human Factors and Ergonomics Society 38th Annual Meeting* (pp. 21-25). Santa Monica,CA: Human Factors and Ergonomics Society.

Selcon, S. J. (1990). Decision support in the cockpit: Probably a good thing? *Proceedings of the Human Factors Society 34th Annual Meeting* (pp. 46-50). Santa Monica, CA: Human Factors Society.

Selcon, S. J., & Taylor, R. M. (1990). Evaluation of the situational awareness rating technique (SART) as a tool for aircrew systems design, *Situational awareness in aerospace operations (AGARD-CP-478)* (pp. 5/1 -5/8). Neuilly Sur Seine, France: NATO - AGARD.

Selcon, S. J., Taylor, R. M., & Koritsas, E. (1991). Workload or situational awareness?: TLX vs SART for aerospace systems design evaluation. *Proceedings of the Human Factors Society 35th Annual Meeting* (pp. 62-66). Santa Monica, CA: Human Factors Society.

Selcon, S. J., Taylor, R. M., & McKenna, F. P. (1995). Integrating multiple information sources: Using redundancy in the design of warnings. *Ergonomics, 38*, 2362-2370.

Selcon, S. J., Taylor, R. M., & Shadrake, R. A. (1992). Multi-modal cockpit warnings: Pictures, words, or both? *Proceedings of the Human Factors and Ergonomics Society 36th Annual Meeting* (pp. 62-66). Santa Monica, CA: Human Factors and Ergonomics Society.

Sexton, G. A. (1988). Cockpit-crew systems design and integration. In E. L. Wiener & D. C. Nagel (Eds), *Human factors in aviation* (pp. 495-526). San Diego: Academic Press.

Sheridan, T. (1981). Understanding human error and aiding human diagnostic behavior in nuclear power plants. In J. Rasmussen & W. B. Rouse (Eds), *Human detection and diagnosis of system failures*. New York: Plenum Press.

Sheridan, T. B., & Verplanck, W. L. (1978). *Human and computer control of undersea teleoperators*. Cambridge, MA: MIT Man-Machine Laboratory.

Sklar, A. E., & Sarter, N. B. (1999). Good vibrations: Tactile feedback in support of attention allocation and human-automation coordination in event-driven domains. *Human Factors, 41*(4), 543-552.

Slamecka, N. J., & Graf, P. (1978). The generation effect: Delineation of a phenomenon. *Journal of Experimental Psychology: Human Learning and Memory, 4*(6), 592-604.

Smith, P. J., McCoy, E., Orasanu, J., Denning, R., Van Horn, A., & Billings, C. (1995). *Cooperative problem solving in the interactions of airline operations control centers with the national aviation system*. Columbus, OH: Cognitive Systems Engineering Laboratory, Ohio State University.

Smolensky, M. W. (1993). Toward the physiological measurement of situation awareness: The case for eye movement measurements. *Proceedings of the Human Factors and Ergonomics Society 37th Annual Meeting* (pp. 41). Santa Monica, CA: Human Factors and Ergonomics Society.

Sorkin, R. (1989). Why are people turning off our alarms? *Human Factors Bulletin, 32*(4), 3-4.

Sorkin, R. D., Mabry, T. R., Weldon, M. S., & Elvers, G. C. (1991). Integration of information from multiple element displays. *Organizational Behavior and Human Decision Processes, 49*, 167-187.

Stanton, N. A., Harrison, D. J., Taylor-Burge, K. L., & Porter, L. J. (2000). Sorting the wheat from the chaff: A study of the detection of alarms. *Cognition, Technology and Work, 2*, 134-141.

Stanton, N. A., & Stammers, R. B. (1998). Alarm-initiated activities: Matching visual formats to alarm handling tasks. *International Journal of Cognitive Ergonomics, 2*(4), 331-348.

Stein, E. S. (1992). *Air traffic control visual scanning* (DOT/FAA/CT-TN92/16). Atlantic City International Airport, NJ: Federal Aviation Administration William J. Hughes Technical Center.

Still, D. L., & Temme, L. A. (2001). Oz: A human-centered computing cockpit display. *Proceedings of the Interservice/Industry Training, Simulation and Education Conference (I/ITSEC)*, Arlington, VA.

Stout, R. J., Cannon-Bowers, J. A., & Salas, E. (1996). The role of shared mental models in developing team situational awareness: Implications for training. *Training Research Journal, 2*, 85-116.

Taylor, J. C., Robertson, M. M., Peck, R., & Stelly, J. W. (1993). Validating the impact of maintenance CRM training. *Proceedings of the Seventh International Symposium on Aviation Psychology* (pp. 538-542). Columbus, OH: Department of Aviation, The Ohio State University.

Taylor, R. M. (1990). Situational awareness rating technique (SART): The development of a tool for aircrew systems design, *Situational awareness in aerospace operations (AGARD-CP-478)* (pp. 3/1 - 3/17). Neuilly Sur Seine, France: NATO - AGARD.

Taylor, R. M., Endsley, M. R., & Henderson, S. (1996). Situational awareness workshop report. In B. J. Hayward & A. R. Lowe (Eds), *Applied aviation psychology: Achievement, change and challenge* (pp. 447-454). Aldershot, UK: Ashgate Publishing Ltd.

Taylor, R. M., & Selcon, S. J. (1991). Subjective measurement of situation awareness. In Y. Queinnec & F. Danniellou (Eds), *Designing for Everyone:*

Proceedings of the 11th Congress of the International Ergonomics Association (pp. 789-791). London: Taylor & Francis.

Tullis, T. (1983). The formatting of alphanumeric displays: A review and analysis. *Human Factors, 25*, 657-682.

Tyler, R. R., Shilling, R. D., & Gilson, R. D. (1995). *False alarms in naval aircraft: A review of Naval Safety Center mishap data* (Special Report 95-003). Orlando, Florida: Naval Air Warfare Center Training Systems Division.

United Kingdom Air Accidents Investigation Branch. (1990). *Report on the accident to Boeing 737-400 G-OBME near Kegworth, Leicestershire on 8 January 1989* (Report No. 4/90 (EW/C1095)). Farnborough, UK: Department for Transport.

Vakil, S. S., & Hansman, R. J. (1997). Predictability as a metric of automation complexity. *Proceedings of the Human Factors and Ergonomics Society 41st Annual Meeting* (pp. 70-74). Santa Monica, CA: Human Factors and Ergonomics Society.

Vakil, S. S., & Hansman, R. J. (1998). Operator directed common conceptual models for advanced automation. *Proceedings of the 17th Digital Avionics Systems Conference*. Seattle, WA: IEEE.

Vakil, S. S., Midkiff, A. H., & Hansman, R. J. (1996). *Development and evaluation of an electronic vertical situation display*. Cambridge, MA: MIT Aeronautical Systems Laboratory.

Verity, J. W. (1991, April 29). I can't work this thing! *Business Week,* 58-66.

Vidulich, M. (2000). Testing the sensitivity of situation awareness metrics in interface evaluations. In M. R. Endsley & D. J. Garland (Eds), *Situation awareness analysis and measurement* (pp. 227-248). Mahwah, NJ: Lawrence Erlbaum.

Vidulich, M. A. (1989). The use of judgment matrices in subjective workload assessment: The subjective workload dominance (SWORD) technique. *Proceedings of the Human Factors Society 33rd Annual Meeting* (pp. 1406-1410). Santa Monica, CA: Human Factors Society.

Vidulich, M. A., & Hughes, E. R. (1991). Testing a subjective metric of situation awareness. *Proceedings of the Human Factors Society 35th Annual Meeting* (pp. 1307-1311). Santa Monica, CA: Human Factors Society.

Waag, W. L., & Houck, M. R. (1994). Tools for assessing situational awareness in an operational fighter environment. *Aviation, Space and Environmental Medicine, 65*(5, supl), A13-A19.

Wenzel, E. M., Wightman, F. L., & Foster, S. H. (1988). A virtual display system for conveying three-dimensional acoustic information. *Proceedings of the Human Factors Society 32nd Annual Meeting* (pp. 86-90). Santa Monica, CA: Human Factors Society.

Wichman, H., & Oyasato, A. (1983). Effects of locus of control and task complexity on prospective remembering. *Human Factors, 25*(5), 583-591.

Wickens, C. D. (1992). *Engineering psychology and human performance* (2nd ed). New York: Harper Collins.

Wickens, C. D. (2000). The tradeoff of design for routine and unexpected performance: Implications of situation awareness. In M. R. Endsley & D. J.

Garland (Eds.), *Situation awareness analysis and measurement* (pp. 211-226). Mahwah, NJ: Lawrence Erlbaum.

Wickens, C. D., Conejo, R., & Gempler, K. (1999). Unreliable automated attention cueing for air-ground targeting and traffic monitoring. *Proceedings of the Human Factors and Ergonomics Society 43rd Annual Meeting* (pp. 21-25). Santa Monica, CA: Human Factors and Ergonomics Society.

Wickens, C. D., & Kessel, C. (1979). The effect of participatory mode and task workload on the detection of dynamic system failures. *SMC-9*(1), 24-34.

Wiegmann, D. A., Rich, A., & Zhang, H. (2001). Automated diagnostic aids: The effects of aid reliability on users' trust and reliance. *Theoretical Issues in Ergonomics Science, 2*(4), 352-367.

Wiener, E. L. (1985). Cockpit automation: In need of a philosophy. *Proceedings of the 1985 Behavioral Engineering Conference* (pp. 369-375). Warrendale, PA: Society of Automotive Engineers.

Wiener, E. L. (1989). *Human factors of advanced technology ("glass cockpit") transport aircraft* (NASA Contractor Report No. 177528). Moffett Field, CA: NASA-Ames Research Center.

Wiener, E. L. (1993). Life in the second decade of the glass cockpit. *Proceedings of the Seventh International Symposium on Aviation Psychology* (pp. 1-11). Columbus, OH: Department of Aviation, The Ohio State University.

Wiener, E. L., & Curry, R. E. (1980). Flight deck automation: Promises and problems. *Ergonomics, 23*(10), 995-1011.

Wilkinson, R. T. (1964). Artificial signals as an aid to an inspection task. *Ergonomics, 7*, 63-72.

Wilson, G. F. (2000). Strategies for psychophysiological assessment of situation awareness. In M. R. Endsley & D. J. Garland (Eds), *Situation awareness analysis and measurement* (pp. 175-188). Mahwah, NJ: Lawrence Erlbaum.

Woods, D. D., & Roth, E. M. (1988). Cognitive engineering: Human problem solving with tools. *Human Factors, 30*(4), 415-430.

Wurman, R. S. (1989). *Information anxiety*. New York: Doubleday.

Xiao, Y., & MacKenzie, C. F. (1997). Uncertainty in trauma patient resuscitation. *Proceedings of the Human Factors and Ergonomics Society 41st Annual Meeting* (pp. 168-171). Santa Monica, CA: Human Factors and Ergonomics Society.

Xiao, Y., Mackenzie, C. F., & Patey, R. (1998). Team coordination and breakdowns in a real-life stressful environment. *Proceedings of the Human Factors and Ergonomics Society 42nd Annual Meeting* (pp. 186-190). Santa Monica, CA: Human Factors and Ergonomics Society.

Yeh, M., & Wickens, C. (2001). Display signaling in augmented reality: Effects on cue reliability and image realism on attention allocation and trust calibration. *Human Factors, 43*(3), 355-365.

Yeh, M., Wickens, C. D., & Seagull, F. J. (1999). Target cueing in visual search: The effects of conformality and display location on the allocation of visual attention. *Human Factors, 41*(4), 524-542.

Young, L. R. A. (1969). On adaptive manual control. *Ergonomics, 12*(4), 635-657.

Zachary, W., Ryder, J., Ross, L., & Weiland, M. Z. (1992). Intelligent computer-human interaction in real-time, multi-tasking process control and monitoring systems. In M. Helander & M. Nagamachi (Eds), *Human factors in design for manufacturability*. New York: Taylor & Francis.

Zachary, W. W. (1998). Decision support systems: Designing to extend the cognitive limits. In M. Helandar (Ed), *Handbook of human-computer interaction*. Amsterdam: Elsevier Science Publishers.

Zlotnik, M. A. (1988). Applying electro-tactile display technology to fighter aircraft. *Proceedings of the IEEE National Aerospace and Electronics Conference* (pp. 191-197). Dayton, OH: IEEE.

Zsambok, C. E., & Klein, G. (Eds.). (1997). *Naturalistic decision making*. Mahwah, NJ: Lawrence Erlbaum.

Index

A

B

C

D

E

F

G

H

P

R

S

T

U

V

W

Z